Springer Series on Cultural Computing

Founding Editor

Ernest Edmonds

Series Editor

Craig Vear, University of Nottingham, Nottingham, UK

Editorial Board

Paul Brown, University of Sussex, Brighton, UK

Nick Bryan-Kinns, Queen Mary University of London, London, UK

David England, Liverpool John Moores University, Liverpool, UK

Sam Ferguson, University of Technology, Sydney, Australia

Bronać Ferran, Birkbeck, University of London, London, UK

Andrew Hugill, University of Leicester, Leicester, UK

Nicholas Lambert, Ravensbourne, London, UK

Jonas Lowgren, Linköping University, Malmo, Sweden

Ellen Yi-Luen Do, University of Colorado Boulder, Boulder, CO, USA

Sean Clark, De Montfort University, Leicester, UK

Nelson Zagalo, Department of Communication & Arts, University of Aveiro, Aveiro, Portugal

Matthias Rauterberg, Eindhoven University of Technology, Eindhoven, The Netherlands

Cultural Computing is an exciting, emerging field of Human Computer Interaction, which covers the cultural impact of computing and the technological influences and requirements for the support of cultural innovation. Using support technologies such as artificial intelligence, machine learning, location-based systems, mixed/virtual/augmented reality, cloud computing, pervasive technologies and human-data interaction, researchers can explore the differences across a variety of cultures and cultural production to provide the knowledge and skills necessary to overcome cultural issues and expand human creativity.

This series presents monographs, edited collections and advanced textbooks on the current research and knowledge of a broad range of topics including creativity support systems, creative computing, digital communities, the interactive arts, cultural heritage, digital culture and intercultural collaboration.

This Series is abstracted/indexed in Scopus.

Belinda J. Dunstan · Jeffrey T. K. V. Koh · Deborah Turnbull Tillman · Scott Andrew Brown
Editors

Cultural Robotics: Social Robots and Their Emergent Cultural Ecologies

Editors
Belinda J. Dunstan
Creative Robotics Lab
University of New South Wales
Paddington, NSW, Australia

Deborah Turnbull Tillman
Creative Robotics Lab
University of New South Wales
Paddington, NSW, Australia

Jeffrey T. K. V. Koh
Singapore Institute of Technology
Singapore, Singapore

Scott Andrew Brown
Creative Robotics Lab
University of New South Wales
Paddington, NSW, Australia

ISSN 2195-9056 ISSN 2195-9064 (electronic)
Springer Series on Cultural Computing
ISBN 978-3-031-28140-2 ISBN 978-3-031-28138-9 (eBook)
https://doi.org/10.1007/978-3-031-28138-9

© The Editor(s) (if applicable) and The Author(s), under exclusive license to Springer Nature Switzerland AG 2023
This work is subject to copyright. All rights are solely and exclusively licensed by the Publisher, whether the whole or part of the material is concerned, specifically the rights of translation, reprinting, reuse of illustrations, recitation, broadcasting, reproduction on microfilms or in any other physical way, and transmission or information storage and retrieval, electronic adaptation, computer software, or by similar or dissimilar methodology now known or hereafter developed.
The use of general descriptive names, registered names, trademarks, service marks, etc. in this publication does not imply, even in the absence of a specific statement, that such names are exempt from the relevant protective laws and regulations and therefore free for general use.
The publisher, the authors, and the editors are safe to assume that the advice and information in this book are believed to be true and accurate at the date of publication. Neither the publisher nor the authors or the editors give a warranty, expressed or implied, with respect to the material contained herein or for any errors or omissions that may have been made. The publisher remains neutral with regard to jurisdictional claims in published maps and institutional affiliations.

This Springer imprint is published by the registered company Springer Nature Switzerland AG
The registered company address is: Gewerbestrasse 11, 6330 Cham, Switzerland

The future is unthinkable. Yet here we are, thinking it.
Coexisting, we are thinking future coexistence. Predicting it and more: keeping the unpredictable one open.

—Timothy Morton, Dark Ecology

Foreword

At the start of the new millennium, the rise of *social robotics*, as a novel branch of Embodied AI and HRI, has marked the beginning of a transdisciplinary undertaking bringing robotics far beyond the boundaries of engineering.

The scope of social robotics is not limited to that of a technological discipline engaged in moving, from the realm of science-fiction to that of contemporary human social contexts, machines that are able to communicate with us through social signals compatible with our own. The success reached by social robotics in building "social partners" for humans is based on innovative scientific research. To create robots that can be effectively integrated into our social contexts, social robotics embodies in these artefacts hypotheses about us, namely hypotheses generated by a wide variety of sciences (from biology to ethology, from anthropology to sociology and psychology, from cognitive sciences to semiology and epistemology…) to describe scientifically how we know, how we communicate, how we perceive each other, how we relate to our environment and our social world—in short: who we are. And, by evaluating the quality of our interactions with these machines, social robotics tests these hypotheses, both in its labs and in the field, and provide feedback on them.

Social robotics is a science. We can conceive it as an emergent form of anthropology, addressing the issue of human self-knowledge based on the research method introduced by cybernetics to study life and cognition synthetically—the "synthetic" or "understanding by building" method. Indeed, social robotics can be seen as one of the most original and comprehensive expressions of the cybernetic project of "synthetic science." A "synthetic anthropology", which, by building robotic models of humans, and introducing them into human social contexts, on one side, generates unprecedented knowledge about us, and, on the other, transforms us in unprecedented ways.

The diffusion of social robots will change us. It will transform us and our world by revealing, amplifying, and reorienting features of our sociability. It will do so in ways that we do not know or understand for now, since these changes will cause the very process by which we will know and understand ourselves and our social universe better. More than a paradox, this is a challenge, which social robotics has been imposing on us for more than two decades now: Creating a generative loop between the process of self-knowledge and the process of self-transformation in which social robots are involving us.

Today, the birth of *cultural robotics* in the form of an autonomous research domain, as announced in this book, reflects the beginning of a new transdisciplinary adventure, in contemporary science, which promises to address this challenge proactively.

Born as a branch of social robotics dedicated to (self-)reflection on its cultural dimensions, now cultural robotics appears as an extremely original new area of study, whose specific contributions are grounded in a profound awareness of the potentialities that social robots can express in our evolution. Cultural robotics emphasizes that the project of social robots is not, nor can it be, merely technological. This new transversal domain recognizes that social robotics, while introducing its social machines as a new technology, deals also, and inseparably, with the introduction of a novel category of social relations—"human-robot" social relations. Based on this acknowledgment, cultural robotics carries a specific perception of the scale of the changes that social robots are likely to produce in our social contexts. It identifies them as complex transformations, developed through a network of circuits of co-determination (robotics-society, mind-technology, humankind-nature, natural-artificial...) which are irreducibly mediated by cultural components. For these kinds of transformations, cultural analysis can in no way be confined to specialized debates of marginal relevance. Concerning human-robot social co-evolution, cultural analysis is destined to play a concrete, decisive role, since, by directly affecting the ways we design, interpret, integrate, and live, in our public and private spaces, with the social actors produced by robotics, it exerts a deep influence both on the imminent and the long-term futures accessible to humanity in this new phase of its evolution.

One of the most promising aspects of cultural robotics' approach to the challenge of social robots is its critical inclination, which avoids the extremes polarizing current debate about these new machines—the sterile alternative between *techno-phobia* and *techno-enthusiasm.*

The complex transformations triggered by social robots—changes in our relationships with technology, with our everyday environments, with other social agents, and, ultimately, with ourselves and our identity—are perceived by cultural robotics as an opportunity.

While a negative perspective on the impacts of deploying robotic social partners tends to prevail in the scientific discourses developed by the human sciences—from philosophy to ethics, from anthropology to sociology—the research lines engaged in cross-fertilizing into cultural robotics, although centered on humans and their cultural specificities, stand for the possibility that we can make social robots means of a positive metamorphosis. Within cultural robotics, human sciences show themselves ready to engage critically and proactively in support of the ambition originally associated with the notion of social robots: building artificial agents able to play for us the role of *social connectors*, and thus facilitate, stimulate, and enhance relationships among us. In other words: making social robot tools that can help us get on the path of positive self-development, directed toward the growth of our self-knowledge, and our moral and cultural growth.

The chapters collected in this volume, in my view, converge in indicating comprehensive and virtuous ways in which we can move in this direction. Together, they represent a "creative foresight" of how we can generatively address the challenge of social robots: Creating an alliance among sciences directed to establishing an alliance with our social machines—a twofold alliance in support of our own generative (co-) evolution.

The emerging field of cultural robotics, as delineated in this volume, seems to express this view, and to implicitly promote, as a frame for our work on building synergic relationships between disciplines, between humans and machines, and among humans, what Francisco Varela, while exploring the profound biological roots of the human mind, brought forward as a "participatory epistemology" for a sustainable future: "**This world is our dance together—not my projection, nor yours; its something we do together, and what we do changes what the world is like.**"

Luisa Damiano
Logic and Philosophy of Science
IULM University
Milan, Italy
luisa.damiano@iulm.it

Acknowledgments

Thank you to the UNSW Creative Robotics Lab and the National Facility for Human-Robot Interaction Research for supporting the editorial team in the development of this publication and to Mari Velonaki for her vision and leadership of the lab. Thank you to David Rye, Jorge Forseck and Caleb Kelly for their advice and support.

Thank you to UNSW School of Art and Design for committing grant funding to support the final stages of this publication through the Publication Support Scheme funded by the Faculty of ADA Research Office. Our deepest thanks to Peter Blamey for his alchemic work on this book.

Thank you to Helen Desmond at Springer for her gracious understanding and guidance throughout this editorial process, particularly during the pandemic and all the difficulties that presented.

We extend our thanks to the international cultural robotics community, who have continued the discourse that began in 2015 and have once again entrusted their research to us in this publication.

Contents

1 **Emergent Cultural Ecologies in Social Robotics** 1
Jeffrey T. K. V. Koh and Belinda J. Dunstan

Part I Human Futures

2 **Social Robot Morphology: Cultural Histories of Robot Design** 13
Belinda J. Dunstan and Guy Hoffman

3 **The Robot Soundscape** 35
Frederic Anthony Robinson, Oliver Bown, and Mari Velonaki

4 **Reimagining Robots** 67
Ingrid Bachmann

5 **Data, Site, Materials: Robotics and Digital Fabrication Within Installation Art** 75
Vaughan Wozniak-O'Connor

6 **The Future of Non-fungible Tokens: PNFTs as a Medium for Programmatic Art Enabling a Fully Realized AI-Driven Art Ecosystem** 89
Jeffrey T. K. V. Koh

Part II Assistive Technology

7 **From Assistive to Adaptive: Can We Bring a Strengths-Based Approach to Designing Disability Technology?** 101
Scott Andrew Brown

8 **The Intersection of Social Impact, Technology and Design: A Catalyst for Cultural Change** 109
Melanie Tran

9 Culture in Social Robots for Education ... 127
Barbara Bruno, Aida Amirova, Anara Sandygulova, Birgit Lugrin, and Wafa Johal

10 Towards an Autistic User Experience (aUX) Design for Assistive Technologies ... 147
Sebastian Trew and Scott Andrew Brown

11 Drone Swarms to Support Search and Rescue Operations: Opportunities and Challenges ... 163
Maria-Theresa Oanh Hoang, Kasper Andreas Rømer Grøntved, Niels van Berkel, Mikael B Skov, Anders Lyhne Christensen, and Timothy Merritt

Part III Creative Platforms and Their Communities

12 Culture and Technology: Curating New Media in Collaborative Ways ... 179
Deborah Turnbull Tillman

13 Soft Robotics Workshops: Supporting Experiential Learning About Design, Movement, and Sustainability ... 189
Anca-Simona Horvath, Elizabeth Jochum, Markus Löchtefeld, Karina Vissonova, and Timothy Merritt

14 Sonic Robotics: Musical Genres as Platforms for Understanding Robotic Performance as Cultural Events ... 219
Wade Marynowsky, Julian Knowles, Oliver Bown, and Sam Ferguson

15 Rouge and Robot: The Disruptive Feminine ... 237
Lian Loke and Dagmar Reinhardt

16 On Display: Robots as Culture ... 257
Deborah Turnbull Tillman and Mari Velonaki

Index ... 275

Contributors

Aida Amirova Department of Robotics and Mechatronics, School of Engineering and Digital Sciences, Nazarbayev University, Nur-Sultan, Kazakhstan

Ingrid Bachmann Concordia University, Montreal, QC, Canada

Oliver Bown Interactive Media Lab, University of New South Wales, Sydney, NSM, Australia

Scott Andrew Brown Creative Robotics Lab, University of New South Wales, Sydney, Australia

Barbara Bruno École Polytechnique Fédérale de Lausanne (EPFL), Lausanne, Switzerland

Anders Lyhne Christensen University of Southern Denmark, Odense, Denmark

Belinda J. Dunstan Creative Robotics Lab, University of New South Wales, Sydney, NSW, Australia

Sam Ferguson Creativity and Cognition Studios, School of Computer Science, Faculty of Engineering and IT, University of Technology Sydney, Sydney, Australia

Kasper Andreas Rømer Grøntved University of Southern Denmark, Odense, Denmark

Maria-Theresa Oanh Hoang Aalborg University, Aalborg, Denmark

Guy Hoffman Cornell University, Ithaca, NY, USA

Anca-Simona Horvath Department of Communication and Psychology, Aalborg University, Aalborg, Denmark

Elizabeth Jochum Department of Communication and Psychology, Aalborg University, Aalborg, Denmark

Wafa Johal School of Computer Science and Information Systems, Faculty of Engineering and Information Technology, University of Melbourne, Melbourne, VIC, Australia

Julian Knowles Macquarie University, Sydney, Australia

Jeffrey T. K. V. Koh Singapore Institute of Technology, Singapore, Singapore

Lian Loke The University of Sydney, Sydney, Australia

Birgit Lugrin Human-Computer Interaction, Institute of Computer Science, Julius-Maximilians-Universität Würzburg, Würzburg, Germany

Markus Löchtefeld Department of Architecture, Design and Media Technology, Aalborg University, Aalborg, Denmark

Wade Marynowsky University of Technology Sydney, Sydney, Australia

Timothy Merritt Department of Computer Science, Aalborg University, Aalborg, Denmark

Dagmar Reinhardt The University of Sydney, Sydney, Australia

Frederic Anthony Robinson Creative Robotics Lab, University of New South Wales, Sydney, NSM, Australia

Anara Sandygulova Department of Robotics and Mechatronics, School of Engineering and Digital Sciences, Nazarbayev University, Nur-Sultan, Kazakhstan

Mikael B Skov Aalborg University, Aalborg, Denmark

Deborah Turnbull Tillman Creative Robotics Lab, University of New South Wales, Sydney, Australia

Melanie Tran Sydney, NSW, Australia

Sebastian Trew Institute of Child Protection Studies, Australian Catholic University, Canberra, Australia

Niels van Berkel Aalborg University, Aalborg, Denmark

Mari Velonaki Creative Robotics Lab, University of New South Wales, Sydney, NSM, Australia

Karina Vissonova Institute for Advanced Design Studies, non-profit, Budapest, Hungary

Vaughan Wozniak-O'Connor ARC Centre of Excellence for Decision-Making and Society, University of New South Wales, Sydney, NSW, Australia

Chapter 1
Emergent Cultural Ecologies in Social Robotics

Jeffrey T. K. V. Koh and Belinda J. Dunstan

Abstract This chapter introduces the edited collection *Cultural Robotics: Social Robots and their Emergent Cultural Ecologies*. We present and describe the three themes that we see as contemporarily emergent within cultural robotics research: human futures, assistive technology, and creative platforms and their communities. With these themes demarcating the publication, we canvas the contributions to each section. We offer a new lens for examining the reach of social robotics, that of *cultural ecology*, where consideration for the broader political, economic, and social factors impacted by this field become inseparable to our evaluation of it. We argue for the development of social robotics to be increasingly informed by community-led transdisciplinary research, to be decentralised and democratised, shaped by teams with a diversity of backgrounds, informed by both experts and non-experts, and tested in both traditional and non-traditional platforms.

1.1 Introduction

In 2015, the authors, together with David Silvera-Tawil, held a workshop at IEEE RO-MAN in Kobe, Japan, which called for contributions attending to a relatively new premise: in what ways are social robots participants in, and creators of, culture? We were both pleased and surprised at the breadth and depth of the contributions from around the world, which signified a growing interest in the influence and contributions of robots and artificial intelligence (AI) to culture, as well as the influence of human cultures on the design and applications of social robots. As we now introduce a new collection of research on cultural robotics, we reflect on the origins and development of this field and offer an overview of what we have identified as emerging themes of research in the cultural sphere of social robotics.

J. T. K. V. Koh (✉)
Singapore Institute of Technology, Singapore, Singapore
e-mail: valino.koh@singaporetech.edu.sg

B. J. Dunstan
Creative Robotics Lab, University of New South Wales, Sydney, NSW, Australia
e-mail: belinda.dunstan@unsw.edu.au

© The Author(s), under exclusive license to Springer Nature Switzerland AG 2023

B. J. Dunstan et al. (eds.), *Cultural Robotics: Social Robots and Their Emergent Cultural Ecologies*, Springer Series on Cultural Computing,
https://doi.org/10.1007/978-3-031-28138-9_1

The research presented at the 2015 workshop was brought together and published as a collection of works entitled *Cultural Robotics* (2016, Springer). In the opening chapter of this book, Dunstan et. al. described how robots could not only be maintainers of and participants in human culture but could also have the potential to develop their own culture, which could quite possibly become completely unrecognisable to humans.

The authors of the contributed papers brought a variety of research; some concerning the morphology and development of social robots, while others discussed topics such as human–robot interaction (HRI) collaborative tasks, and methods for how one could evaluate HRI from sociological perspectives. More directly, chapters such as Gemeinboeck et al. discussed the socialisation of non-anthropomorphic robots via harnessing the kinaesthetic awareness of dancers; Chesher described how robots participate in the development of cinema, television, and digital media; and Marynowsky et al. shared case studies on operatic works by robotic systems. Discussions on robotic-supported food experiences were presented via Laursen et al., and Davies and Crosby wrote about the potential advent of robot-generated culture via the musical performances of the all-robot band Compressorhead. Six years later, we attempt to gauge the expanding state of the field of cultural robotics in this new collection.

With the benefit of hindsight, we have reflected on some definitions posited in the original publication, which we feel have expanded since. Although we previously defined social robotics as the foundation to cultural robotics, through our observations of these fields over the last six years, we now view these fields as engaged in a symbiotic relationship, where cultural robotics can be used as a lens to look deeper into the impact of social robotics. While our 2016 publication engaged an exciting breadth of "cultural" robotic participation, much of the content was focused on surface-level signifiers of culture, such as dance, traditional dress, music, and food. In this edition, we seek to delve deeper into what the term "culture" encompasses, across topics such as geopolitical boarders, creative community building, challenges to gender normativity, neurodiverse engagement with technology, sonic communication, personal grooming rituals, human and robotic agency, the impact and authenticity of data tracking, the historical origins and ethical implications of robot morphology typologies, and diversity-led technology design. An extensive and thoughtful discussion on all that the term "culture" encompasses can be found in the chapter presented by Bruno et al., "Culture in Social Robots for Education". The reader will also note a variety of robotic-adjacent technology included within this publication, such as biodata, NFTs, film, and sound, as we seek to understand the broadening technological ecologies that come to impact upon robotics development, and that which has a hand in shaping culture in tandem with social robotics.

1.2 Emergent Thematics

Reviewing the deeper cultural integration and impact of social robotics that has occurred since our first publication, we have summarised the emergent thematics that have provided impetus and given shape to this publication under three key streams: human futures, assistive technology, and creative platforms and their communities.

The three themes of this publication align with the key research streams of the Creative Robotics Lab (University of New South Wales, Australia) but also speak to the deeper, diverse, and generative applications for which social robots are being used.

1.2.1 Human Futures

Human futures encompass the aesthetic and ethical touchpoints between humans and social robots, including the history and future of social robot morphology design, movement planning for affective expression, sensory and sonic interaction with robots, technology ethics, material explorations of embodiment, and robotic performed sentience.

The opening chapter from Belinda J. Dunstan and Guy Hoffman traces the historical origins and cultural influences on the prevailing dominant social robot morphological typologies and issues a call to action for roboticists to engage in the aesthetic design of robots in a more informed and knowing manner. Following this, Frederic Robinson, Oliver Bown, and Mari Velonaki survey sonic robotic communication and the sonification of human actions, questioning how sound can be used to enrich human–robot interactions.

Artists Ingrid Bachmann and Vaughan Wozniak-O'Connor each discuss the use of their artworks to challenge and critique cultural assumptions. Bachmann's robots are messy, furry, "breathing" and without application, questioning the notion of creating machine life that is not necessarily productive. Wozniak-O'Connor's work renders self-tracking data as installation artworks, highlighting the disruption that technology and its shortcomings can have on traditional cultural notions such as the definition of "installation art" as well as the "white cube" of the gallery space. At the centre of the emergent robotic experience, Jeffery T. K. V. Koh discusses the notion of the AI robotic art ecosystem, where art as non-fungible tokens (NFTs) is created, traded, stored, and owned, all by robots.

1.2.2 Assistive Technologies

Assistive technologies acknowledge that technology, including robotics, has largely been designed for the "middle of the middle" and instead looks to design technology for and by marginalised populations. This research stream encapsulates those working in community-led teams, adopting a strengths-based approach to designing assistive technologies for those with disability or neurodiversity. In the realm of social robotics, assistive technologies can also include robots as household and workplace collaborators, co-workers, and assistants, as well as the design of assistive robotic objects.

Scott Andrew Brown opens this section by exploring the capacity of assistive technology to augment and empower the user. He argues for a social model of disability, where a community-led approach to technology design places the user at the centre of the design process. Melanie Tran offers insights into designing user experience (UX) and disability-focused social enterprises that redefine the concept of inclusion, and Sebastian Trew and Scott Andrew Brown offer assistive technologies as an approach for addressing the social and sensory challenges faced by autistic individuals.

Barbara Bruno and colleagues contribute a survey of the literature on social robotics for education, examining its cultural impact with focus on cultural sensitivity and adaptation. They provide guidelines for designing cross-cultural robots and systems that are culturally adaptive. Maria-Theresa Oanh Hoang and colleagues bring insights to the future use of drone swarms to assist in emergency events, with the aim of minimising distress and harm, and highlighting the opportunities of using swarms in search and rescue operations.

1.2.3 Creative Platforms and Their Communities

Creative platforms and their communities look to the creative cross-disciplinary researchers adopting robotics within their practices, those contributing creatively to more traditional robotics research, and the testing of robotics in non-traditional platforms such as museum and gallery spaces.

Deborah Turnbull Tillman brings new media and new methods of collaboration to the forefront of this section in her introductory chapter. Highlighting the disruptive and interdisciplinary nature of the technologies used in contemporary media art, she positions collaborative relationships as an effective facilitator for extending cultural experiences beyond the gallery and into the public sphere.

Within this section, authors Anca-Simona Horvath and colleagues present methods for increased accessibility and a focus on sustainability in robotics through their documented workshops for soft robotics. Their studio-based courses support transdisciplinary teaching and act as a non-traditional entry point to learning robotics. Artist Wade Marynowsky and collaborators describe sonic robotic performances that use known musical genres to position social robots as producers of culture, from an all-robot opera to disco dancing roller skates, questioning notions of robotic agency. Performer and architect Lian Loke and Dagmar Reinhardt present work that integrates a robotic arm with intimate feminine personal-care rituals, questioning traditional boundaries of the subject and object in HRI, introducing notions of "collaborative care", and providing keen critical insight into the use of robots for cultural practices.

Concluding this section, Deborah Turnbull Tillman and Mari Velonaki explore the display of robots as cultural objects within the context of museum and galleries settings, where these settings act to both reinforce and de-silo historical taxonomies and constructs, particularly those of research disciplines. These notions are illustrated through case studies of contemporary exhibitions.

1.2.4 Platforming with Purpose

Reflecting our desire to platform non-traditional robotics research as well as essential work being conducted and communicated by neurodivergent people and people with a disability, some of the chapters within this publication may be presented in a way that is outside a traditional academic context. We wish to share the research of all our contributors in a way that allows their voices to be heard and thoughts to remain authentically structured, without being constrained by the academic tradition. We invite you to approach these chapters with curiosity; we all have much to learn from one another.

1.3 Cultural Ecologies

Social robotics has grown to include a wide range of applications, with deepening cultural implications. Beyond defining what delineates the current state of the art in cultural robotics, we wish to describe an approach to research that we envisage as beneficial to the future of robotics. Where architects can no longer afford to simply think of timber as a "sustainable material" without asking the deeper questions concerning its origin, land clearing, plantations, personnel, transport, durability, and waste disposal, social roboticists must inquire more deeply into the social, political, and cultural reaches of social robotics. Within this publication, we entitle this process *cultural ecology*.

The origins of this term stem from the term "political ecology", which has been defined by Watts (2015), and earlier by Robbins (2019), as:

> the study of the relationships between political, economic and social factors with environmental issues and changes. Political ecology differs from apolitical ecological studies by politicizing environmental issues and phenomena.

In the context of social robotics, the "environmental issues and changes" to be considered include those of the social and cultural environment. The definition of political ecology has more recently been expanded upon by Morton (2016) with broader application. In *Dark Ecology*, Morton acknowledges the complexity, nonlinearity, and interconnectedness of the socio-ecological system and calls for adaptations to be made.

> Through ecological awareness, differences between R2-D2-like beings and humans become far less pronounced; everything gains a spectral quality (p. 138).

Morton adds that,

> Coexisting with these non-humans is ecological thought, art, ethics and politics (p. 159).

In the following section, we suggest potential "adaptations" for approaching robotics research with an "ecological" awareness, to promote a more diverse and ethically engaged approach to the design and applications of social robots.

1.4 Case Studies

Technology is deeply rooted in political ecology. As current technologies develop and new technologies emerge, governments and other types of organisations seek out new ways to engage with their citizens, patrons, customers, and users. With technology, some people are included in the discourse of their society, while others become estranged for a variety of reasons–not for lack of want, but simply via a lack of access. While social robotics looks to address the user at the centre as the primary driver for the development of robotic applications, we can look to adjacent technologies that may indicate and inform future developments in the field of cultural robotics.

Rooted in maker culture, prototyping platforms such as Arduino and Raspberry Pi have allowed for a kind of democratisation across the Internet of things (IoT). This enables many new people a means to experiment with ubiquitous and pervasive technologies. Ospanova et al. (2021) discussed how IoT devices such as the Raspberry Pi have allowed students and educators a means to actively participate in prototyping, increasing engagement and positive student perception in regard to technology. This phenomenon has extended accessibility to more people to participate in the development of human–computer interaction, including social robotics. Practitioners of the fine arts, for example, are now able to develop robots and AI for cultural applications. This was the original impetus for the defining of cultural robotics (Koh et al. 2016).

Prototyping platforms and their communities, such as the one that has coalesced around the Raspberry Pi, have developed into an ecosystem based on the principles of open source and accessibility, allowing for a variety of communities a means to engage in engineering and computer science prototyping. From education to wildlife conservation, the affordances of open hardware and software have made this possible.

Regarding cultural robotics, a proprietary approach to the development and maintenance of culture would not only be self-defeating, but also impossible to govern. We feel it important to highlight that for the development of cultural robotics to flourish and be valid, an open, non-proprietary, and non-confidential approach is required. The need for these deeper issues of democratisation, access, and designing "for and with" the user is explored in Part II, "Assistive Technologies".

In this book's chapter, titled "The Future of Non-Fungible Tokens: pNFTs as a Medium for Programmatic Art Enabling a Fully Realized AI-Driven Art Ecosystem", Koh discusses the how artificial intelligence geared towards the creation of artworks can have their artworks exchanged and collected by fully autonomous artificial intelligence collectors, operating within a fully automated digital marketplace. It questions the notion of not only art making (cultural artefact development), but also the notion of what it is to buy, sell, and collect artworks in an age of artificial intelligence.

Further to non-proprietary approaches to technology, distributed ledger technologies (DLTs) such as blockchain allow for decentralised communities to form around open standards and transparency. While there has been much criticism on the environmental impacts of technologies such as Bitcoin, advances in this space have quickly moved to address some of these concerns. Incumbents such as Ethereum are quickly moving towards proof-of-stake algorithms versus energy inefficient models such as Bitcoin's proof-of-work, and others such as Cardano have fully adopted low-energy models such as proof-of-stake from the onset, utilising exponentially less energy than previous generations of DLTs. These decisions have not been driven by a centralised institution but by fully distributed and autonomous organisations. For cultural robotics to gain a significant foothold in the zeitgeist of social robotics, it must adopt decentralised approaches to the encoding of cultural norms in order to best serve the communities these robots are being made for.

As we move towards digitising culture via cultural robotics, a transformation in the economy of cultural goods will occur. Cultural robotics has much to adopt from the technologies mentioned above, not in terms of their techniques but in the way that their communities and design principles are formed, to enable a rich cultural robotics ecosystem to emerge.

1.5 Conclusion

This introduction summarises the three themes that we see as contemporarily emergent within cultural robotics research: human futures, assistive technology, and creative platforms and their communities. We offer a new lens for examining the reach of social robotics, that of a cultural ecology, with consideration for the political, economic, and social factors that impact the development of the field.

Within this introduction and the chapters supported in this publication, we argue for the development of social robotics to be increasingly informed by community-led transdisciplinary research, to be decentralised and democratised, shaped by teams with a diversity of backgrounds, informed by both experts and non-experts, and tested in both traditional and non-traditional platforms. In this way, we posit the field of cultural robotics as an ecological approach to encompassing the widest possible spectrum of human experience in the development of social robotics.

We are honoured by the deep cultural and disciplinary diversity of the authors who have contributed their research to this publication. The curiosity and critical examination evident in their work truly offers a cultural ecology of the deeper implications of social robotics in the present day, as well as speculating on the near future. The chapters herein incidentally approach common themes within social robotics from various perspectives, often challenging or compounding the positions of the others. We welcome this robust discourse as being vital to the future development of social robots. While in no way exhaustive of the reach of robotics, this collection cements the role of social robots as independent contributors to and producers of a vast array of culture, worthy of ongoing critical examination.

List of Terms

Social Robotics

A social robot is an autonomous robot that interacts and communicates with humans or other autonomous physical agents by following social behaviours and rules attached to its role. Like other robots, a social robot is physically embodied (avatars or on-screen synthetic social characters are not embodied and thus distinct). (Henschel, A., Laban, G., Cross, E. S. [2021]. "What Makes a Robot Social? A Review of Social Robots from Science Fiction to a Home or Hospital Near You". Current Robotics Reports. Springer Nature: 9–19. https://doi.org/10.1007/s43154-020-00035-0.)

Political Ecology

Political ecology is the study of the relationships between political, economic, and social factors with environmental issues and changes. Political ecology differs from apolitical ecological studies by politicising environmental issues and phenomena (Robbins 2019).

Distributed Ledger Technology (DLT)

A distributed ledger (also called a shared ledger or distributed ledger technology or DLT) is the consensus of replicated, shared, and synchronised digital data that is geographically spread (distributed) across many sites, countries, or institutions (Distributed Ledger Technology: beyond block chain).

Blockchain

A blockchain is a type of distributed ledger technology (DLT) that consists of growing list of records, called blocks, that are securely linked together using cryptography (Narayanan, A., Bonneau, J., Felten, E., Miller, A., Goldfeder, S. [2016]. Bitcoin and cryptocurrency technologies: a comprehensive introduction. Princeton: Princeton University Press. ISBN 978-0-691-17169-2).

Non-Fungible Token (NFT)
A non-fungible token (NFT) is a unique digital identifier that cannot be copied, substituted, or subdivided, that is recorded in a blockchain, and that is used to certify authenticity and ownership (https://www.merriam-webster.com/dictionary/NFT).

Internet of Things
The Internet of things (IoT) describes physical objects (or groups of such objects) with sensors, processing ability, software, and other technologies that connect and exchange data with other devices and systems over the Internet or other communications network ("Internet of Things Global Standards Initiative". ITU. Retrieved 26 June 2015).

User Experience (UX)
User experience (UX) is the experience that products create for the people who use them in the real world. It is about how a product works on the outside, when people come into contact with it (Garrett, J. J., 2011. The elements of user experience: user-centred design for the web and beyond (voices that matter). *New riders, 2.*

References

Koh JTKV, Dunstan BJ, Silvera-Tawil D, Velonaki M (eds) (2016) Cultural robotics. CR 2015. Lecture notes in computer science, vol 9549. Springer, Cham
Morton T (2016) Dark ecology. Columbia University Press, New York
Ospanova A, Tuleuov B, Zharkimbekova A, Kussepova L, Mangmuryn M (2021) Mobile devices and portative classroom based on Raspberry Pi computers. In: 12th National conference with international participation (ELECTRONICA). IEEE, pp 1–4
Robbins P (2019) Political ecology: a critical introduction. Wiley, New York
Watts MJ (2015) Now and then: the origins of political ecology and the rebirth of adaptation as a form of thought. In: Perreault T, Bridge G, McCarthy J (eds) The Routledge handbook of political ecology. Routledge, Abingdon, pp 41–72

Part I
Human Futures

Chapter 2
Social Robot Morphology: Cultural Histories of Robot Design

Belinda J. Dunstan and Guy Hoffman

Abstract Social robot morphologies are not conceived in a void but build on cultural trajectories of artifact design that precede them. We suggest three design tropes that are predominant in many robots morphological design choices: the human replica, the futuristic machine, and the cute companion. We discuss the first two of these tropes in the context of their historical origins, and the third from a contemporary lens. For all three, we present cultural implications of the aesthetic typologies to emphasize the critical importance of conscious engagement with these contexts when designing social robots.

2.1 Introduction

The physical appearance of a robot does not suddenly materialize from the imagination of its designer but exists within a cultural history of artifact design, drawing on this history and its traditions. Yet many designers of social robots do not recognize or acknowledge their design's position as part of a lineage of cultural traditions, instead citing interaction requirements, user preferences, or pure inspiration as the basis for their design choices (see: the "motivation" column in Dunstan (2019), *Twenty-Five Robots in Twenty-Five Years*).

This lack of acknowledgment can limit or complicate the reception and treatment of robots and the subsequent success in interaction with social robots. In her consideration of *Robots in Society, Society in Robots*, Šabanović (2010) identified that the design of social robots had been primarily developed in a unidirectional, technologically determinist manner, where technology is developed in a linear fashion of continual progress and society fulfills a passive role by accepting and adapting to the results of technical innovation. Due to the highly social contexts for which

B. J. Dunstan (✉)
Creative Robotics Lab, UNSW, Sydney, NSW, Australia
e-mail: belinda.dunstan@unsw.edu.au

G. Hoffman
Cornell University, Ithaca, NY, USA
e-mail: hoffman@cornell.edu

© The Author(s), under exclusive license to Springer Nature Switzerland AG 2023

B. J. Dunstan et al. (eds.), *Cultural Robotics: Social Robots and Their Emergent Cultural Ecologies*, Springer Series on Cultural Computing,
https://doi.org/10.1007/978-3-031-28138-9_2

social robots are designed, Šabanović called for a move away from the technocentric forward-march of social robot development and instead proposed a "bidirectional shaping" between society and robots that

> paves the way for approaching design in a value-centred manner, consciously incorporating social and cultural meaning-making into design.

Šabanović proposed that it was not sufficient to consider the social and cultural impacts of a robot in post-production user testing, but rather that

> the meaning of various technological choices ... should be questioned throughout the process of technology design (2010, p. 445).

Šabanović also notes that the integration of robots into broader society should incorporate the study of both the social and technical aspects of the technology. While there is a significant body of contemporary critical theory concerning aesthetic trends in technological product design and the emergence and acceptance of social robots, the identification and analysis of aesthetic trends and their origins, specifically in social robot morphology, is necessary to foster a more conscious incorporation of social values and cultural meaning into these artifacts that are being designed to share social spaces with humans.

In this chapter, we aim to map several cultural trajectories of artifact design that lead up to contemporary social robot morphologies. We suggest three design tropes that are present in many robots morphological design choices. The first is that of the human replica, a wish to artificially recreate with mechanical means the naturalistic structure of a human. The second is that of the futuristic machine, a neutrally designed, streamlined device that is often represented through clean lines and neutral color palettes, suggesting a better-than-nature efficiency. The third is that of the cute toy or companion, emphasizing child-like or pet-like features and suggesting a certain naïvité, helplessness, vulnerability, and loyalty. We discuss the first two of these tropes in the context of their historical origins, and the third from a contemporary lens. For all three, we present the cultural implications of these aesthetic typologies for robots.

Social robot morphology has been surveyed several times in the existing HRI literature. However, these surveys are often in the form of categorical classifications of robot forms with limited cultural or historical analysis (Hegel et al. 2009; Diana and Thomaz 2011; Mahdi et al. 2022). We start to fill this gap by tracing some historical and cultural origins of social robot design, which reveal underlying notions about the function of technology that echo in contemporary applications and contexts.

In Sect. 2.2, we trace the origins of the human replica from antiquity through the design of clockwork automata to the "steam men" of the Victorian age. In Sect. 2.3, we discuss the transition of design from the naturally inspired to the machine-centric in the twentieth century. We particularly emphasize the evolution of the streamlined aesthetic through post-Industrial Revolution Italian Futurism, suggesting an ideology of speed, efficiency, and hygiene.

We then move to a more contemporary lens. Section 2.4 presents the development of a cute aesthetic for social robots, exemplified early on in robots such as Kismet,

Leo, and iCat, and persistent in the quarter-century since in contemporary research robots such as Blossom and commercial robots such as Pepper, Astro, and Otto. We detail the development and persistence of this typology in social robot design, along with its ethical implications, through the lens of Sianne Ngai's aesthetic theory.

To create robots for effective and affective interaction, social roboticists must design morphologies with an awareness of the cultural origins and social implications of their chosen aesthetics, just as designers in any other discipline would.

2.2 The Human Replica

We start our historical analysis with perhaps the most obvious and uninspired robot design: an attempted replica of the human form. This design trope can be traced back to prehistoric human figurines and mechanically actuated puppets in antiquity but began to take on a more decidedly robot-like form in the sixteenth century through the development of clockwork automata.

The urge to recreate a semblance of intelligent life via artificial means was never culturally neutral and has often been consciously related to questions of control, be it control over nature, over death, or over other living creatures. For example, sophisticated clockwork mechanisms were built as scientific tools to give humans control over the seasons, seas, and crops, and simultaneously inspired attempts to recreate living creatures, including humans, via mechanical means. The relationship between anthropomorphic machines and control is also evident in the robot-like designs of "steam men" in the nineteenth century, where the imagining of such steam men was closely linked to racism. We argue that the cultural associations between human form robots and control over other humans linger today, as does the notion that the design of a human-like machine, can help its designer both overcome the limitations of nature and reveal important truths about the mystery of humanness.

Humans have been creating replicas of the human form since prehistoric times, with human-like figurines dating back 35,000–40,000 years. These early sculptures indicate the long-standing interest of humans in creating artificial versions of themselves. Sculptures led to articulated and jointed masks and dolls, for example, those found in Egypt as early as the 2nd millennium BC. Some of these figures are described as being augmented with hidden voice boxes for dramatic effect. There is also written evidence of Roman wax figures that were actuated with complex mechanisms, including an attempt by Mark Antony to "revive" the dead Julius Caesar to shock a crowd of observers. Derek J. de Solla Price (1964) presents a clear and concise history of such pre-modern automata, and the reader can find an extensive presentation of ancient-to-modern automata in Chapuis (1958).

Along with the creation of these figurative representations, there is also long-documented contemplation about the possible aliveness and humanness of artificially created human figures. Two well-known examples are the Greek myth of Pygmalion from the eighth century, and the Golem, an animated anthropomorphic creature of Jewish folklore, dating back at least to the Middle Ages. In many of the treatments of possibly-alive artificial creatures, the theme of control arises, be it over nature, death, gods, or other humans. These mythologies also usually come with moral warnings related to the hubris of control and the inevitable disaster that it brings. These questions and warnings remain to this day in the context of robotics.

The link between man-made mechanisms, artificial creatures, and control over nature clarifies during the early modern era, starting in the late fifteenth century CE and continuing throughout the sixteenth and seventeenth centuries. This period is marked by three simultaneous and interleaved developments: first, sophisticated metalwork leading to the ability to build complex spring-driven clockworks; second, the seeds of a pre-Enlightenment mechanistic and secular scientific thinking moving away from a sacred idea of humanity; and third, the expansion of European colonial empires along with a culture of exploitation.

At the same time, as European monarchies expanded their control over distant regions, including exercising control over the humans who lived in these colonies, metalwork improved to enable, among others, the construction of sophisticated measurement machines, used for long-distance ocean navigation and timekeeping. Some examples are depicted in Fig. 2.1. The increased precision of these machines—precursors of automata, then calculators, and eventually computers and robots—must have given their owners a heightened sense of control over complex natural processes. The leap from mastering the stars and seasons to mastering other living creatures and viewing them as nothing but sophisticated machines was short. De Solla Price cites St. Thomas Aquinas as stating that,

> [...] animals show regular and orderly behavior and must therefore be regarded as machines. (de Solla Price 1964)

De Solla Price adds that

> [s]urely, such a near-Cartesian concept could only become possible and convincing when the art of automaton-making had reached the point where it was felt that all orderly movement could be reproduced, in principle at least, by a sufficiently complex machine.

Fig. 2.1 Astronomical clock, circa 1568; mirror clock, ca. 1565–1570; clock-watch with sundial, ca. 1605–10

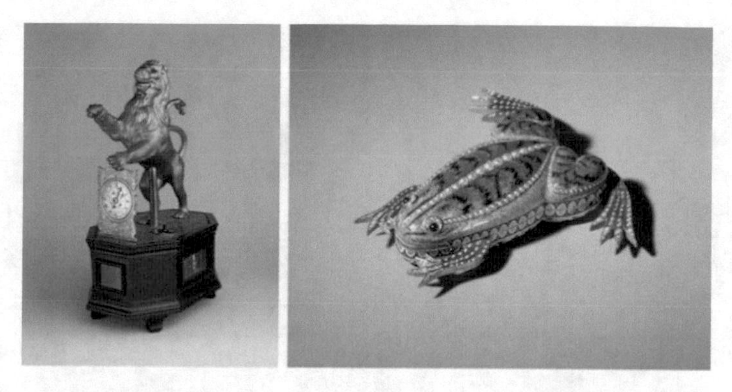

Fig. 2.2 Automaton clock in the form of a lion (1620–35), and automaton in the form of a frog (c.1820)

By the end of the eighteenth century, the courts of Europe were awash with automata, celestial prediction machines, and calculators, all made of precious metals and stones often harvested from faraway slave colonies. True to the idea that even animals and perhaps humans are nothing but complex machines, many of these clockwork devices were in the shape of natural creatures (see Fig. 2.2) and attempted to recreate the behaviors of animals and humans. This rise of these clockwork machines was wonderfully documented in the exhibition "Making Marvels" at the Metropolitan Museum of Art in 2019–2020, and in the accompanying catalog (Koeppe 2019).

Fast-forward into the Victorian era, as steam engines become the height of technology, but the dream of artificial humans is still alive and well. In the nineteenth century, spring-loaded clockwork animals and human figures are supplemented by so-called steam men. These mostly imaginary constructions appear both in Europe and across the Atlantic in the USA. One example is Edward S. Ellis's early science fiction dime novel *The Steam Man of the Prairies* (1868), which triggered a range of copycat and sequel novels published throughout the second half of the nineteenth century. In it, an anthropomorphic steam machine, its face painted black, carries white frontier-people across the American prairie through a variety of nineteenth-century adventures (see Fig. 2.3, bottom). Another notable example is Zodoc Dederick's 1868 US patent application for a steam carriage shaped like a human pulling a cart, thought to be the real-life inspiration for the above-mentioned series of novels. Other steam men of the nineteenth century (Fig. 2.3, top), which quite clearly resembling modern-day robots, were imagined for a variety of labor tasks, from hauling carriages of cargo and passengers to agricultural chores.

It is not difficult to see the imagination of slave owners in both the fictional and the engineering versions of American steam men. The machines' human appearance was often designed with stereotypical features of Black men, and the narratives of the steam man novels are laden with racist tropes. The tasks given to these steam men were, at the time, associated with labor of enslaved people, including agriculture or—in the dime novel series—capturing Native Americans. If the automata of the sixteenth to eighteenth century were set in a cultural moment of the scientific conquest

Fig. 2.3 Fictional steam men from the late nineteenth century: steam plowmen depicted in a British journal (top), the Steam Man of the Prairies (bottom left), detail from the 1868 android-drawn Steam Carriage patent drawing by Dederick and Grass (bottom right)

of nature and its inhabitants, the robots of the nineteenth century—especially in the United States—were set in the recent historical context of the normalization of owning other humans for labor. Evans (2018) reads these inventions as being distinctly shaped by the historical context following the emancipation of enslaved people:

> Coming on the heels of the Civil War, the steam man stories, like Dederick's patent, drew on extant black caricatures to explicitly racialize their central invention, vividly illustrating the afterlife of slavery at the birth of America's machine culture.

According to Evans, these design speculations mark

> a moment when, in the wake of emancipation and the early promises of Reconstruction, the former economic base built on enslaved labor slowly transformed into what became nominally free forms of wage labor, supplemented by an increasing adoption of mechanical labor… [These] cultural artifacts index a shift happening across the country as old forms of mastery, structured by the institution of chattel slavery, were co-opted by and transformed into new narratives of mastery over machines.

The shift from dependence on wage labor to mastery over machines is still a central theme in the justification of robotics today, and the aesthetic trope of a human replica robot is the continuation of the artificial human designs of past centuries.

2.2.1 Implications

A continuous cultural thread runs from articulated figurines, through mechanically reincarnated emperors, semi-alive Golems and Pygmalions, to clockwork animals and humans, steam servants, and eventually today's humanoid robots. It is a thread of a mechanistic attempt to understand, capture, and control nature and its inhabitants. No matter what the current high technology was, humans made the leap from its precision to its ability to model nature and its creatures and, subsequently, to control them. Especially if a technology—be it a clock or a machine learning model—can accurately predict events, such as the movements of stars, the seasons, or animal behavior, engineers were tempted to use it to recreate and control natural and social phenomena. This recreation was often accompanied by a sense of deep understanding of the natural phenomenon simply due to the creation of a superficial replica.

The desire and sense of ability to control nature can lead to a similar desire to control humans. The nineteenth-century comfort with "using humans" of a perceived lower rung set the groundwork for a fantasy of creating artificial lower-rung humans, to be placed in jobs roboticists now call *dangerous, dull, and dirty*. Today's human-like robots suggest a similar exploitative connotation. They are being seen as less-than human but still imagined as being capable of replacing humans in a variety of tasks. This tension between the sense of approaching humanness but being placed in roles that humans shy away from, challenges contemporary social mores and may cause some of the unease felt in relation to robots in human society.

In addition, the construction of human-like robots and their suggested intelligence also uncovers a hubris, running millennia-long, that if some remaining engineering issues could just be resolved (and they always will be, "soon"), humans would not only get better tools, but also a better understanding of themselves. It is not hard to see a foreshadowing of the current technological hubris of being able to represent "all orderly movement" of people, both online and offline, by "a sufficiently complex" machine learning model, to paraphrase de Solla Price.

Beyond the well-known "uncanny valley" of robot design, we suggest that the cultural heritage delineated in this chapter explains some of the issues with the humanoid form, just as many of the predecessors of today's humanoid robots were viewed as dangers, moral warnings, and uncanny cohabitants.

2.3 The Futuristic Machine

Design in the twentieth century moved away from the cultural and political values of the past and with it discarded the aesthetic trappings of prior generations. Minimalist, abstract, and futurist ideas emerged around the turn of the century, took root in product design in the period between the World Wars, and became ubiquitous in consumer homes in the mid-century and beyond. This history of what a “futuristic machine” looks like is the second design tradition that we trace to contemporary robot design.

Several principles drove these “aesthetics of the future,” the primary of which was a desire to escape from the ornamental, naturalistic, rich, and decadent images and ideologies of the past. Instead, abstraction, minimalism, and streamlining were considered to be the height of design in the early to mid-twentieth century. This attempt to discard the history of detailed arts and crafts from the sixteenth to nineteenth centuries was already apparent in the minimalist works of the Vienna Workshop around the turn of the century. But the drive toward a different future was highly accelerated after the collective trauma of World War I, ushering in a new era in politics, be it fascism or socialism, and with it, new aesthetic principles.

Politics was not the only dramatic driver of change in design. The twentieth century also became an era of mass manufacturing, which replaced handcraft and one-off designs. New processes promised access to technology for people beyond the aristocracy, and the aesthetic that accompanied this promise spoke to new materials, methods, and production speed. The design language that grew out of these technological novelties was one of simple geometric shapes, smooth surfaces, and an increased use of stamped metals and molded plastics.

The aesthetic of mass manufacturing stood in dialogue with contemporaneous currents in the visual arts, which also moved away from realism toward clean lines, abstraction, curves and non-representation. These qualities that came to represent the “futuristic” aesthetic in the early 1900s can be seen in the sculptures of the Italian Futurist artists, who mimicked the form and finishes of mass manufacturing technologies, including broad smooth surfaces and basic geometric shapes that appear to meet at a seamed line or fold. The Futurists also emphasized dynamism, the difficult-to-capture but omnipresent movement that was absent from the static representational art of prior periods.

Futurist aesthetics bled into the minimalism of European avant-garde art and the mass-produced aesthetic of design schools like the German Bauhaus, and eventually inspired a new movement in US American product design, known today as “Streamline Moderne” (Kowalik 2017). As its name implies, this style is inspired by the aerodynamic lines of fast-moving vehicles, from automobiles to aircraft. The adoption of streamlined designs into other realms, most notably household appliances, came to symbolize the intangible qualities of the machine age: speed, efficiency, predictability, reliability, and strength. These qualities remain associated with the streamlined designs of robots, even when these robots do not need to worry about “gliding efficiently through space”—to paraphrase Marshall (2012)—any more than the streamlined toasters of the 1950s.

The clean aesthetics also aligned with the twentieth-century political ideal of equality. Instead of the mechanical replica of a human servant, as described in the previous section, the machines of the future were neutral, modern ways to live a life of affluence without exploiting others. This promise carries over to today's "intelligent machines," from smart speakers through to mobile devices and social robots: clean lines represent simplicity and efficiency, while masking the still-existing human toll that enables their existence.

2.3.1 Futurism, Dynamism, and the Simultaneity of Human and Machine

The beginning of the twentieth century was marked by an acceleration across many aspects of European life, along with an erasure of boundaries due to the speed of new modes of transportation and communication. This acceleration led to an artistic focus on the future, perhaps brought on by the sense that technology was accelerating faster than traditional styles could capture. In 1909, Filippo Tommaso Marinetti published the "Futurist Manifesto" (Marinetti 1909), initializing an ideology that would be expressed though a multitude of media, including poetry, film, drawing, painting, and sculpture. Two notable ideas were permeating in Futuristic art: The first was *dynamism*, the wish to capture qualities of movement and change, even in static artforms. The second is that of *simultaneity*, both of times and places, and of the human body and machines. These principles were evident in an obsession with the aerodynamic and functional shapes of vehicles and other machines, which later led to the association of certain machine-like elements with speed, efficiency, and a hopeful future.

Between the years of 1900 and 1916, the erasure of boundaries across many aspects of European life was immense. The reader can find these changes insightfully documented by Marjorie Perloff (2003). Perloff notes how the appearance of advertising posters saw "art" and "life" blurring together in every shop window and towering above the streets. In 1905, the Trans-Siberian Railway was completed, linking western Russia to the Pacific coast, together with the Trans-African Railway and the Trans-Andean Railway, dissolving nation boarders and previously impossible distances. Between 1909 and 1914, the corners of the globe were tugged tightly together as the world witnessed the first successful expeditions to both the North and South Poles, together with the first extended airplane flight (Wilbur Wright 1909), and the first flight across the English Channel (Louis Blériot 1909); humankind now traversed the globe somewhere between the land and the sky. One's thoughts and words could be spoken in one place and heard miles away in another with the increasing availability of the telegraph and telephone. In any of the nearly 200 new cinemas that had opened in Paris by 1913, audiences could sit in one place and be transported entirely to another through the advent of film. The increasing production of automobiles, together with the expansion of steam train lines, saw the body and self-move with speed and dynamism through space, observable as a blur: simultaneity.

Among the slippages and states of simultaneity, Futurists experienced blurred boundaries between the space of the human body and the qualities of the machine age, as is detailed in the "Futurist Manifesto":

> 4. We affirm that the world's magnificence has been enriched by a new beauty: the beauty of speed… a roaring car that seems to ride on grapeshot is more beautiful than the Victory of Samothrace. (Marinetti 1909)

Similarly, Boccioni et al. wrote in the "Technical Manifesto of Futurist Painting" (1914):

> Our bodies penetrate the sofas upon which we sit, and the sofas penetrate our bodies. The motor bus rushes into the houses which it passes, and in their turn the houses throw themselves upon the motor bus and are blended with it.

The speed, efficiency, mobilization, and dynamism of the mechanical world would be expressed not simply as something they experienced, but rather, something they felt they were becoming, and would become, as

> motion and light destroy the materiality of solid bodies (Lista 1986)

In his "Technical Manifesto of Futurist Sculpture," Boccioni states,

> [O]bjects will not be placed alongside the statue, [...] they will be embedded in the muscular lines of a body. We will see, for example, the wheel of a motor projecting from the armpit of a machinist (Apollonio et al. 1973)

and urges,

> Let us open up the figure like a window and enclose within it the [mechanical] environment in which it lives. (Apollonio et al. 1973)

This Futurist philosophy resulted in a dynamic vehicle-inspired aesthetic typified by sleek lines, simplification of surfaces, elongation, a reduction to basic geometric shapes, and a neutral color pallet. Aerodynamic lines were only broken by tubes and other machine components, merging human bodies and mechanisms, and setting the tone for what is viewed as "futuristic" and "advanced" for a century to come. When met with the true power of mass manufacturing, this aesthetic of the future would permeate not only art galleries and catalogs, but the design of everyday objects and, eventually, robots.

2.3.2 *Streamlined Design*

The dynamism and affinity for fast-moving vehicles embodied in Italian Futurism found its way into the consumer market through a design movement known today as Streamline Moderne. Kowalik (2017) provides an excellent survey of this movement, its influences, and the principles it embodies, and much of this section is based on his discussion. While previous twentieth century art forms, including Art Deco, Futurism, and Art Nouveau, all similarly rejected the opulence of the aristocratic art of previous centuries, Streamline offered a

> further abstraction [...], rounding the angular edges, making the style less harsh and more approachable, and shaped by the speed of progress. (Kowalik 2017)

Streamlined design took on the forms of trains and cars, such as their rounded, smooth, and teardrop-shaped lines that enabled them to move quickly through space. Just as Futuristic sculptures and paintings connected vehicular shapes with the speed of a new century, the streamlined design of consumer objects offered connotations of speed, efficiency, and modernity, creating a completely new ideal of elegance that did not rely on intricate craftsmanship and ornamentation. This promise of a bold future embodied by the fast-moving machines of the new century was especially necessary in the years following the trauma of World War I and the subsequent economic depression, both of which could be viewed as the complete failure of the "old ways." The promise of fast-paced modernity soon flowed over beyond transportation, as many household items were designed using the language of aerodynamics, even though this was not a functional requirement of these objects. Contemporaneous industrial designers such as Raymond Loewy progressed directly from designing steam engines and automobiles to designing pencil sharpeners, hair dryers and Dutch ovens (Loewy 1999) (Fig. 2.4).

As a result, the mid-century kitchen was replete with curved, smooth surfaces that were meant to remind the modern housekeeper of the efficient and dynamic life she was part of. Nickles (2002) presents an excellent case study of the mid-century streamlined refrigerator as a functional machine, a mass-appeal technology, and an icon of the clean and hygienic future. She positions the technology as part and parcel of a new sociology of the middle class and an ideology of progress.

Ideologically, streamlining was futuristic and hopeful. New mass manufacturing methods were more than just technologies; they were objects that promised participation in a brave new world, full of hope, and convenience for all. Consequently, parts stamped or extruded from steel, or molded in new materials such as polystyrene and polyvinyl chloride (PVC) (Roser 2016) and joined through fastener-less methods

Fig. 2.4 Streamlined designs by Raymond Loewy: the Pennsylvania Railroad K4-Class 4–6-2 3768 train known as "The Torpedo," 1938, and Pencil Sharpener, 1934

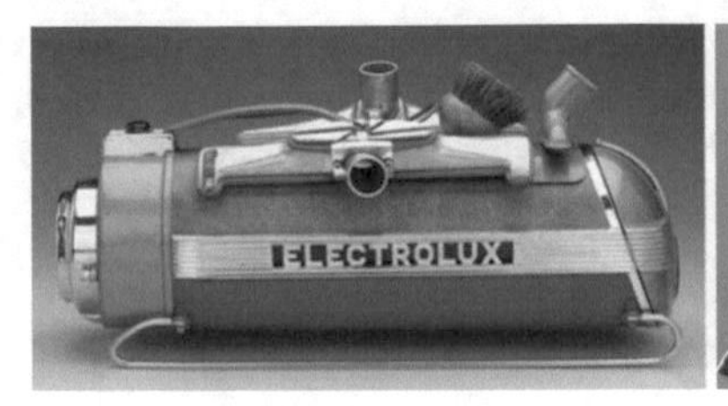

Fig. 2.5 Streamlined designs of household objects, designed in the intra-war and post-war period. From left to right: Electrolux vacuum cleaner (designed in 1937), electric iron (ca. 1930s), electric mixer (1945–1955)

such as gluing, welding, or heat sealing, realized both an aesthetic and an ideology of simplicity and efficiency. These techniques could lead to quickly mass-manufactured objects and therefore could democratize luxury. The smooth lines and lack of visible fasteners also indicated a fetish of simplicity, ease, health, and cleanliness, which was part of the hopeful future (Fig. 2.5).

The streamlined aesthetic claimed an adherence to and celebration of "functionality." According to this idea, morphology is driven by its function and therefore it *is* its aesthetic, along the lines of Sullivan (1896): "form ever follows function." The catalog of the 1934 exhibition "Machine Art" at the Museum of Modern Art devotes a section to "Function." The exhibition was contemporary with many of the streamlined designs and celebrated mass-produced household objects as "art." About the beauty of functional design, the curators write:

> A knowledge of function may be of considerable importance in the visual enjoyment of machine art [...] Whoever understands the dynamics of pitch in propeller blades (No. 41) or the distribution of forces in a ball bearing (No. 50) [...] is likely to find that this knowledge enhances the beauty of the objects.

Ironically, the diffusion of streamlined design from vehicles of transportation to household objects goes against this form-follows-function principle. In fact, the streamlined design of non-moving objects often concealed the mechanics of their operation. As Marshall (2012) notes regarding Loewy's streamlined pencil sharpener (also in Kowalik 2017):

> That pencil sharpener might have *looked* [emphasis in original] smoothly aerodynamic, but when was the last time a pencil sharpener had to glide quickly through space? [...] Here the streamlined shape was a metal casing put over an old-fashioned machine, not only obscuring the turning, spiraled blades of its working interior, but pretending to a 'look' of functionalism at cross purposes with a function!

This role of hiding a device's mechanism also appears in Nickels's analysis of refrigerator design. She especially points out the design decision to move the cooling element from the top of the appliance to a more hidden position. Quoting a complaint communicated inside the General Electric company,

> These people would rather *not* [emphasis in original] have the mechanical part of the device in evidence ... [W]hat interests them in such a product, that is, the machine itself, is the very thing that the woman buying it wants kept out of sight and out of mind. (Nickles 2002)

Finally, it is important to note the social context of luxury without servants that Streamline Moderne represented, namely an attempt to break from the feudal and colonial traditions of servants and, subsequently, of automata and steam men. Nickles speaks of the mid-century refrigerator as a technology aimed at the "servantless housewife." Ensconced in her house, which architect Le Corbusier termed a "machine for living" (Cohen 2014), this servantless housewife was now required to take on many more of the domestic chores, aided by the new mass-produced technology that promises to relieve her from the difficulties associated with this new situation. Nickles also speaks of the transition from marketing to "class" to focusing on "mass," enabled, of course, by the innovative manufacturing techniques of the new century.

That said, amidst all of this technological innovation hid a highly regressive social program. The ultimate goal of the new appliances of the twentieth century was to maintain women in the role of homemakers instead of encouraging them to join the workforce,

> as a way to stabilize not only the economic crisis but also a social one that threatened to undermine traditional gender roles. (Nickles 2002)

2.3.3 Implications

The aesthetic of the futuristic machine that dominated the first half of the twentieth century lives on into the digital age, from "smart" appliances all the way to robot aesthetics. Robot designs employing smooth, simple curves and made from plastics and metal remain abundant, even as mass manufacturing techniques give way to the on-demand "mass customization" of rapid prototyping.

The design trope of the futuristic machine aims to suggest neutrality, and form based purely on function. Robot designers draw on these non-realistic abstractions that originated in the early twentieth century in an attempt to make robots look "futuristic". It also suggests a robot's non-human objectness, along with its objectivity and anonymity.

Tracing the social history of Futurism and streamlined design shows that, again, its cultural baggage is not empty. The design of today's robots is tightly linked with the merging of human and machine, and are, in a sense, a Futuristic sculptural attempt to enclose the mechanical and artificial within an organic, human, or animal-like form. The result is a mechanical object which has been made to look like a natural creature and perform reconstitutions of behaviors and emotion. For these robots, the dissonance between appearance and behavior has often proven to be objectionable and bothersome.

The robots of the twenty-first century *do* present several technological revolutions, in both hardware and software engineering. This technological breakthrough could have sparked a disruptive innovation in aesthetics and design philosophy, as the machines of the early twentieth century did in Italian art. Instead, the robot designs of today fail to be an expression of our time, as their designers are—perhaps unknowingly—reproducing exterior aspects of the stereotypical 100-year-old "futuristic look." This retrospective quality can be attributed to a limited critical and cultural consideration in robot design, and a lack of knowledge of the strong aesthetic links to the original work of the Futurists. Sadly, robot morphologies from the last twenty years might just have likely been designed by Futurist sculptors working in the early twentieth century.

Boccioni asserted:

> It is not simply by reproducing the exterior aspects of life that art becomes the expression of its time. (Apollonio et al. 1973)

But the exterior aspects of Italian Futurist sculpture have endured, and contemporary social robotics design appears to be caught in a 100-year loop. An example can be seen in the visual comparison between artist Henry Gaudier-Brzeska's sculpture *Dog* (1914) and Sony's AIBO robot:

Gaudier-Brzeska's *Dog* could be interpreted as a dachshund, but without a full set of breed-defining features, it may simply be Gaudier-Brzeska's approximation of "dog". Yet, of all the possible dog-like appearances, it shares remarkable aesthetic similarity to Sony's AIBO, designed and produced 85 years later. Both "dogs" sport a conical and almost featureless snout, with only an indication of eyes, and floppy ears that are rendered motionless at the side of the head, widening at the bottom. For both forms, the body is distinctly separate from the head, with one smooth form arching at the lower back to meet the raised hind legs. AIBO's "forearms" are formed in two key sections, distinguished by their points of articulation. Both the snout and the chest are finished at a flat angle. The mimicking of this 100-year-old "futuristic" aesthetic is preventing contemporary social robots from becoming a true and unique aesthetic expression of their time (Fig. 2.6).

In addition, rather than celebrating the merger of human and machine in robot design, many robots participate in the concealment of their reality, a fallacy already critiqued of the streamlined design of consumer goods, as discussed above. While streamlined design emerged in part as a result of the mass manufacturing techniques that promised to democratize luxury and convenience, it was quickly adopted as a veneer aesthetic, hiding the complexities of the machine that does the work. Robots today exploit the same metaphorical connotations to suggest a futuristic fantasy of

Fig. 2.6 Henry Gaudier-Brzeska, *Dog*, 1914 (left) and SONY's, AIBO (ERS-110), 1999 (right)

a "smart device" or "intelligent companion." As with the General Electric refrigerator of the streamlined pencil sharpener, the smooth cladding of the robot hides its complexity and frailness. Similarly, the round edges of a smart speaker hide the messy natural and human resources that work behind the scenes to enable the fantasy of intelligent devices (Crawford 2021).

The mass manufacturing of robots themselves poses additional social and ethical concerns. Since robots are often considered autonomous, the mass manufacturing of an autonomous agent raises the issue of the value of uniqueness, as described in Hoffman (2020). Moreover, the objectness, neutrality, and anonymity suggested by a streamlined robot design also promises the same values that Streamline Moderne promoted: efficiency, speed, hygiene, and low friction. However, this fantasy of clean and low friction interaction with a robot can also come at a cost. As Hoffman suggests, building on Turkle (2007), the cleanliness of the robot interaction fantasy could lead to a world in which

> People will accept the obviously designed social behavior of a machine in place of real relatedness. (Hoffman 2020)

In other words, if robot designers promote the concept of an easy-to-interact companion, we might run the risk of people ultimately finding human interaction too difficult (Turkle 2007). Finally, robot designers must ask themselves what kind of social structure their design represents. The futuristic design of the mid-twentieth century spoke of a brave new world of materials and technologies in a servantless society, but also represented a conservative view of the family and the gender roles within it. What kind of society does a robotic device suggest when it nostalgically feeds on the design tropes of the early and mid-twentieth century?

2.4 The Cute Companion

The third significant trend in robot morphology is that of the cute robotic companion. From the earliest days of social robotics, Breazeal and Foerst (1999) identified that "[p]eople tend to react emotionally to someone or something 'cute'" in a "tender and caring way." They designed the robot Kismet using cute features to encourage users to "treat it like an infant, and to modify their own behavior to play the role of the caregiver." One of the motivations for doing so was to help the robot better read facial expressions and voice cues, as they would be exaggerated. This idea is also reflected in the design of the robot Simon by Diana and Thomaz (2011). The authors decide to incorporate "cute" features for "people to immediately perceive the robot's lack of initial knowledge, or 'innocence'," which would in turn encourage them to teach the robot. The robot's cute appearance can also help to set realistic expectations about a robot's capacity and thus avoid high user expectations in human–robot interaction, which are easily disappointed.

For reasons that may be cultural (Lee and Sabanović 2014), and also, as discussed above, in response to the complications that can accompany designing robots that are too hyper-realistic, an observable aesthetic trend of cuteness has emerged in robot morphologies. Typified by the use of large bright eyes, pastel colors, soft or shiny surface treatment, use of a high-pitched voice, and an overall small stature, cute robots can be seen in many examples, from research prototypes like Blossom (Suguitan and Hoffman 2019) and DragonBot (Westlund 2015) to industry-manufactured robots such as SoftBank's Pepper robot (Lafaye et al. 2014), Amazon's Astro, and Samsung's Otto. While cuteness may be used to increase technology acceptance and invite interaction that is more appropriate for a robot's technology readiness level, further analysis of this aesthetic is needed to understand its cultural function and to encourage a more informed usage of cute morphologies in social robotics.

In her seminal work *Our Aesthetic Categories*, Ngai (2010) discusses the aesthetic categories of the zany, the cute and the interesting, and argues for the contemporary centrality of these aesthetic categories. She examines the category of cute with the same philosophical seriousness that has previously been afforded to beauty and the sublime in literature. Ngai describes the capacity for the aesthetics of cute, when applied to an object or product, to *commodify*, *domesticate* and *pacify*, with the further potential to render an object charming, irrelevant, vulnerable, and inconsequential. We explore these three consequences of a cute aesthetic in this section.

By way of illustration, Ngai (2012) describes a cute frog-shaped sponge that shares remarkable similarities with many social robot morphologies. It has "an enormous face (it is, in fact, nothing but a face), and exaggerated gaze (but interestingly no mouth)," which she says emphasizes the way that cuteness is often achieved by moving away from realism. In Ngai (2010), the author explains that

> realist verisimilitude or even formal precision tend to work against or even nullify cuteness. (2010, p. 64)

Ngai argues that the move away from detail and realism toward cute is best sought in objects with round contours and little to no detail, which suggest a certain pliancy and responsiveness, either materially or metaphorically, where "the less formally articulated… the cuter" (Ngai 2012). This rounded blob of detail-less mass is best accompanied by the qualities of "smallness, compactness, formal simplicity, softness and pliancy." While it is not a functionally plausible option for many social robots to be pliable or soft, some are, including The Hug (2003), Keepon (2007), PARO (2014), and Blossom (2019). Others imply their social pliancy and compliance through rounded contours, an exaggerated face, smallness, and simplicity of form.

In the remainder of this section, we discuss how cuteness aids in commodifying, domesticating, and pacifying a design, and ask what the implications are for robot appearances.

2.4.1 Commodify

Cuteness, according to Ngai (2012), while passive and pliable, is also seductive and "capable of making surprising demands," such as the demand to be purchased and thus turned into a commodity. Ngai (2012) quotes Johnson (2010), who says that the purchaser of an object that uses a cute aesthetic is often

> seduced into feeling that buying the product is, in fact, carrying out the wishes of the product itself.

One must only consider the cute images of farm animals on the packages of animal food products as an example of this fallacy of transferring the wish of the consumer to the imagined wish of the product. In this manner, cuteness allows robots to transcend the category of "expensive research prototype" or "out-of-reach home appliance" and be trivialized, appealing to the consumerist tug to be purchased.

Cuteness also invites interaction. In the case of a cute small object, the aesthetic invites the "subject to handle it physically" (Ngai 2012). In robotics, cuteness also promotes interaction, inviting a viewer to move closer, and to participate and interact with a robot. However, cuteness has the power to go further than mere interaction. Ngai explains that consumption is conflated with identity, where "wanting to have" becomes "wanting to be like," and cuteness "thus produces what Doane (1987) describes as a "strange constriction of the gap between consumer and commodity," where "commodity and consumer share the same attributes."

In this symbiotic chicken-and-egg relationship, cuteness thus not only appeals to those wanting to be like the cute object, but cuteness is a mimetic aesthetic that generates more cuteness; those interacting with a cute object tend to get smaller and lower to the ground, the pitch of their voice getting higher, and they use "small sized adjectives." This blurring of boundaries is exploited by designers of cute artifacts, leading consumers to want to not only "own" but "adopt" objects. Adoption, however, brings with it an additional layer of object characteristics. Harris (1992) argues

> advertisers have learned that consumers will 'adopt' products that create … an aura of motherlessness, ostracism and melancholy.

2.4.2 Domesticate

A robot's cute aesthetic contributes not only to a robot's commercial appeal but also to its domestication, taming, and introduction into the sphere of the home. Young et al. (2009) argue that cuteness fosters the passage of the "Other" into domestic environments and is indicative of powerlessness, and therefore the implied safety of the consumer. In their study *Toward Acceptable Domestic Robots: Applying Insights from Social Psychology* (2009), they posit that

> one of the most important and unique barriers to the widespread adoption of robotics is an especially complex socialisation process,

and that for social robotics

> the problems of technology acceptance are far more significant in a domestic environment than an industrial one.

Young et al. (2009) argue that the domestic socialization or 'absorption' of social robots into domestic settings will be largely dependent on upon

> domestic consumer perceptions of what robots are [...] and what exactly they are and are not capable of doing.

The study contains guidelines for the acceptance of social robots, where the authors outline the importance of robot design methodologies, and how designers may choose to leverage user tendencies to anthropomorphize robot appearances to influence the user's perception of the harmlessness or safety of the robot. One example they cite is that consumers referred to Roomba robotic vacuum cleaners as "cute," thus

> [The] Roomba can become a social part of the home and in a sense, a social participant in the family, not that different from, say, a pet hamster.

To become a pet, however, an animal must be controlled, made submissive, and rendered harmless. In other words, it must be stripped of any surplus power or dominance. Cuteness is a design mechanism that aides in this process, as it

> solicits a regard of the commodity as an anthropomorphic being less powerful than the aesthetic subject, appealing specifically to us for protection and care. (Ngai 2012)

In this way, consumer concerns about the harmlessness and safety of domestic social robots might be assuaged through cuteness, which Lori Merish argues can transform "transgressive subjects into beloved objects" (Lori Merish in Ngai 2012, p. 60).

2.4.3 Pacify

As noted in the previous sections, Ngai (2005) argues that the aesthetic of cute is undeniably trivializing, and as noted above, this design choice embodies a sense of vulnerability that can evoke a desire in us to protect, potentially pacifying any concerns we may have about these "transgressive subjects." However, Ngai also asserts that the cute (just like the interesting and zany) can have an ambivalent nature that evokes contradictory effects. Specifically, cuteness can evoke both empathy and aversion (Ngai 2012), as well as both tenderness and aggression (Ngai 2005).

In *Cuteness*, Harris (1992) warns that cuteness

> disempowers its objects, making them appear more ignorant and vulnerable than they really are

and, as summarized by Ngai (2012), cuteness

> excites the consumer's sadism or desire for mastery as much as her desire to protect and cuddle.

The contradictory affective implications of cuteness can be seen in the treatment of a number of contemporary robots. HitchBOT, the hitch-hiking robot, was designed with several cute features. It has a digital smiley face, pool-noodle arms, yellow rain boots, and is completely dependent on the goodwill of the public to hitch-hike to its destination. HitchBOT successfully hitch-hiked across Canada and Europe, gaining a popular online following, but later met a violent end in Philadelphia, USA, where civilian pranksters fabricated a surveillance tape of them repeatedly kicking and eventually "destroying" the robot.

In a study of the cute, small, yellow robot named Keepon, Kozima et al. (2009) reported a variety of behaviors toward Keepon, including

> Violent versus Protective behaviour: ... a boy ... beat Keepon several times, and a girl stopped him, 'No! No!'. When [the boy] hit Keepon's head several times, [the girl] stopped him by saying 'It hurts! It hurts!'. [The boy] hit Keepon's head a couple of times ... observing this, [the girl] approached Keepon and checked if it had been injured ... stroking it gently."

In 2015, a man was reportedly arrested in Japan for allegedly "kicking a Pepper robot in a fit of rage," although the man admitted he was "frustrated with a store clerk." The robot reportedly now moves more slowly and has been permanently damaged. While the imbalance of power that cuteness affords robots may pacify them and render them unthreatening to humans, it evidently has the capacity to give rise to aggressive and exploitative behavior in their human counterparts.

The deliberate pacification of robots through a cute aesthetic may also carry other social undertones. Ngai explains that the act of "'giving face' to an object," particularly an expressive face to a dumb object with no mouth, is to "phantasmically make it lose face," which is categorically "an act of humiliation." Ngai describes the use of overly large eyes as "perversely literalizing the gaze (as described by Walter Benjamin)," enabling the robot to "empathetically return our gaze" while "other facial features—the mouth in particular—tend to be simplified to the point of barely being there" (Ngai 2012). It is interesting in this context to see how many cute robot designs have increased the eyes and either left out the mouth or made it unnaturally small (see examples such as Robotics Today's PaPeRo R500, 2001, and Mayfield Robotics' Kuri, 2017). This trend is also observable in the documented design stages for the robot Simon, where in the prototyping phase, the robot was modeled to include lips and a mouth, which were later removed, together with the decision to make the head smaller (Diana and Thomaz 2011). Ngai asserts that much like the appearance of Hello Kitty, the large eyes and lack of any mouth seem to "amount to denying speech," establishing and maintaining a strong power differential.

2.4.4 Implications

What are the implications of mass-producing and populating our homes with cute robotic agents that adoringly gaze at us and are unable to speak for themselves? Will this morphological subjugation be something we are called to account for?

Considering the symbiotic and mimetic powers of cuteness, is this belittling, helplessness, and subjugation something we also willingly bring upon ourselves? Social robot designers may unwittingly be designing cute robots in order to address interaction difficulties, aid in their complex socialization process, and boost commercial appeal. However, by pacifying, domesticating, and commodifying robots, they also run the risk of engaging with the politically and socially ambivalent nature of this aesthetic category, provoking elements of both engaging and potentially problematic responses from the human counterparts interacting with them. Furthermore, in light of the above, when we consider not only the process of humans designing and making robots, but also in defining robots as "making humans," this exploitation becomes particularly concerning.

2.5 Conclusion

This chapter traces historical origins and influences along with cultural implications on the most dominant robot morphological typologies. We specifically focus on the design tropes of cute companions, futuristic machines and human replicas. We posit that most robot designs today draw on these typologies and that almost all social robots today are located somewhere along the human replica–futuristic machine–cute companion spectrum.

Design is never "original," and just like any artifact design, the design of social robots is dependent on the cultural lineages that preceded them, and converse with the political and ethical implications that these lineages provide. Designing a social robot requires an informed awareness of the history and implications of the aesthetic that one is drawing upon and implementing. We issue a call to action for robot designers to explicitly choose and appreciate a certain aesthetic cultural lineage to build upon. This also calls for more scholarship into cultural topics as a basis for robot design and collaboration with those outside of traditional robotics research to contribute historical, cultural, and aesthetic knowledge to the design process.

In her discussion of Judith Butler's *Bodies That Matter* (1993), Suchman (2009) likens the "gendering" of human bodies over time through "the reiteration of norms" to technology construction, "as a process of materialization through reiterative forms." She argues that these reiterative forms can come to represent more or less uncontested "normative identifications of matter." It is not argued here that these normative forms of human replica, futuristic machines, or cute companion robots are inherently problematic, or always inappropriate. Rather, it is argued that these forms have materialized through multiple iterations and have resulted in normative and largely uncontested robotic typologies. An acknowledgment of responsibility for this technology and its agential capacity must be paired with a commitment to the critique and continual contestation of these human-like machines, for, as contended by Dumouchel et al. (2017),

[H]ow we live with robots, and what kinds of robots we live with, reflects our own moral character.

Šabanović (2010) similarly identifies that robot design contributes to the construction of "technoscientific imaginaries" or "narratives about social order, human behavior and psychology, and common norms." The integration of robots into human environs will impact social norms and values, and the manner of this impact may be shaped, in the first instance, by design. It is therefore the responsibility of roboticists to be aware of their agency and responsibility in shaping social and cultural norms through design, and to seek to integrate aesthetic choices in an ethical and informed way.

References

Apollonio U, Brain R, Flint R, Higgitt J Tisdall C (1973) Futurist manifestos. Thames and Hudson London (1973)

Breazeal C, Foerst A (1999) Schmoozing with robots: exploring the boundary of the original wireless network. In: Proceedings of the 3rd international cognitive technology conference (CT-99), pp 375–390. San Francisco, CA. http://www.ai.mit.edu/projects/ntt/projects/NTT9904-01/NTT9904-01.html

Butler J (1993) Bodies that matter: on the discursive limits of "sex." Routledge, New York

Chapuis A (1958) Automata: a historical and technological study. Central Book Company, New York

Cohen JL (2014) Le corbusier's modulor and the debate on proportion in France. Architectural Histories 2(1)

Crawford K (2021) The atlas of AI: power, politics, and the planetary costs of artificial intelligence. Yale University Press

de Solla Price DJ (1964) Automata and the origins of mechanism and mechanistic philosophy. Technol Cult 5(1):9. https://doi.org/10.2307/3101119

Diana C, Thomaz AL (2011) The shape of Simon: creative design of a humanoid robot shell. In: CHI '11 extended abstracts on human factors in computing systems, CHI EA '11. Association for Computing Machinery, New York, pp 283–298. https://doi.org/10.1145/1979742.1979648

Doane MA (1987) The desire to desire, vol 27. Indiana University Press, Bloomington

Dumouchel P, Damiano L, DeBevoise M (2017) Living with robots. Harvard University Press. https://doi.org/10.4159/9780674982840

Dunstan BJ (2019) The plastic dynamism of the human aesthetic: employing futurist methodologies in the cross-disciplinary design of social robot morphologies. PhD thesis, UNSW Sydney. https://doi.org/10.26190/UNSWORKS/3869

Evans T (2018) The race of machines: blackness and prosthetics in early American science fiction. Am Lit 90(3):553–584. https://doi.org/10.1215/00029831-6994805

Harris D (1992) Cuteness. Salmagundi 96:177–186

Hegel F, Lohse M, Wrede B (2009) Effects of visual appearance on the attribution of applications in social robotics. In: RO-MAN 2009-The 18th IEEE international symposium on robot and human interactive communication. IEEE, pp 64–71

Hoffman G (2020) The social uncanniness of robotic companions. In: Frontiers in artificial intelligence and applications. IOS Press. https://doi.org/10.3233/faia200953

Koeppe W (ed) (2019) Making marvels. Metropolitan Museum of Art, New York

Kowalik W (2017) Streamline modern design in consumer culture and transportation infrastructure: design for the twentieth century. New Errands 5(1). https://doi.org/10.18113/P8NE5160475

Kozima H, Michalowski MP, Nakagawa C (2009) Keepon: A playful robot for research, therapy, and entertainment. Int J Soc Robot 1(1):3–18

Lafaye J, Gouaillier D, Wieber PB (2014) Linear model predictive control of the locomotion of Pepper, a humanoid robot with omnidirectional wheels. In: 2014 IEEE-RAS international conference on humanoid robots. IEEE. https://doi.org/10.1109/humanoids.2014.7041381

Lee HR, Sabanović S (2014) Culturally variable preferences for robot design and use in South Korea, Turkey, and the United States. In: Proceedings of the 2014 ACM/IEEE international conference on human-robot interaction, pp 17–24

Lista G (1986) Futurism. Art Data, London

Loewy R (1999) Raymond Loewy and streamlined design. Universe Pub (1999)

Mahdi H, Akgun SA, Saleh S, Dautenhahn K (2022) A survey on the design and evolution of social robots—past, present and future. Robot Auton Syst 156:104193. https://doi.org/10.1016/j.robot.2022.104193

Marinetti FT (1909) The futurist manifesto. Le Figaro 20:39–44

Marshall JJ (2012) Machine art, 1934. University of Chicago Press, Chicago

Ngai S (2005) The cuteness of the avant-garde. Crit Inq 31(4):811–847

Ngai S (2010) Our aesthetic categories. PMLA 125(4):948–958

Ngai S (2012) Our aesthetic categories: zany, cute, interesting. Harvard University Press, Cambridge

Nickles S (2002) Preserving women: refrigerator design as social process in the 1930s. Technol Cult 43(4):693–727

Perloff M (2003) The futurist moment: avant-garde, Avant guerre, and the language of rupture. University of Chicago Press, Chicago

Roser C (2016) Faster, better, cheaper in the history of manufacturing: from the stone age to lean manufacturing and beyond. Productivity Press, Routledge

Šabanović S (2010) Robots in society, society in robots. Int J Soc Robot 2(4):439–450

Suchman L (2009) Agency in technology design: Feminist reconfigurations. In: Digital cultures: participation–empowerment–diverstity. Online proceedings of the 5th European symposium on gender & ICT

Suguitan M, Hoffman G (2019) Blossom: a handcrafted open-source robot. J Hum Robot Interact 8(1). https://doi.org/10.1145/3310356

Sullivan LH (1896) The tall office building artistically considered. In: Lippincott's Magazine. J. B. Lippincott, Pennsylvania

Turkle S (2007) Authenticity in the age of digital companions. Interact Stud 8(3):501–517. https://doi.org/10.1075/is.8.3.11tur

Westlund JK (2015) Telling stories with green the DragonBot. In: Proceedings of the tenth annual ACM/IEEE international conference on human-robot interaction extended abstracts. ACM. https://doi.org/10.1145/2701973.2702089

Young JE, Hawkins R, Sharlin E, Igarashi T (2009) Toward acceptable domestic robots: applying insights from social psychology. Int J Soc Robot 1(1):95–108

Chapter 3
The Robot Soundscape

Frederic Anthony Robinson, Oliver Bown, and Mari Velonaki

Abstract As social robots make their way into human environments, they need to communicate with the humans around them in rich and engaging ways. Sound is one of the core modalities of communication and, beyond speech, affects and engages people across cultures and language barriers. While a growing body of work in human–robot interaction (HRI) investigates the various ways it affects interactions, a comprehensive map of the many approaches to sound has yet to be created. In this chapter, we therefore ask "What are the ways robotic agents can communicate with us through sound?", "How does it affect the listener?" and "What goals should researchers, practitioners and designers have when creating these languages?" These questions are examined with reference to HRI studies, and robotic agents developed in commercial, artistic and academic contexts. The resulting map provides an overview of how sound can be used to enrich human–robot interactions, including sound uttered by robots, sound performed by robots, sound as background to HRI scenarios, sound associated with robot movement, and sound responsive to human actions. We aim to provide researchers and designers with a visual tool that summarises the role sound can play in creating rich and engaging human–robot interactions and hope to establish a common framework for thinking about robot sound, encouraging robot makers to engage with sound as a serious part of the robot interface.

F. A. Robinson (✉) · M. Velonaki
Creative Robotics Lab, University of New South Wales, Sydney, NSM, Australia
e-mail: frederic.robinson@unsw.edu.au

M. Velonaki
e-mail: mari.velonaki@unsw.edu.au

O. Bown
Interactive Media Lab, University of New South Wales, Sydney, NSM, Australia
e-mail: o.bown@unsw.edu.au

© The Author(s), under exclusive license to Springer Nature Switzerland AG 2023

B. J. Dunstan et al. (eds.), *Cultural Robotics: Social Robots and Their Emergent Cultural Ecologies*, Springer Series on Cultural Computing,
https://doi.org/10.1007/978-3-031-28138-9_3

3.1 Introduction

As social robots make their way into human environments, they need to communicate with the humans around them in rich and engaging ways (Breazeal et al. 2016). To do so, robots have a variety of modalities at their disposal, including gesture, light and sound. While the latter is one of the core modalities of communication, it remains comparatively underexplored in human–robot interaction research. Bethel and colleagues highlighted its potential as a cross-cultural form of communication, which does not rely on semantic content and can therefore transcend language barriers (Bethel and Murphy 2006). An illustrative example of how sound can work across cultures is prosody, the pattern of melody, stress and timing inherent to human speech (Pinker 1989). Fernald and Mazzie investigated how the characteristics of mothers' prosody when talking to their child differ across different languages. They found that some characteristics are shared across cultures and may even be universal to human communication (Fernald and Mazzie 1991; Fernald et al. 1989).

A growing body of work in human–robot interaction (HRI) investigates the various ways robots can use sound to influence and engage the humans around them. Some areas, such as robotic musicianship (Bretan and Weinberg 2016) and semantic-free utterances (Yilmazyildiz et al. 2016), are well established, while others, such as motor sound (Moore et al. 2017) or robot movement sonification (Frid et al. 2018), have only recently received significant attention. However, a comprehensive map that charts the many existing approaches to robot sound, and how these might benefit interactions, has yet to be created. In this chapter, we therefore ask "What are the ways robotic agents communicate with us through sound?", "What goals do researchers, practitioners and designers have when creating these languages?" and "How does it affect the listener?" These questions are examined with reference to HRI studies and to robotic agents developed in commercial, artistic and academic contexts, resulting in a map of machinic languages that provides a comprehensive picture of the robot soundscape, or, in Kadish's terms, the Robophony (Kadish 2019). While some of the approaches are firmly established amongst HRI researchers, others have been explored in adjacent fields, allowing us to highlight areas that may provide a fruitful source of insight and inspiration for future HRI research endeavours.

A map of the robot soundscape is depicted in Fig. 3.1. It broadly distributes robot sound across five categories: sound uttered by robots, sound and music performed by robots, sound as background to HRI scenarios, sound associated with robot movement, and sound responsive to human actions. The definition of these top-level categories has been guided by two decisions: (1) considering all types of sound equally, without differentiating between music, speech, sound effects or other; and (2) considering the *context* in which a sound occurs, rather than the sound's characteristics or the intention behind it. It should be noted that these categorisations are not strict and some overlap is unavoidable.

The following sections are structured as follows: Sect. 3.2 examines the use of speech and semantic-free utterances. The focus in speech is on timbre and tonal qualities, rather than semantic content. Semantic-free utterances are discussed

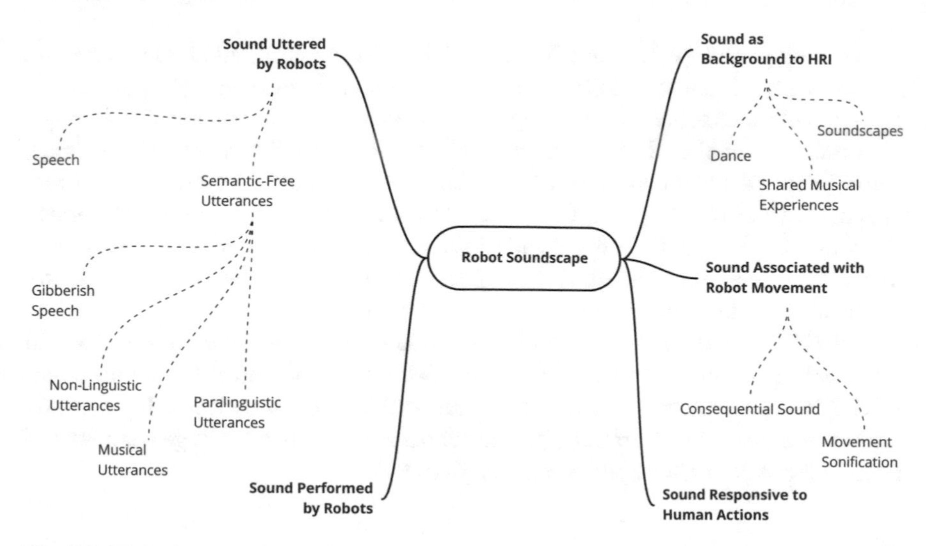

Fig. 3.1 The robot soundscape, a map of the various contexts sound occurs in human–robot interaction

both with regards to their creation and their effect on human–robot interactions. Section 3.3 discusses sound created by or with robotic agents in a performative context. Section 3.4 looks at sound and music used concurrent with human–robot interaction. It summarises the contexts in which this occurs and what effect this has on the interactions. Section 3.5 focuses on sound associated with robot movement. This includes sound as a by-product of robot movement as well as the deliberate sonification of robot movement. Finally, Sect. 3.6 focuses on sound emitted by robots that is responsive to human actions. As interactions with robots are in large parts determined by the user, this category can be found in all contexts mentioned above. This section will look at instances where sound responsive to human actions manifests as a separate sound category. With this chapter, we aim to provide researchers and designers with a comprehensive picture of the role of sound can play in creating rich and engaging human–robot interactions. In doing so, we hope to establish a common framework for thinking about robot sound, and a call to arms for robot makers to engage with sound as a serious part of the robot interface.

3.2 Sound Uttered by Robots

Much of the sound-related studies in HRI have, unsurprisingly, focused on the voice, as natural language is one of the primary modalities of social interactions with robots (Belpaeme et al. 2013). Seamless natural language interaction, however, remains a challenge in HRI (Goodrich and Schultz 2008). Despite much research in natural language processing (NLP) aiming at improving language recognition, understanding, and generation (Connell 1807; D'Mello et al. 2005; Gorostiza and Salichs 2011; Mozos et al. 2007), it still remains significantly flawed, which negatively affects interactions by causing uncertainty or miscommunication, amongst other issues (Mubin et al. 2009). Even small errors can lead to breakdowns in interaction (Holzapfel and Gieselmann 2004; Shiwa et al. 2009). For example, a slight error in a service robot's perception can lead to it missing an instruction by a human, who may then feel like the robot is not helpful or not willing to assist. These potential breakdowns need to be mitigated by expectancy setting strategies, such as clearly communicating the robot's limitations, or recovery strategies, such as apologies (Lee et al. 2010). The average human's ability to understand and adapt to most real-world contexts is still superior to any algorithm (Moore 2014) and the state-of-the-art in NLP is not advanced enough to facilitate seamless, open-ended dialog between humans and machines. Currently, "social robots are unable to leverage the power of natural language" (Yilmazyildiz et al. 2016, p. 64).

3.2.1 Speech

Natural language and speech are a large enough topic to be considered separately to the wider issue of robot sound design. This chapter focuses instead on every aspect of sound design besides language. Beyond its semantics, however, the voice can affect HRI in various ways. Eyssel et al. showed how the gender of a robot's voice, as well as how closely the voice resembled that of a human, impacted the impressions formed by study participants (Eyssel et al. 2012). The impact of natural or synthesised voices was also shown in a study by Walters et al., where participants chose various approach distances when meeting a robot, depending on its vocal qualities. An artificially generated voice resulted in the largest distances (Walters et al. 2008). Nakagawa et al. showed how whispering robots were more successful in motivating participants to perform certain tasks (Nakagawa et al. 2013), and Chang et al. modulated the pitch, speech rate and intonation of their robot and subsequently showed that participants attributed different personality traits to the robot (Chang et al. 2018). Focusing on subtle differences in intonation, Aarestrup et al. found how various versions of a simple greeting would lead to significant difference in users' perception of a robot's friendliness, assertiveness and engagement (Aarestrup et al. 2015). Fischer et al. analysed the prosody of famous public speakers and used them to inform the speech melody of the synthesised speech of various social robots. This led to the robots being perceived as more engaging, passionate and charming (Fischer et al. 2019). Introducing the notion of 'appropriate' voices for robotic agents, Moore showed how a robot's embodiment and application context require a tailored approach when designing its voice, in order to be deemed appropriate by a user. He further notes that "ubiquitous deployment of inappropriate human-like voices for non-living artefacts might deceive users into overestimating their capabilities" (Moore 2017, p. 1). These studies show that a robot's speech communicates more than semantic content. Researchers are therefore investigating semantic-free utterances as a way to facilitate communication between robots and humans outside of language.

3.2.2 Semantic-Free Utterances

Yilmazyildiz and colleagues coined the term *semantic-free utterances* (SFUs), defining them as "auditory communication or interaction means for machines that allow emotion and intent expression, composed of vocalizations and sounds without semantic content." (Yilmazyildiz et al. 2016, p. 64). These utterances have been described as a powerful channel of communication across cultures and language barriers when used within a multimodal communication system (and Other Non-verbal Expressions of Affect for Robots 2006). The creation of SFUs has been informed by a wide range of disciplines, such as sound design for film and cartoons, music composition, and linguistics. Recent work has also drawn inspiration from biological rules derived from basic communication patterns found in animals to convey valence and arousal (Korcsok et al. 2020). Part of the motivation for this is the argument that an artificial agent should be regarded as a separate species, which needs only basic communicational skills that should be distinct from human communication (Miklósi et al. 2017). According to Yilmazyildiz et al., SFUs can be broadly categorised as gibberish speech, musical utterances, non-linguistic utterances and paralinguistic utterances (see Fig. 3.2) (Yilmazyildiz et al. 2016).

Gibberish speech contains many of the characteristics of language but has no actual meaning to the listener, because essential elements of it are masked (Remez et al. 1981; Scherer 1985) or modified (Burkhardt and Sendlmeier 2000; Cahn 1990), or because the words used are nonsensical in combination (Chomsky 1956). It can be produced by altering audio recordings of speech through filtering (Knoll et al. 2009; Scherer et al. 1972; Snel and Cullen 2013) or restructuring (Friend 2000; Johnson et al. 1986; Scherer 1971, 1982; Teshigawara et al. 2007) various elements of the speech signal, while aiming to leave affective and prosodic cues intact. Another approach is to create artificial speech using synthesisers. This allows precise control

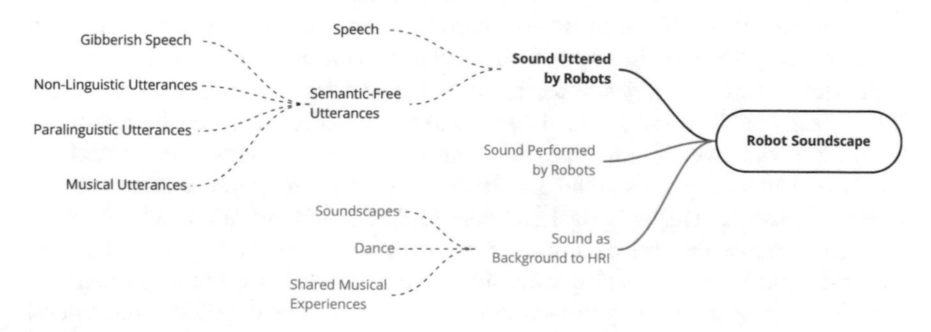

Fig. 3.2 Sound uttered by robots

of the signal's individual parameters, such as intonation, timbre and rhythm (Schroder et al. 2010).

Gibberish speech has long been used in the context of HRI. Breazeal's robot Kismet (Breazeal 2000, 2004) used childlike speech comprised of random vowel and consonant combinations, created by a speech synthesiser (Hallahan 1995). By altering parameters such as speech rate, intensity and pitch, the robot produced affective utterances based on the vocal affect parameter mapping by Cahn (1990). In studies evaluating the accuracy of affect recognition, participants correctly identified emotions about two thirds of the time.

Oudeyer generated words with predetermined syllable combinations. He created exaggerated affective speech by modifying parameters, such as the accentuation, duration and pitch of these syllables, which was used to convey the basic emotions–happiness, calmness, anger, sadness, and comfort (Oudeyer 2003). In a subsequent evaluation of the system, participants were able to correctly distinguish the utterances in unsupervised experiment conditions in around half of the cases. The difficulties encountered stemmed mainly from mixing up utterances with high arousal (e.g., anger and happiness) and from correctly recognising more neutral emotions, such as calmness. Nemeth et al. later introduced the term *Spemoticons* to describe simple emotional cues based on speech (Németh et al. 2011). This practise is related to the established earcons (Blattner et al. 1989) and auditory icons (Gaver 1986) from the field of human–computer interaction (HCI). For the communication of emotion and intention in their representation system, they created synthetic gibberish speech using a speech synthesiser Profivox (Olaszy et al. 2000) and subsequently modified the prosody by adjusting rhythm, pitch and intensity. They identified the most effective of their designs in perceptual studies, asking participants to categorise sounds into seven emotional archetypes, but did not validate the designs with absolute benchmarks.

Musical utterances use compositional techniques to inform their content and structure. They aim to convey affect or information through the expressive language of music. The communication of affect is an integral part of music, and listeners have a well-established listening experience to draw associations from, thereby potentially making the utterances highly intuitive. The field of music also has an extensive body of research investigating how music conveys emotion (Meyer 2008), providing HRI researchers with a broad variety of tools to create utterances.

Various studies in HRI have involved musical utterances. Johannsen implemented basic musical utterances in a service robot to communicate functional information, such as planned motion trajectories, internal state, and degree of urgency (Johannsen 2001, 2002, 2004), using pitch, rhythm, and timbre. For example, the directions 'up' and 'down' were communicated using a melody with either an upwards or downwards trajectory (this could also be seen as conforming to Gaver's notion of auditory icons with a metaphorical association). The directions 'left' and 'right' were encoded using rhythmical patterns. Several user studies involving musicians and non-musicians indicated that the latter significantly outperformed other participants after they were given the opportunity to learn the meaning of the various utterances. While sample sizes were too small to draw any conclusion, this indicates that musical utterances can be intuitive and easy to learn for musically inclined listeners, but are less effective for the wider population.

In 2005, Esnaola and Smithers developed a musical language for their robot MiReLa (Esnaola and Smithers 2005). The robot's name is derived from solmisation, the practise of assigning syllables to the notes in a scale (McNaught 1892) and, similarly, their language assigns letters to ten different pitches that are combined to form predefined words that contain information and instructions. The notes were meant to be whistled, which placed them above the frequency range of human speech and various types of ambient sounds. Even though this vocabulary needed to be learned by humans to understand and communicate with the robot, the researchers argued that this approach provided an easy-to-learn method of communication, which was resilient to background noise. Their validation studies focused on the robot's ability to detect and recognise utterances from the human. However, with the current state of voice recognition technology, this focus is perhaps less relevant than it was at the time.

Investigating the potential of musical utterances to convey affect, Jee et al. used musical theory and notation to develop a synthetic sound vocabulary. In an experiment, participants had to categorise utterances into four basic affective categories: happiness, sadness, fear, and dislike (Jee et al. 2007), and the results showed that their musical utterances reliably conveyed the intended emotion. In further work, they added additional affective categories, such as pride and expectation, and modified the tempo, pitch and volume of these utterances to additionally communicate the intensity of the emotion (Jee et al. 2009). They also analysed musical utterances from popular science fiction films and tested intention statements such as *encouragement*, *affirmation*, and *questions* (Jee et al. 2010). The latter provides an especially interesting intersection of musical material (ascending melodic line) and prosodic content (utterance with rising pitch). Results from that study supported the effectiveness of music for archetypal emotions, with 80% of participants reporting that the utterances are sufficient to express emotion. The communication of intention was less successful, with around half of participants reporting them to be sufficient. Combining insights from musicology and natural language behaviour in animals, Ayesh developed what he termed a "Musical Language for Emotional Interaction" (Ayesh 2006, 2009). He created basic building blocks of behavioural states called 'atoms', which could then be combined to create emotions. For example, excitement and stress combine to create anger. However, no user studies were performed to assess the effectiveness of the approach.

Recent approaches to musical utterances apply machine learning techniques to aid in their creation. The aims are to either make them more intuitive to understand, or to personalise them for the listener. Savery and colleagues created a data set with improvised emotional phrases that were based on the Geneva Emotion Wheel, an instrument for assessing emotional responses (Savery et al. 2020). They then used that data set to generate affective utterances through non-linguistic musical prosody (Savery et al. 2009). Early results indicate that machine learning may provide robots with a powerful way to reliably communicate a range of emotions during interactions. Ritschel and colleagues presented an approach where the pitch sequences of musical utterances—the score, so to speak—were determined by the researcher, but the tempo and timbre were adjusted to listener preference through a training process with comparative human feedback (Ritschel et al. 2019). The majority of the 27 participants of the study preferred the personalised version. It remains to be discussed whether this is due to changes to the actual sound characteristics—meaning each of the participants had their own particular set of synthesis parameters that they favoured—or whether sound that is 'made for me' simply feels better to listen to. In either case, personalised sound appears to be a promising direction for this modality.

Non-linguistic utterances are based on the long and rich history of auditory communication in human–computer interaction, particularly sonification, auditory icons, and earcons. Sonification is the "technique of rendering sound in response to data and interactions" (Hermann et al. 2011). It uses sound as an alternative to visual displays to convey the information contained in data streams. Perhaps the most famous example of sonification is the Geiger counter, which represents radioactivity levels through the frequency of clicks. The notion of auditory icons was proposed by William Gaver in 1986 in the context of sound design for computer interfaces (Gaver 1986). Auditory icons use a listener's associations to the possible source of a sound to convey information, rather than objective parameters like timbre and volume. When presented with an auditory icon, listeners are "hearing the world, not the sound" itself (Gaver 1986, p. 169). A listener's associations can be drawn from symbols established through cultural conventions (e.g., police siren), metaphors (e.g., connecting descending pitch with falling) or physical causation. Examples for the latter are the crumpling of paper when moving a file into the trash or the stylised 'swoosh' resembling a passing plane engine when sending off an email. Earcons use the language of sonificiation within the context of computer interfaces. Blattner et al. distinguish between representational earcons (which are synonymous with Gaver's auditory icons) and abstract earcons (Blattner et al. 1989). Through short and modular digital sound patterns, earcons are meant to convey information about a system's state in a flexible and easily extendable way. Despite them being unable to build on a listener's associations and therefore being less memorable and intuitive (Dingler et al. 2008), they have become widespread in a wide variety of interfaces. Examples range from the interaction feedback on today's smartphone apps to the startup sounds of desktop computers. Earcons also have a variation that is based on processed speech. Spearcons are spoken phrases that are sped up to the point of intelligibility (Walker et al. 2013). The benefits of these are an improved learnability in comparison to more abstract earcons.

What separates non-linguistic utterances from the approaches above is their use as social cues by a robotic agent (Yilmazyildiz et al. 2016). Beyond the context of HRI, these would be called user experience (UX) sounds. The presence of an artificial agent acting in a social context makes these sounds utterances, rather than notifications. In HRI, non-linguistic utterances have been implemented in a wide variety of robotic agents with the aim of investigating how machines can convey positive or negative sentiments using the most basic audio cues possible (Cha and Mataric 2016; Komatsu 2005; Komatsu et al. 1941; Korcsok et al. 2020; Song and Yamada 2017). A study by Read and Belpaeme explored how children interpret affect in non-linguistic utterances and found that rhythmic patterns played an important role in the interpretation. These interpretations, however, were not consistent, with the participants assigning different meanings to the same sounds throughout the study. The study also found indicators that the perception of affect is categorical, meaning subtle changes in the utterances did not lead to subtle changes in the interpretation (Read and Belpaeme 2012). This was confirmed in a follow-up study involving adults (Read and Belpaeme 2016). While the listener's tendency to focus on and categorically interpret particular prosodic elements of non-linguistic utterances simplifies the design of sounds for this particular purpose, it may also push other potentially relevant sound content into the background. If humans naturally latch on to prosodic elements present in a robot's utterances and focus on deciphering their affective meaning, other sound characteristics which do not represent aspects of human speech may be filtered out.

Paralinguistic utterances are, by Schuller and Batliner's definition, the vocal elements of speech that do not contain words (Schuller and Batliner 2013). These include factors such as intonation and voice quality, which are already an important aspect of other types of SFUs mentioned above. However, these also contain non-verbal vocal cues, such as back-channel feedback (e.g., an approving hum) (Ward 1996), and affect bursts like gasps and sighs (Scherer 1994), or laughter (Becker-Asano and Ishiguro 2009; Becker-Asano et al. 2011). These utterances have been shown to be effective and reliable at communicating various emotions (Schroder 2003). Prendiger et al. used grunts and moans to support an artificial agent's facial expressions conveying various emotions in a game scenario (Prendinger et al. 2006). Their findings indicated that a robot's affective display could cause or increase a similar response in the human participant. In their sensitive artificial listener (SAL) system, Schröder et al. implemented back-channel feedback and found it increased engagement in participants (Schroder et al. 2011). Kobayashi and Fujie presented a system that integrated paralinguistic cues into various robots to communicate the robots' conversational state with the aim of achieving more seamless communication (Kobayashi and Fujie 2013). No user studies were performed to evaluate their implementation. Investigating children's interpretation of affective sounds, Rossi et al. implemented a range of paralinguistic utterances into the social robot NAO (Gouaillier et al. 2009). The robot used vocalisations such as 'Grrrr', 'Yuck', and 'Ahh ahh' to communicate anger, disgust and relief, respectively. Their results showed that children clearly perceived the valence (positive or negative notion) of the utterances (Rossi et al. 2019).

In summary, robotic agents use speech and semantic-free utterances to convey affect, positive and negative attitudes, attention, intent, and states. They are further used to structure conversations and generate interest and engagement. In many of their forms, they draw from the prosodic content in utterances, based on our long established experience of using the voice to communicate with other living beings (Korcsok et al. 2020). In the case of musical utterances, they further draw from culture-specific musical experiences to facilitate clear categorical emotion communication.

There are various arguments supporting the use of SFUs as an alternative to speech: (1) They avoid creating expectations of an agent's intelligence by circumventing the many pitfalls of verbal human–machine communication; (2) Their ambiguity makes them a valuable alternative to concrete semantic content, as the task of interpretation lies with the human counterpart—the intelligent other—who can take context into account and project more complex traits onto the robot than intended by the designer; (3) It avoids a language barrier both in intercultural contexts and in contexts where verbal communication might not be possible, such as rescue missions. However, the perceived meaning and intensity of SFUs can be affected by various other factors. Context, for example, can dramatically change how utterances are interpreted. A study by Read and Belpaeme showed this by coupling videos of various actions performed on a robot (e.g., hitting, stroking and kissing) with various utterances. The utterances were interpreted in the same way when combined with certain videos, meaning the content of the SFU was overridden by its context (Read and Belpaeme 2014). Embodiment also influences how people interpret SFUs. A study by Komatsu and Yamada, for example, showed that the same set of utterances was identified with different success rates when uttered by different robotic agents, as well as by a computer (Komatsu and Yamada 2011). Interestingly, the identification was most successful when the agent was a computer, suggesting that robot embodiment introduces factors that interfere with how SFUs are perceived when isolated. With the dominant area of sound-related research in HRI covered, the next section covers an activity which is equally natural to humans: making music.

3.3 Sound Performed by Robots

Developments in the fields of mechatronics and machine perception have given rise to the research area of robotic musicianship. When looking at the various contexts in which robots emit sound besides speech and utterances, the notion of robotic musicians is an obvious next step. While the primary aim in this context is not to convey functional information, musical robots do need to be able to give social cues in order to effectively communicate with human co-performers and the audience (Cosentino and Takanishi 2021). In the words of Bretan and Weinberg, "robotic musicianship focuses on the construction of machines capable of producing sound, analysing music, and generating music in such a way that allows them to showcase musicality and interact with human musicians" (Bretan and Weinberg 2016, p. 100). With the aim of enriching musical culture and expanding human creativity, robotic musicians are made to play various instruments either alone or in ensembles with other machines or humans. This process entails both the creation of robots that are physically capable of creating sound (Kapur 2005), as well as developing cognitive models to a point where the robots know when and what to play (Weinberg and Driscoll 2006). (For a discussion around the cultural impact of robotic performance, see this book's chapter on sonic robotics by Marynowsky and colleagues.) (Fig. 3.3).

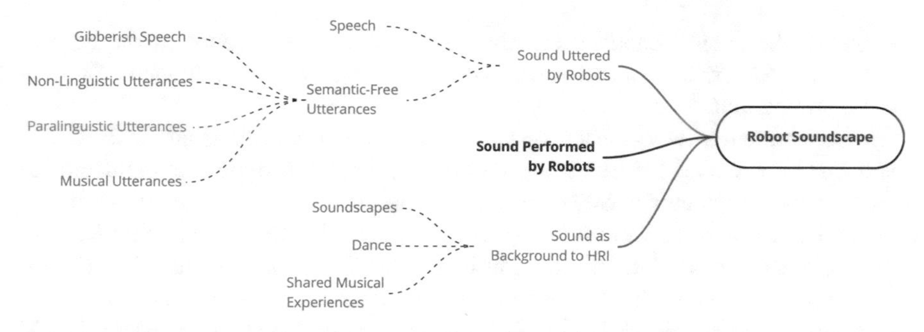

Fig. 3.3 Sound performed by robots

The interdisciplinary nature of this field, involving engineering, computation, music and psychology, gives it the potential to contribute to other fields, including cognitive sciences, human anatomy, and human–robot interaction. In the context of the latter, researchers aim to use the universal and intercultural language of music to find ways to better integrate robots into offices, care facilities and domestic environments. This approach can be seen in the wide range of social robots that include some kind of musical functionality, most often in connection with dance. This phenomenon is expanded upon in Sect. 3.4.

Researchers have developed robots able to perform on a wide range of acoustic and electronic instrument. These include the robot percussionists Cog (Williamson 1999), DB (Atkeson et al. 2000), the modular, non-anthropomorphic LEMUR robots (Singer et al. 2004), and Georgia Tech's Shimon (Savery et al. 2021; Weinberg and Driscoll 2007), as well as pianists WABOT (Kato et al. 1987), and ACT (Zhang et al. 2011), and animatronic robot rock band Compressorhead, whose cultural impact was investigated by Davies and Crosby in 2006s publication on Cultural Robotics (Davies et al. 2016). Robot string players include Sergi Jorda's Afasia (Jordà 2002), a harp-playing robotic hand by Chadefaux et al. (2012), an anthropomorphic violin-playing robot by Shibuya et al. (2012), and the recent Strum-Bot by Vindris and Carnegie (Vindriis and Carnegie 2016). Wind instruments include the robotic bagpipe player McBlare (Dannenberg et al. 2005), the Autosax (Maes et al. 2011), a blowing machine modelling the human mouth by Ferrand and Vergez (Ferrand and Vergez 2008), and a saxophone-playing robot by Solis et al. (2010).

Many robotic musicians play together with humans and take cues from their human counterparts by using machine vision to detect human performer gestures (Bretan et al. 2012; Cicconet et al. 2013; Solis et al. 2005) and head nods (Pan et al. 2010), along with using various machine listening techniques to estimate tempo (Barton 2013; Kapur 2011; Weinberg and Driscoll 2007) and recognise different playing styles (Kapur et al. 2012). As the focus is mostly on how robotic agents technically perform, either alone or in the presence of humans, human behaviour studies are rare, and the research's contributions to HRI are often implicit. Little is therefore known about the effect of robotic musicianship–or, more specifically, the sound of a robot musician–on aspects such as acceptance in a domestic environment or trust in the agent's capabilities. The entertainment value amongst the general public is, however, well documented, with robots such as Georgia Tech's Shimon even going on tour.

Robotic sound artworks often do not involve humanoid robots and instead take the form of augmented instruments or modular systems. Examples for augmented or entirely new instruments played by robotic agents are Trimpin's Contraption Instant Prepared Piano and the *Conloninpurple* installation (Trimpin. 2011), and Richard Logan-Greene's musical robotic systems Brainstem Cello and ActionclarInet (Weinberg et al. 2020) and mechanical instrument Submersible (Logan-Greene 2011). The audiovisual robotic artwork *Looks Like Music* by Japanese media artist Yuri Suzuki comprises several differently shaped miniature robots that can identify and follow lines drawn on a surface. Along these lines, colour markings are then picked up by the robots and translated into musical melodies and rhythms (Suzuki 2014). An ensemble of robots can thereby create musical textures designed and guided by the audience. *Cicadas*, a sound installation by artist Bob Meanza, consists of a large number of small robotic insects, which each have a microcontroller and a sound-producing element, such as a piezo loudspeaker or buzzer. As implied in the name, the robots blend into the environmental sound, creating dense and spatially rich nature-like soundscapes through artificial means (Meanza 2013). Hoyer et al. propose a kinetoacoustic work involving the real-time sonification of the movements of soccer-playing robots (Hoyer et al. 2013). Robots are augmented with various pieces of audio equipment, creating audio feedback loops between the individual agents. The resulting soundscape is combined with audio material derived from the robot's control signals. Consequently, the combination of robot location and movement results in a complex aleatoric soundscape. A similar focus on the interplay between several musical robots listening and responding to each other can be found in work by Krzyz˙aniak, who designed swarms of autonomous musical robots and used them as a platform for exploring human–robot and robot-robot musical interactions (Krzyžaniak 2021). Media artist and musician Moritz Simon Geist creates Sonic Robots that are combined to create augmented musical instruments and sound installations. His Glitch Robot uses various small actuators to hit various materials, such as drums and recycled computer hardware, combining the precision and predictability of electronic sound creation with the organic unpredictability of physical materials (Geist 2014). His MR-808 Drum Robot reimagines the iconic Roland TR-808 drum machine by emulating its sounds through mechanical reproduction (Geist 2013).

3.4 Sound as Background to HRI

Robotic musicianship is not the only HRI context that involves music. Another prominent application is dance, a feature standard to most commercial robotic agents by now. In this case, however, robots are not involved in the music creation process, but instead music provides a backdrop for interactions to take place. While the creation of robotic dance, similar to robotic musicianship, is a complex endeavour that involves various disciplines that go beyond the scope of this chapter (see (Peng et al. 2015) for a systematic review of dance in robotics). However, its aims in the context of human–robot interaction can be summarised as (1) creating rich and interesting social behaviour (Kozima et al. 2009; Michalowski 2010; Vircíková et al. 2011; Vircíková and Sinčák 2010; Vircikova and Sincak 2010); (2) non-verbally expressing emotion and intention (Beck et al. 2010; McColl and Nejat 2014); and (3) increasing entertainment, novelty, and commercial value (Fujita et al. 2005; Kac 1997; Yoshida et al. 2012). A prominent example of a dancing embodied agent is the small creature-like robot Keepon by Michalowsky et al., which extracts rhythm from auditory and visual stimuli in its environment (Michalowski et al. 2007). Through analysis of videotaped interactions with children, they showed that a synchronisation of the robot's movement with music playing in its surroundings affected children's engagement, suggesting dance as a promising application for working towards rhythmically synchronised social interactions. Another implementation of robotic dance was done by Seo et al. for the robot platform DARwIn-OP. They presented a movement generation system combining real-time rhythm detection with a gesture vocabulary based on Laban movement analysis (Seo et al. 2013). Lin et al. created Disco Lamp, an interactive robotic lamp inspired by the famous lamp from the logo of animation studio Pixar. When in performance mode, the lamp analyses the frequency distribution in environmental sound and uses this to control servo motors across its body and head (Lin et al. 2014).

Robotic dance's inherent power to engage through shared musical experience was expanded upon in studies investigating shared listening experiences. Hoffman and Vanunu developed the robotic companion Travis to investigate how a robotic listening companion might influence people's music experiences (Fig. 3.4).

They found that the rhythmical head nodding of the robot influenced the participant's enjoyment of the song being played. Additionally, it also positively influenced their impressions of the robot. It should be pointed out that this effect operated subconsciously, as participants were not aware of the difference between on-beat and off-beat robot movement, yet they were influenced by it (Hoffman and Vanunu 2013). In later work, they introduced the notion of robotic experience companionship, extending their studies to include shared video watching (Hoffman et al. 2016).

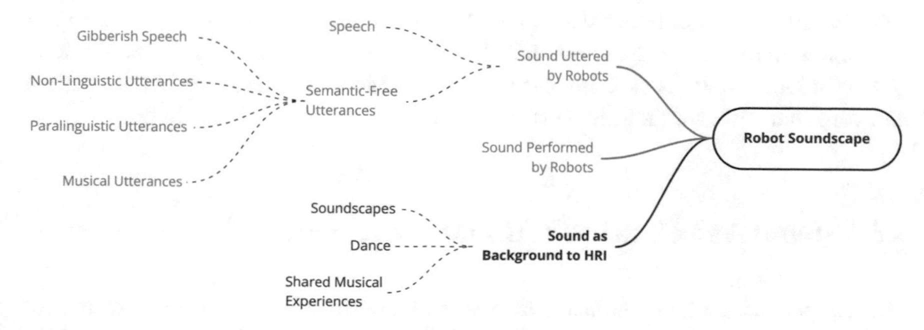

Fig. 3.4 Sound as background to HRI scenarios

Their studies indicated that these shared experiences increased the number of positive human character traits attributed to the robot. A similar shared listening setting, but in this instance controlled by the robot, was presented by Zhang et al. Robotic Prototype ROMO detects emotional cues in children with autism spectrum disorder and uses that information to generate context-appropriate real-time sonification (Zhang et al. 2016, 2015). By generating musical material as background for assistive HRI contexts, its creators aim to promote emotional and social communication. However, no validation experiments were performed to assess this claim. This process of user-responsive music generation was later implemented as part of a multi-modal adaptive framework for emotional engagement in child-robot interaction (Javed et al. 2018).

Little work in HRI has focused on using non-musical sound as a backdrop to HRI scenarios, although the effects that have been demonstrated are intriguing. Thiessen et al. investigated the effect the infrasound on a NAO robot's affective communication (Thiessen et al. 2019). Infrasound is inaudible, low-frequency vibration in the range of around 16–20 Hz (Berglund et al. 1996) and is impossible to clearly localise, due to its large wavelengths (Middlebrooks and Green 1991). When this type of sound coincided with a robot's affective gestures, they found that the basic emotion happiness was perceived more strongly without participants being aware of the infrasound's presence. While the underlying mechanism behind this is unclear, the authors noted that the "presence of infrasound will impact how people interact with robots" (Thiessen 2019, p. 17). Komatsu et al. used what they call "vibrational artificial subtle expressions" to convey a system's confidence level (Komatsu et al. 2018). They found that it is possible to communicate system information through variations in the vibrations frequency. While the communicative potential of pitch contours, prosody and intonation is well established in various forms of utterances, this showed that these notions stay effective even when applied to the related, but distinct, modality of vibration. While both of these cases show some resemblance to utterances presented in Sect. 3.2, especially with regards to their aim of communicating affect and intention, their subtle nature makes it difficult to define them as social cues. While the use of subtle, non-musical sound is still largely unexplored as a modality in human–robot interaction, it has shown to influence humans in various ways in other fields. Infrasound, for example, can affect blood pressure (Danielsson and Landstrom 1985), influence concentration and cognitive performance (Harris and Johnson 1978), and impact balance and situational awareness (Evans and Tempest 1972). Ambient soundscapes can make listeners feel more or less safe (Sayin et al. 2015), induce emotions of calm and vibrancy (Cain et al. 2013), or impact patient recovery in healthcare facilities (Loupa 2020).

3.5 Sound Associated with Robot Movement

This chapter has so far examined sound as a deliberate utterance, sound created for musical purposes, and sound as backdrop to HRI scenarios. The next two sections will focus on sound associated with robot movement and sound dependant on user

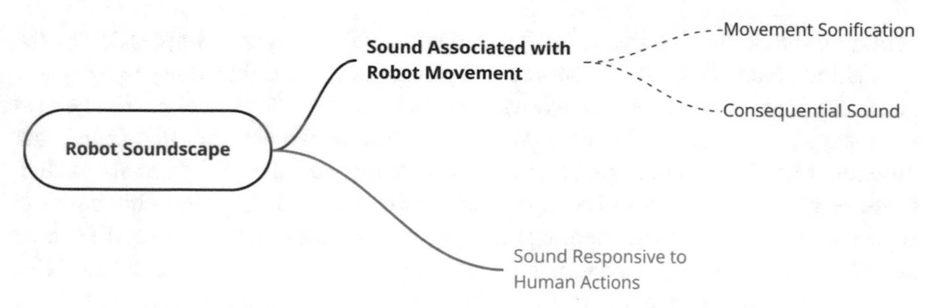

Fig. 3.5 Sound associated with robot movement

behaviour (see Fig. 3.5). Even if sound is not an intentional part of a robot's communication, it always accompanies interactions and always affects the listener. The saying that true silence exists only in space remains true to any interaction involving robots in motion.

One perspective on sound associated with robot movement is the categorisation of *internal* and *external auditory cues* proposed by Cha et al. (2018b) in their survey of non-verbal signalling methods in HRI. Differentiating these cues by their source and method of production, Cha et al. define internal auditory cues as sounds "generated entirely by the human body," such as sighs or groans. External auditory cues, in contrast, arise from physical interactions with the environment, such as footsteps (Cha 2018b, p. 271). Internal auditory cues have clear parallels to the various forms of utterances mentioned in this chapter, and Cha et al. mention how they tend to communicate information about internal states like mood or intention. External auditory cues are said to provide more functional information.

Another perspective on sound associated to robot movement can be found in HRI studies examining *consequential sound*, defined as "a consequence of the moving parts of products" (Van Egmond 2008, p. 69). In the context of robotics, this could be the sound of a motor, for example. Studies on consequential sound have shown that it influences a human's perception of the robot as a whole (Moore et al. 2017, 2019; Tennent et al. 2017). Human–robot interaction, therefore, appears to be affected by a robot's sound in an implicit way that so far has received little attention. Non-auditory forms of implicit communication have been shown to have a positive effect on human–robot interaction (Breazeal et al. 2005), indicating that implicit auditory communication could be a valuable tool in shaping human–robot interactions.

3.5.1 Consequential Sound

A range of studies have shown that consequential sound can influence human–robot interactions. Participants of a study by Trovato and colleagues were concerned about their safety when hugging a robot, because the noise emitted from the robot's hand was threatening (Trovato et al. 2016). In a study involving robotic seal Paro, Inoue et al. found that the motor sound left participants with a negative impression of the interaction (Inoue et al. 2008). Moore and colleagues found that study participants generally disliked sound associated with a robot's motors (Moore et al. 2017). Frid et al. explored the servo sounds associated with affective gestures and found that "sounds inherent to robot movement can communicate other affective states than those originally intended to be expressed" (Frid et al. 2018, p. 9). This claim is supported by various studies demonstrating that consequential sound can affect how humans perceive robot characteristics such as trustworthiness (Moore et al. 2019), precision and strength (Moore et al. 2017), or competence (Tennent et al. 2017). An interesting manifestation of consequential sound in the arts can be found in Woolf and Bech's robotic artwork *Boundless in Space*. This robot consists of a textured organic surface and a mobile base. Ultrasonic proximity sensors are attached to the robot's sides, giving it the possibility to flee across the installation space if spectators come too close. The electronics facilitating movement contain a large number of relay switches, meaning a sudden change of direction and onset of movement is accompanied by "sharp rhythmic clicking sounds" that are a byproduct of the switches but "resemble panicked warning signals or please for help" (Woolf and Bech 2002, p. 32). This is an intriguing example of consequential sound that fulfils a clear communicative purpose. These challenges have also been identified by industry, with Microsoft's researchers applying active noise cancellation–using artificial sound to cancel out or lower the volume of consequential sound–to an unspecified robot (Ikeuchi et al. 2020).

The ability of sound to communicate subtext and influence how we perceive the world around us is well known in other disciplines, where this communication channel is actively used. In the domain of product sound design, for example, the user rarely notices the designed sound elements, but instead subconsciously perceives the product characteristics the sound designer aims to convey. For example, the sound of shutting a car door conveys that the vehicle is difficult to break into (Özcan and van Egmond 2006), the subtle click of an expensive lighter communicates the quality of its materials (Lageat et al. 2003), and the fizz when opening a carbonated drink emphasises its freshness (Spence and Wang 2015). Using a process called active sound design (ASD), car manufacturers add artificial engine noise to their vehicles to enhance the experience of the driver. In the case of electric vehicles like the Jaguar I-Pace, ASD no longer augments engine sound but replaces it entirely. In some instances, sound directly impacts what we perceive through other senses. Jousmäki and Hari conducted an experiment where participants were asked to rub their hands together. This process was then captured with a microphone and amplified. Participants who had a brighter sound accompanying this action reported having dryer hands (Jousmäki and Hari 1998). Watanabe and Shimojo demonstrated how the motion of two objects on a screen would be described as either "passing through" or "bouncing off" each other, depending on which sound accompanied the movement (Watanabe and Shimojo 2001).

As part of their work towards defining a design space for the targeted use of consequential sound in HRI, Frid and colleagues propose the masking or enhancing of consequential robot sound (Frid and Bresin 2021; Frid et al. 2018). They reference the process of *blended sonification*, originally defined by Tünnermann and colleagues as changing a sound's perception by combining it with artificial audio elements that convey additional information while preserving the coherence of the original sound event (Tünnermann et al. 2013). Blended sonification was somewhat validated by Trovato and colleagues, who showed that the negative response to a simulated robot motor sound could be mitigated by combining it with a soundscape (Trovato et al. 2018). Their experiment explored whether blended sonification on a static robot with a running motor would impact how close participants approached the robot. While they identified no significant effect on participant distance, they found that participants rated the robot more favourably when the motor sound was masked with additional sound material. A considerably earlier study involving designed motor sound can be seen in work by Johannsen (Johannsen 2001, 2002, 2004). They mixed together musical utterances with a range of robot motion audio recordings to let the user localise the robot even if it is out of view. Additionally, the pitch and timbre of the audio recordings was adjusted to better blend with the musical elements. A more recent study involving designed consequential sound was conducted by Cha et al., who used modified sound recordings of servo motors belonging to the TurtleBot robot, and played them through a different robot while it was performing a collaborative task with participants (Cha et al. 2018a). Their results showed that combining a robot's consequential sound with masking noise made it more localisable, thereby aiding task performance, and was rated less annoying by participants. Cha et al. describe the sounds in this study as auditory icons, designed to resemble and enhance a robot's natural sound cues in a way that makes them intuitive for listeners to understand. While the content of the communication–the location of the robot–is rather focused in this case, the link to Gaver's terminology is clear. Another parallel can be drawn to the representational earcons defined by Blattner and colleagues (Blattner et al. 1989). One of the defining features that differentiate robotic agents from virtual agents is their embodiment. All robotic motion is a physical process involving both the robot's environment and its internal mechanics. As a result, the use of auditory icons does not only draw connections to the listener's prior experience with physical processes but can directly represent the physicality of the agent.

While previous studies have demonstrated the impact of consequential sound and argued for using this channel of communication as a design space in HRI, researchers have only recently begun applying and investigating this modality. Several studies indicate that the addition of sound to a motion could alter how robotic gesture is perceived. In a paper discussing the potential of robotic gesture as a communication channel, Hoffman and Ju note how "well-designed robot motion can communicate, engage, and offer dynamic possibilities beyond the machines' surface appearance or pragmatic motion paths" (Hoffman and Ju 2014, p. 89). Sound could potentially be used as an additional modality for this context, making robotic gesture appear more controlled, elegant, or expressive. Adding sound to robot motion–movement sonification–will be the focus of the next section.

3.5.2 Movement Sonification

Even though the term *movement sonification* usually describes the practise of adding sound to human movement, in this chapter we will define it as designing sound to coincide with robotic movement. This area has only recently received attention in the HRI community, with studies examining the sonification of expressive robot gestures. The SONAO project, for example, explored this area of research as a way to enrich interactions with a NAO robot (Frid et al. 2018). Dahl and colleagues explored the relationship between movement quality and sound characteristics by having professional musicians sonify robot motion (Dahl et al. 2017), which then informed their conceptual framework for robot sound synthesis (Bellona et al. 2017). Latupeirissa and colleagues investigated the sound emissions of several robots in science fiction and argued that fictional robot sound characteristics could provide a basis for conveying internal states and sonifying robot gesture (Latupeirissa et al. 2019). None of these implementations were validated using perceptual studies involving non-experts. Recent work by Zahray and colleagues, however, explored mapping robot motion data to various sound parameters and asked study participants to rate enjoyment and appeal, and to describe the perceived movement information (Zahray et al. 2020). They found that sounds with emotional subtext and musical elements were generally favoured by the listeners. Recent work by the authors explored sonifying robot gesture using various sound sources and found that accompanying robotic movement with musical sound affects how humans rate movement characteristics (Robinson et al. 2021). Sonifying a single robot motion sequence with different sound accompaniments made participants describe the motion as more or less elegant, precise, uncontrolled, or jerky, amongst others.

In the domain of utterances, researchers have synchronised affective expressions with robot motion, which could also be considered as movement sonification. Aiming to provide more realistic speech, Otsuka et al. presented a voice manipulation method that takes robot head movement into account (Otsuka et al. 2009). Based on an analysis of human vocalisations, their implementation modulates the timbre of a robot's voice depending on its head's pitch-axis rotation. Jee et al. aimed to synchronise their musical utterances with a robot's motion trajectories to make the emotion expression more "effective." To this end, they designed the musical structure of their utterances in a way that allowed for repeating short sections of the music. Utterance length could then be adjusted to coincide with various robot movements without running out or being cut off (Jee et al. 2009). With the aim of increasing the emotional impact of affective gestures, Bramas and colleagues built a system which synchronised various utterances such as sound effects, voice recording snippets or musical extracts with a robot's motion (Bramas et al. 2008). Their system modulated sound characteristics like prosody and volume and applied various audio processing techniques. They did not aim to communicate particular affective states, but rather to investigate the temporal progressing of utterances and their interplay with other modalities. This work did not include an evaluation study.

This notion of increasing the impact of what is being conveyed by synchronising multiple modalities can also be found in the infrasound work by Thiessen and colleagues, which was described in Sect. 3.4 (Thiessen et al. 2019). In that particular case, robotic motion coincided with diffuse vibrations in the environment and did in fact impact how expressions were perceived. A clear increase of impact was, however, not shown. A technical implementation by Schwenk and Arras used their robot Daryl's movement data to created paralinguistic utterances in real time (Schwenk and Arras 2014). Using what they call "motion modulation," they changed timbre and pitch of the robot's utterances. This allowed them to combine expressive movements of ears, head and body with expressive non-linguistic utterances. This case provides an interesting example of blending sound categories. While Schwenk and Arras's sonic language is clearly based in utterances, their direct link of motion and sound makes the implementation a clear example of movement sonification. Furthermore, these utterances directly blend with the robotic agent's consequential sound. This could potentially mask or enhance Daryl's natural sound profile. The authors did not report on the consequential sound element and removed it in their video demonstration of the system. Savery et al. explored the interplay between robot gestures and emotional prosody, using robot movement that was tightly synced with musical phrases to successfully convey a range of emotions to the listener (Savery et al. 2019). An example of a commercial social robot blending non-linguistic utterances with movement is robot Vector, initially created by Anki, a consumer robot manufacturer based in California. In a patent filing, the company describes enhancing Vector's consequential sound with artificial sound material (Wolford et al. 2019). Using what they call "condition-based audio techniques," they synchronise audio material to the robot's gestures. The sound is additionally modified based on the robot's mood at any given time, meaning the motion of a sad robot would sound different from that of a joyful one. A detailed investigation into the motivation behind, and implementation of these audio techniques can be found in our recent interview with Anki's lead audio designer (Robinson et al. 2022).

3.6 Sound Responsive to Human Actions

We now move to the final shape that sound can take in human–robot interactions: sound responsive to human actions. Bartneck and Forlizzi define interactivity in social robots as "having the potential to exhibit causal behaviour that is to respond in reaction to interaction with a human" (Bartneck and Forlizzi 2004, p. 593). Given that the fundamental purpose of social robots is to interact with the humans around them, it can be argued that all sound emitted by a robot and discussed in this chapter is in some way dependant on user behaviour. This final section focuses on changes in sound that are *directly* linked to user behaviour (e.g., adjusting speech volume according to user distance), rather than part of a multimodal response by the robot (e.g., answering a question) (Fig. 3.6).

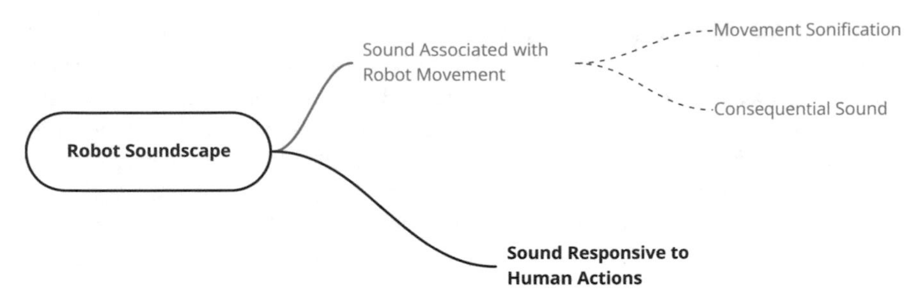

Fig. 3.6 Sound responsive to human actions

In perhaps the most functional and straightforward example of making robotic sound responsive to user behaviour, Brock and Martinson propose the notion of "auditory perspective taking" to optimise a robot's speech intelligibility. They suggest four measures adaptive systems might take: facing the listener, adjusting speaking volume, pausing when environmental sound is too loud, and moving to another location when environmental sound persists (Brock and Martinson 2006). While these measures are natural aspects of human-to-human communication, they involve a range of relevant audio concepts that are rarely applied in human–robot interaction. Facing the listener, for example, builds on the dispersion characteristics of the human voice: high-frequency content containing consonants vital to intelligibility is most clearly audible in front of the speaker. NAO and Pepper, two of the most widely used social robots, have speakers on the side of their head, meaning the action of facing the listener does not enhance speech intelligibility. Another example of user-reactive sonic behaviour in HRI is the sonic interaction implementation by Schwenk and Arras mentioned in Sect. 5.2. Next to the already discussed utterances linked to robot motion, they additionally implemented behaviour that continuously sonified the distance between a human and the robot (Schwenk and Arras 2014). Their goal was to create a "reactive sonic feedback on people's proximity to the robot" (Schwenk and Arras 2014, p. 166).

Other types of responsive sound in HRI are notably removed from the notions of speech and utterances. Consequently, few studies explore their effects on the listener. However, with this shift of focus from functional communication to engaging and novel interactive experiences, robotic artworks become a valuable source of insight. The robotic artwork *Echidna* by Woolf and Bech is a fist-sized tangle of wires, placed on top of a podium that conceals the underlying electronics. The wire is the robot's body as well as its sensor, as it functions as an antenna that captures variations in the electromagnetic field around it. Visitor movement (gesture and proximity) causes these variations, which are subsequently sonified, giving a voice to the artwork. The chaotic nature of the sensor input leads to a wide range of sonic behaviour, which, in Woolf and Bech's words, "reflects not the sophistication of the underlying electronics, but the complexity of the environment in which the sculpture is situated" (Woolf and Bech 2002, p. 32). The works of artist Peter Vogel can in some way be seen as responsive robotic artworks. The skilful circuit designer creates interactive sound objects that are entirely made of openly visible analogue electronics (Martin and Gleeson 2011). In works such as his *Sound Wall*, light sensors pick up movement and behaviour from gallery visitors and translates them into complex, entirely non-digital electronic soundscapes (Vogel 1979). Consequently, the artwork is a combination of the sculpture itself and its response to the presence of the audience.

Given the fact that responsive auditory feedback is rare in non-artistic human–robot interaction contexts, one might ask whether this type of sound even has a place in social robotics. Writing about the potential futures of interaction between humans and artificial life, Rinaldo envisions a "cybernetic ballet of experience, with the computer/machine and viewer/participant involved in a grand dance of one sensing and responding to the other" (Rinaldo 1998, p. 375). Having a robot respond to a human's proximity or gestures with sonic cues may non-verbally convey that their presence is being perceived and acknowledged, providing a subtle and continuous answer to the question "Is this robot aware of my presence?" By linking the agent's sound to parameters like a person's proximity or gestural information, users might get an indicator that their presence is perceived and acknowledged by the robot, providing a continuous and subtle answer to the question "Does this robot know I am here?".

Woolf and Beck state, that "reactive robots often behave in ways that belie their apparent simplicity" (Woolf and Bech 2002, p. 33). One might argue that this can facilitate the creation of rich and engaging experiences that are not determined or limited by a robot's technical capabilities. A similar argument is made for semantic-free utterances, which provide a channel of communication that does not bring with it the expectation of fully realised natural human–robot conversation.

Forlizzi and Battarbee view social robots as products that can facilitate co-experience and social interaction (Forlizzi and Battarbee 2004). This notion of co-experience allows the connection to more general ideas around interactivity in media art. When looking at robotic agents as mobile interactive media installations, one might imagine various other applications for them. For example, interactive environments encourage exploration and appreciation of spaces (Hespanhol and Tomitsch 2012). In museum contexts, they increase engagement (Hornecker and Stifter 2006) and a sense of social connectedness amongst visitors (Hu and Le 2013; Roussou 1999). In the words of Urbanowicz and Nyka, "such art form is used to create lively spaces, where people may interact with the installation itself or with one another" (Urbanowicz and Nyka 2016, p. 591). These case studies may provide interesting perspectives on interactions between robots and groups of people, and we hope to see more diverse applications of interactive sound in human–robot interaction in the future.

3.7 Conclusion

The contribution of this chapter is a map of the robot soundscape. We identified the various contexts in which sound occurs in HRI: sound uttered by robots, sound performed by robots, sound as background to HRI scenarios, sound associated with robot movement, and sound responsive to human actions. In doing so, we additionally identified some trends regarding what types of sound HRI researchers predominantly focus on, and which are comparatively underexplored, despite being well established in other disciplines and potentially beneficial to HRI. While utterances and robot musicianship have both received considerable attention over the years, consequential sound and movement sonificiation have only recently been more thoroughly examined. Few studies have looked into sound as a background to HRI scenarios, and sonifications of human actions and application examples are more likely found in other areas, such as ambient soundscapes and interactive media art. Through bringing all these contexts together in a shared design space, we hope to motivate designers and researchers to more thoroughly explore sound over the coming years, working towards providing robots with a broad sonic palette that moves beyond speech to engage and affect people across cultures and languages.

References

Aarestrup M, Jensen LC, Fischer K (2015) The sound makes the greeting: interpersonal functions of intonation in human-robot interaction. In: 2015 AAAI spring symposium series

Atkeson CG, Hale JG, Pollick F, Riley M, Kotosaka S, Schaul S, Shibata T, Tevatia G, Ude A, Vijayakumar S et al (2000) Using humanoid robots to study human behavior. IEEE Intell Syst Appl 15(4):46–56

Ayesh A (2006) Structured sound based language for emotional robotic communicative interaction. In: ROMAN 2006—The 15th IEEE international symposium on robot and human interactive communication. IEEE, pp 135–140

Ayesh A (2009) Emotionally expressive music based interaction language for social robots. ICGST Int J Autom Robot Auton Syst 9(1):1–10

Bartneck C, Forlizzi J (2004) A design-centred framework for social human-robot interaction. In: RO-MAN 2004. 13th IEEE international workshop on robot and human interactive communication. IEEE, Kurashiki, Okayama, Japan, pp 591–594. https://doi.org/10.1109/ROMAN.2004.1374827

Barton S (2013) The human, the mechanical, and the spaces in between: explorations in human-robotic musical improvisation. In: Ninth artificial intelligence and interactive digital entertainment conference

Beck A, Hiolle A, Mazel A, Cañamero L (2010) Interpretation of emotional body language displayed by robots. In: Proceedings of the 3rd international workshop on affective interaction in natural environments, pp 37–42

Becker-Asano C, Kanda T, Ishi C, Ishiguro H (2011) Studying laughter in combination with two humanoid robots. AI & Soc 26(3):291–300

Becker-Asano C, Ishiguro H (2009) Laughter in social robotics-no laughing matter. In: Intl. workshop on social intelligence design. Citeseer, pp 287–300

Bellona J, Bai L, Dahl L, LaViers A (2017) Empirically informed sound synthesis application for enhancing the perception of expressive robotic movement. In: Proceedings of the 23rd international conference on auditory display—ICAD 2017. The International Community for Auditory Display, University Park Campus, pp 73–80. https://doi.org/10.21785/icad2017.049

Belpaeme T, Baxter P, Read R, Wood R, Cuayáhuitl H, Kiefer B, Racioppa S, Kruijff-Korbayová I, Athanasopoulos G, Enescu V et al (2013) Multimodal child-robot interaction: building social bonds. J Hum Robot Interact 1(2):33–53

Berglund B, Hassmen P, Job RS (1996) Sources and effects of low frequency noise. J Acoust Soc Am 99(5):2985–3002
Bethel CL, Murphy RR (2006) Auditory and other non-verbal expressions of affect for robots. In: AAAI fall symposium: aurally informed performance, pp 1–5
Blattner M, Sumikawa D, Greenberg R (1989) Earcons and icons: their structure and common design principles. Hum Comput Interact 4(1):11–44 (1989). https://doi.org/10.1207/s15327051hci0401_1. Accessed 24 Sept 2019
Bramas B, Kim YM, Kwon, DS (2008) Design of a sound system to increase emotional expression impact in human-robot interaction. In: 2008 international conference on control, automation and systems. IEEE, pp 2732–2737
Breazeal CL (2000) Sociable machines: expressive social exchange between humans and robots. PhD Thesis, Massachusetts Institute of Technology
Breazeal CL (2004) Designing sociable robots. MIT Press
Breazeal C, Kidd CD, Thomaz AL, Hoffman G, Berlin M (2005) Effects of nonverbal communication on efficiency and robustness in human-robot teamwork. In: 2005 IEEE/RSJ international conference on intelligent robots and systems. IEEE, Edmonton, Alta., Canada, pp 708–713. https://doi.org/10.1109/IROS.2005.1545011
Breazeal C, Dautenhahn K, Kanda T (2016) Social robotics. In: Siciliano B, Khatib O (eds) Springer handbook of robotics. Springer, Cham, pp 1935–1972. https://doi.org/10.1007/978-3-319-32552-172
Bretan M, Cicconet M, Nikolaidis R, Weinberg G (2012) Developing and composing for a robotic musician using different modes of interaction. In: ICMC
Bretan M, Weinberg G (2016) A survey of robotic musicianship. Commun ACM 59(5):100–109
Brock DP, Martinson E (2006) Using the concept of auditory perspective taking to improve robotic speech presentations for individual human listeners. In: AAAI fall symposium: aurally informed performance, pp 11–15
Burkhardt F, Sendlmeier WF (2000) Verification of acoustical correlates of emotional speech using formant-synthesis. In: ISCA tutorial and research workshop (ITRW) on speech and emotion
Cahn JE (1990) The generation of affect in synthesized speech. J Am Voice I/o Soc 8(1):1–1
Cain R, Jennings P, Poxon J (2013) The development and application of the emotional dimensions of a soundscape. Appl Acoust 74(2):232–239. https://doi.org/10.1016/j.apacoust.2011.11.006
Cha E, Cha E, Kim Y, Fong T, Matarić MJ (2018b) A survey of nonverbal signaling methods for non-humanoid Robots. Found Trends in Robot 6(4):211–323. https://doi.org/10.1561/2300000057
Cha E, Mataric M (2016) Using nonverbal signals to request help during human-robot collaboration. In: 2016 IEEE/RSJ international conference on intelligent robots and systems (IROS). IEEE, Daejeon, South Korea, pp 5070–5076. https://doi.org/10.1109/IROS.2016.7759744
Cha E, Fitter NT, Kim Y, Fong T, Matarić MJ (2018a) Effects of robot sound on auditory localization in human-robot collaboration. In: Proceedings of the 2018a ACM/IEEE international conference on human-robot interaction—HRI '18. ACM Press, Chicago, IL, USA, pp 434–442. https://doi.org/10.1145/3171221.3171285
Chadefaux D, Le Carrou JL, Vitrani MA, Billout S, Quartier L (2012) Harp plucking robotic finger. In: 2012 IEEE/RSJ international conference on intelligent robots and systems. IEEE, pp 4886–4891
Chang RC-S, Lu H-P, Yang P (2018) Stereotypes or golden rules? Exploring likable voice traits of social robots as active aging companions for tech-savvy baby boomers in Taiwan. Comput Hum Behav 84:194–210. https://doi.org/10.1016/j.chb.2018.02.025
Chomsky N (1956) Three models for the description of language. IRE Trans Inf Theory 2(3):113–124
Cicconet M, Bretan M, Weinberg G (2013) Human-robot percussion ensemble: Anticipation on the basis of visual cues. IEEE Robot Autom Mag 20(4):105–110
Connell J (2018) Extensible grounding of speech for robot instruction. arXiv preprint arXiv:1807.11838

Cosentino S, Takanishi A (2021) Human–robot musical interaction. In: handbook of artificial intelligence for music. Springer, Heidelberg, pp 799–822

D'Mello S, McCauley L, Markham J (2005) A mechanism for human-robot interaction through informal voice commands. In: ROMAN 2005. IEEE international workshop on robot and human interactive communication. IEEE, pp 184–189

Dahl L, Bellona J, Bai L, LaViers A (2017) Data-driven design of sound for enhancing the perception of expressive robotic movement. In: Proceedings of the 4th international conference on movement computing—MOCO '17. ACM Press, London, United Kingdom, pp 1–8. https://doi.org/10.1145/3077981.3078047

Danielsson A, Landstrom Ü (1985) Blood pressure changes in man during infrasonic exposure: an experimental study. Acta Med Scand 217(5):531–535

Dannenberg RB, Brown B, Zeglin G, Lupish R (2005) McBlare: a robotic bagpipe player. In: Proceedings of the 2005 conference on new interfaces for musical expression. National University of Singapore, pp 80–84

Davies A, Crosby A (2016) Compressorhead: the robot band and its transmedia storyworld. In: Koh JTKV, Dunstan BJ, Silvera-Tawil D, Velonaki M (eds) Cultural robotics. Springer, Cham, pp 175–189

Dingler T, Lindsay J, Walker BN (2008) Learnabiltiy of sound cues for environmental features: auditory icons, earcons, spearcons, and speech. In: Proceedings of the 14th international conference on auditory display. Paris, France, June 24–27

Esnaola U, Smithers T (2005) MiReLa: A musical robot. In: 2005 International symposium on computational intelligence in robotics and automation. IEEE, pp 67–72

Evans MJ, Tempest W (1972) Some effects of infrasonic noise in transportation. J Sound Vib 22(1):19–24

Eyssel F, Kuchenbrandt D, Bobinger S (2012) 'If you sound like me, you must be more human': on the interplay of robot and user features on human-robot acceptance and anthropomorphism, vol 2

Fernald A, Mazzie C (1991) Prosody and focus in speech to infants and adults. Dev Psychol 27(2):209

Fernald A, Taeschner T, Dunn J, Papousek M, de Boysson Bardies B, Fukui I (1989) A cross-language study of prosodic modifications in mothers' and fathers' speech to preverbal infants. J Child Lang 16(3):477–501

Ferrand D, Vergez C (2008) Blowing machine for wind musical instrument: toward a real-time control of the blowing pressure. In: 2008 16th Mediterranean conference on control and automation. IEEE, pp 1562–1567

Fischer K, Niebuhr O, Jensen LC, Bodenhagen L (2019) Speech melody matters—how robots profit from using charismatic speech. ACM Trans Hum-Robot Interact (THRI) 9(1):1–21

Forlizzi J, Battarbee K (2004) Understanding experience in interactive systems. In: Proceedings of the 2004 conference on designing interactive systems processes, practices, methods, and techniques—DIS '04. ACM Press, Cambridge, MA, USA, p 261. https://doi.org/10.1145/1013115.1013152

Frid E, Bresin R (2021) Perceptual evaluation of blended sonification of mechanical robot sounds produced by emotionally expressive gestures: augmenting consequential sounds to improve non-verbal robot communication. Int J Soc Robot, pp 1–16

Frid E, Bresin R, Alexanderson S (2018) Perception of mechanical sounds inherent to expressive gestures of a nao robot-implications for movement sonification of humanoids. In: Sound and music computing

Friend M (2000) Developmental changes in sensitivity to vocal paralanguage. Dev Sci 3(2):148–162

Fujita M, Sabe K, Kuroki Y, Ishida T, Doi TT (2005) SDR-4X II: A small humanoid as an entertainer in home environment. In: robotics research. The eleventh international symposium. Springer, Heidelberg, pp 355–364

Gaver W (1986) Auditory icons: using sound in computer interfaces. Hum-Comput Interact 2(2):167–177. https://doi.org/10.1207/s15327051hci0202

Geist MS (2013) MR-808 drum robot
Geist MS (2014) Glitch robot
Goodrich MA, Schultz AC et al (2008) Human-robot interaction: a survey. Found Trends Hum Comput Interact 1(3):203–275
Gorostiza JF, Salichs MA (2011) End-user programming of a social robot by dialog. Robot Auton Syst 59(12):1102–1114
Gouaillier D, Hugel V, Blazevic P Kilner C, Monceaux J, Lafourcade P, Marnier B, Serre J, Maisonnier B (2009) Mechatronic design of NAO humanoid. In: 2009 IEEE international conference on robotics and automation. IEEE, pp 769–774
Hallahan WI (1995) DECtalk software: text-to-speech technology and implementation. Digit Tech J 7(4):5–19
Harris CS, Johnson DL (1978) Effects of infrasound on cognitive performance. Aviation, space, and environmental medicine
Hermann T, Hunt A, Neuhoff JG (2011) The sonification handbook. Logos Verlag Berlin
Hespanhol L, Tomitsch M (2012) Designing for collective participation with media installations in public spaces. In: Proceedings of the 4th media architecture biennale conference on participation—MAB '12. ACM Press, Aarhus, Denmark, pp 33–42. https://doi.org/10.1145/2421076.2421082
Hoffman G, Bauman S, Vanunu K (2016) Robotic experience companionship in music listening and video watching. Pers Ubiquit Comput 20(1):51–63. https://doi.org/10.1007/s00779-015-0897-1
Hoffman G, Ju W (2014) Designing robots with movement in mind. J Hum Robot Interact 3(1):89. https://doi.org/10.5898/JHRI.3.1.Hoffman
Hoffman G, Vanunu K (2013) Effects of robotic companionship on music enjoyment and agent perception. In: 2013 8th ACM/IEEE international conference on human-robot interaction (HRI). IEEE, Tokyo, Japan, pp317–324. https://doi.org/10.1109/HRI.2013.6483605
Holzapfel H, Gieselmann P (2004) A way out of dead end situations in dialogue systems for human-robot interaction. In: 4th IEEE/RAS international conference on humanoid robots, vol. 1. IEEE, pp 184–195
Hornecker E, Stifter M (2006) Learning from interactive museum installations about interaction design for public settings. In: Proceedings of the 20th conference of the computer-human interaction special interest group (CHISIG) of Australia on computer-human interaction: design: activities, artefacts and environments—OZCHI '06. ACM Press, Sydney, Australia, p 135. https://doi.org/10.1145/1228175.1228201
Hoyer R, Bartetzki A, Kirchner D, Witsch A, van de Molengraft M, Geihs K (2013) Giving robots a voice: a kineto-acoustic project. In: International conference on arts and technology. Springer, Heidelberg, pp 41–48
Hu J, Le D, Funk M, Wang F, Rauterberg M (2013) Attractiveness of an interactive public art installation. In: International conference on distributed, ambient, and pervasive interactions. Springer, Heidelberg, pp 430–438
Ikeuchi K, Fukumoto M, Lee JH, Kravitz JL, Baumert DW (2020) Noise reduction in robot human communication. Google Patents
Inoue K, Wada K, Ito Y (2008) Effective application of Paro: Seal type robots for disabled people in according to ideas of occupational therapists. In: International conference on computers for handicapped persons. Springer, Heidelberg, pp 1321–1324
Javed H, Jeon M, Park CH (2018) Adaptive framework for emotional engagement in child-robot interactions for autism interventions. In: 2018 15th International conference on ubiquitous robots (UR). IEEE, pp 396–400
Jee E-S, Jeong Y-J, Kim CH, Kobayashi H (2010) Sound design for emotion and intention expression of socially interactive robots. Intel Serv Robot 3(3):199–206. https://doi.org/10.1007/s11370-010-0070-7

Jee ES, Park SY, Kim CH, Kobayashi H (2009) Composition of musical sound to express robot's emotion with intensity and synchronized expression with robot's behavior. In: RO-MAN 2009—The 18th IEEE international symposium on robot and human interactive communication. IEEE, Toyama, Japan, pp 369–374. https://doi.org/10.1109/ROMAN.2009.5326258
Jee ES, Kim CH, Park SY, Lee KW (2007) Composition of musical sound expressing an emotion of robot based on musical factors. In: RO-MAN 2007—The 16th IEEE international symposium on robot and human interactive communication. IEEE, Jeju, South Korea, pp 637–641. https://doi.org/10.1109/ROMAN.2007.4415161
Johannsen G (2001) Auditory displays in human-machine interfaces of mobile robots for non-speech communication with humans. J Intell Rob Syst 32(2):161–169
Johannsen G (2004) Auditory displays in human-machine interfaces. Proc IEEE 92(4):742–758
Johannsen G (2002) Auditory display of directions and states for mobile systems. Georgia Institute of Technology
Johnson WF, Emde RN, Scherer KR, Klinnert MD (1986) Recognition of emotion from vocal cues. Arch Gen Psychiatry 43(3):280–283
Jordà S (2002) Afasia: The ultimate homeric one-man-multimedia-band. In: NIME, pp 132–137
Jousmäki V, Hari R (1998) Parchment-skin illusion: sound-biased touch. Curr Biol 8(6):190–191
Kac E (1997) Foundation and development of robotic art. Art Journal 56(3):60–67
Kadish D (2019) Robophony: A new voice in the soundscape. In: RE: SOUND 2019–8th international conference on media art, science, and technology 8, pp 243–252
Kapur A (2005) A history of robotic musical instruments. In: ICMC. Citeseer
Kapur A (2011) Multimodal techniques for human/robot interaction. In: Musical robots and interactive multimodal systems. Springer, Heidelberg, pp 215–232
Kapur A, Murphy JW, Carnegie DA (2012) Kritaanjali: a robotic harmonium for performance, pedogogy and research. In: NIME
Kato I, Ohteru S, Shirai K, Matsushima T, Narita S, Sugano S, Kobayashi T, Fujisawa E (1987) The robot musician 'wabot-2'(waseda robot-2). Robotics 3(2):143–155
Knoll MA, Uther M, Costall A (2009) Effects of low-pass filtering on the judgment of vocal affect in speech directed to infants, adults and foreigners. Speech Commun 51(3):210–216
Kobayashi T, Fujie S (2013) Conversational robots: an approach to conversation protocol issues that utilizes the paralinguistic information available in a robot-human setting. Acoust Sci Technol 34(2):64–72
Komatsu T (2005) Toward making humans empathize with artificial agents by means of subtle expressions. In: International conference on affective computing and intelligent interaction. Springer, Heidelberg, pp 458–465
Komatsu T, Yamada S, Kobayashi K, Funakoshi K, Nakano M (2010) Artificial subtle expressions: intuitive notification methodology of artifacts. In: Proceedings of the 28th international conference on human factors in computing systems—CHI '10. ACM Press, Atlanta, Georgia, USA, p 1941. https://doi.org/10.1145/1753326.1753619
Komatsu T, Kobayashi K, Yamada S, Funakoshi K, Nakano M (2018) Vibrational artificial subtle expressions: conveying system's confidence level to users by means of smartphone vibration. In: Proceedings of the 2018 CHI conference on human factors in computing systems—CHI '18. ACM Press, Montreal QC, Canada, pp 1–9. https://doi.org/10.1145/3173574.3174052
Komatsu T, Yamada S (2011) How does the agents' appearance affect users' interpretation of the agents' attitudes: Experimental investigation on expressing the same artificial sounds from agents with different appearances. Int J Hum Comput Interact 27(3):260–279. https://doi.org/10.1080/10447318.2011.537209
Korcsok B, Faragó T, Ferdinandy B, Miklósi, Korondi P, Gácsi M (2020) Artificial sounds following biological rules: a novel approach for non-verbal communication in HRI. Sci Rep 10(1):1–13
Kozima H, Michalowski MP, Nakagawa C (2009) Keepon. Int J Soc Robot 1(1):3–18
Krzyżaniak M (20121) Musical robot swarms, timing, and equilibria. J New Music Res:1–19
Lageat T, Czellar S, Laurent G (2003) Engineering hedonic attributes to generate perceptions of luxury: consumer perception of an everyday sound. Mark Lett 14(2):97–109

Latupeirissa AB, Frid E, Bresin R (2019) Sonic characteristics of robots in films. In: Sound and music computing conference, pp 1–6
Lee MK, Kiesler S, Forlizzi J, Srinivasa S, Rybski P (2010) Gracefully mitigating breakdowns in robotic services. In: 2010 5th ACM/IEEE international conference on human-robot interaction (HRI). IEEE, pp 203– 210
Lin HS, Shen YT, Lin TH, Lin PC (2014) Disco lamp: An interactive robot lamp. In: 2014 IEEE international conference on automation science and engineering (CASE). IEEE, pp 1214–1219
Logan-Greene R (2011) Submersions I, University of Washington
Loupa G (2020) Influence of noise on patient recovery. Curr Pollut Rep:1–7
Maes L, Raes G-W, Rogers T (2011) The man and machine robot orchestra at logos. Comput Music J 35(4):28–48
Martin J, Gleeson C (2011) The sound of shadows: Peter Vogel. In: Sounding out the museum: Peter Vogel retrospective exhibition
McColl D, Nejat G (2014) Recognizing emotional body language displayed by a human-like social robot. Int J Soc Robot 6(2):261–280
McNaught W (1892) The history and uses of the sol-fa syllables. Proc Music Assoc 19:35–51
Meanza B (2013) Cicadas
Meyer LB (2008) Emotion and meaning in music. University of Chicago Press
Michalowski MP (2010) Rhythmic human-robot social interaction. Carnegie Mellon University
Michalowski MP, Sabanovic S, Kozima H (2007) A dancing robot for rhythmic social interaction. In: Proceedings of the ACM/IEEE international conference on human-robot interaction, pp 89–96
Middlebrooks JC, Green DM (1991) Sound localization by human listeners. Annu Rev Psychol 42(1):135–159
Miklósi A, Korondi P, Matellán V, Gácsi M (2017) Ethorobotics: a new approach to human-robot relationship. Front Psychol 8:958
Moore RK (2014) Spoken language processing: time to look outside? In: International conference on statistical language and speech processing. Springer, Heidelberg, pp 21–36
Moore RK (2017) Appropriate voices for artefacts: some key insights. http://vihar-2017.vihar.org/assets/papers/VIHAR-2017_paper_8.pdf
Moore D, Tennent H, Martelaro N, Ju W (2017) Making noise intentional: a study of servo sound perception. In: Proceedings of the 2017 ACM/IEEE international conference on human-robot interaction—HRI '17. ACM Press, Vienna, Austria, pp 12–21
Moore D, Dahl T, Varela P, Ju W, Næs T, Berget I (2019) Unintended consonances: methods to understand robot motor sound perception. In: Proceedings of the 2019 CHI conference on human factors in computing systems—CHI '19. ACM Press, Glasgow, Scotland UK, pp 1–12. https://doi.org/10.1145/3290605.330073
Mozos OM, Jensfelt P, Zender H, Kruijff GJM, Burgard W (2007) From labels to semantics: an integrated system for conceptual spatial representations of indoor environments for mobile robots. In: ICRA workshop: semantic information in robotics. Citeseer
Mubin O, Bartneck C, Feijs L (2009) What you say is not what you get: arguing for artificial languages instead of natural languages in human robot speech interaction. In: Proceedings of the spoken dialogue and human-robot interaction workshop at IEEE RoMan 2009, Toyama
Nakagawa K, Shiomi M, Shinozawa K, Matsumura R, Ishiguro H, Hagita N (2013) Effect of robot's whispering behavior on people's motivation. Int J Soc Robot 5(1):5–16
Németh G, Olaszy G, Csapó TG (2011) Spemoticons: text to speech based emotional auditory cues. Int Commun Auditory Display
Olaszy G, Németh G, Olaszi P, Kiss G, Zainkó C, Gordos G (2000) Profivox—A Hungarian text-to-speech system for telecommunications applications. Int J Speech Technol 3(3–4):201–215
Otsuka T, Nakadai K, Takahashi T, Komatani K, Ogata T, Okuno HG (2009) Voice quality manipulation for humanoid robots consistent with their head movements. In: 2009 9th IEEE-RAS international conference on humanoid robots. IEEE, Paris, France, pp 405–410. https://doi.org/10.1109/ICHR.2009.5379569T

Oudeyer P-Y (2003) The production and recognition of emotions in speech: features and algorithms. Int J Hum Comput Stud 59(1–2):157–183

Özcan E, van Egmond R (2006) Product sound design and application: an overview. In: Proceedings of the fifth international conference on design and emotion, Gothenburg

Pan Y, Kim MG, Suzuki K (2010) A robot musician interacting with a human partner through initiative exchange. In: NIME, pp 166–169

Peng H, Zhou C, Hu H, Chao F, Li J (2015) Robotic dance in social robotics—a taxonomy. IEEE Trans Hum Mach Syst 45(3):281–293

Pinker S (1989) Language acquisition. Found Cognitive Sci:359–400

Prendinger H, Becker C, Ishizuka M (2006) A study in users' physiological response to an empathic interface agent. Int J Humanoid Rob 3(03):371–391

Read R, Belpaeme T (2012) How to use non-linguistic utterances to convey emotion in child-robot interaction. In: Proceedings of the seventh annual ACM/IEEE international conference on human robot interaction—HRI '12. ACM Press, Boston, Massachusetts, USA, p 219 https://doi.org/10.1145/2157689.2157764

Read R, Belpaeme T (2014) Situational context directs how people affectively interpret robotic non-linguistic utterances. In: Proceedings of the 2014 ACM/IEEE international conference on human-robot interaction—HRI '14. ACM Press, Bielefeld, Germany, pp 41–48. https://doi.org/10.1145/2559636.2559680

Read R, Belpaeme T (2016) People interpret robotic non-linguistic utterances categorically. Int J Soc Robot 8(1):31–50. https://doi.org/10.1007/s12369-015-0304-0

Remez RE, Rubin PE, Pisoni DB, Carrell TD (1981) Speech perception without traditional speech cues. Science 212(4497):947–949

Rinaldo KE (1998) Technology recapitulates phylogeny: artificial life art. Leonardo:371–376

Ritschel H, Aslan I, Mertes S, Seiderer A, André E (2019) Personalized synthesis of intentional and emotional non-verbal sounds for social robots. In: 2019 8th international conference on affective computing and intelligent interaction (ACII). IEEE, pp 1–7

Robinson FA, Bown O, Velonaki M (2022) Designing sound for social robots: candidate design principles. Int J Soc Robot. https://doi.org/10.1007/s12369-022-00891-0

Robinson FA, Velonaki M, Bown O (2021) Smooth operator: tuning robot perception through artificial movement sound. In: Proceedings of the 2021 ACM/IEEE international conference on human-robot interaction, pp 53–62

Rossi S, Dell'Aquila E, Bucci B (2019) Evaluating the emotional valence of affective sounds for child-robot interaction. In: International conference on social robotics. Springer, Heidelberg, pp 505–514

Roussou M (1999) High-end interactive media in the museum. In: ACM SIGGRAPH 99 conference abstracts and applications on—SIGGRAPH '99. ACM Press, Los Angeles, California, USA, pp 59–62. https://doi.org/10.1145/311625.311682

Savery R, Zahray L, Weinberg G (2020) Emotional musical prosody for the enhancement of trust in robotic arm communication. arXiv preprint arXiv:2009.09048

Savery R, Rose R, Weinberg G (2019) Establishing human-robot trust through music-driven robotic emotion prosody and gesture. In: 2019 28th IEEE international conference on robot and human interactive communication (RO-MAN). IEEE, pp 1–7

Savery R, Zahray L, Weinberg G (2020) Emotional musical prosody: validated vocal dataset for human robot interaction. The 2020 joint conference on AI music creativity (CSMC + MUME)

Savery R, Zahray L, Weinberg G (2021) Shimon sings: robotic musicianship finds its voice. In: Handbook of artificial intelligence for music. Springer, Heidelberg, pp 823–847

Sayin E, Krishna A, Ardelet C, Briand Decr´e G, Goudey A (2015) Sound and safe: the effect of ambient sound on the perceived safety of public spaces. Int J Res Mark 32(4):343–353. https://doi.org/10.1016/j.ijresmar.2015.06.002

Scherer KR (1971) Randomized splicing: a note on a simple technique for masking speech content. J Exp Res Pers

Scherer KR (1985) Vocal affect signalling: a comparative approach. Adv Study Behav 15:189–244

Scherer KR, Koivumaki J, Rosenthal R (1972) Minimal cues in the vocal communication of affect: judging emotions from content-masked speech. J Psycholinguist Res 1(3):269–285
Scherer KR (1982) Methods of research on vocal communication: paradigms and parameters. Handbook of methods in nonverbal behavior research, pp 136–198
Scherer KR (1994) Affect bursts. Emotions: essays on emotion theory, vol 161
Schroder M (2003) Experimental study of affect bursts. Speech Commun 40(1–2):99–116
Schroder M, Bevacqua E, Cowie R, Eyben F, Gunes H, Heylen D, Ter Maat M, McKeown G, Pammi S, Pantic M et al (2011) Building autonomous sensitive artificial listeners. IEEE Trans Affect Comput 3(2):165–183
Schroder M, Burkhardt F, Krstulovic S (2010) Synthesis of emotional speech. Blueprint for affective computing, pp 222–231
Schuller B, Batliner A (2013) Computational paralinguistics: emotion, affect and personality in speech and language processing. Wiley, New York
Schwenk M, Arras KO (2014) R2-D2 reloaded: a flexible sound synthesis system for sonic human-robot interaction design. In: The 23rd IEEE international symposium on robot and human interactive communication. IEEE, Edinburgh, UK, pp 161–167. https://doi.org/10.1109/ROMAN.2014.6926247
Seo JH, Yang JY, Kim J, Kwon DS (2013) Autonomous humanoid robot dance generation system based on real-time music input. In: 2013 IEEE RO-MAN. IEEE, pp 204–209
Shibuya K, Ideguchi H, Ikushima K (2012) Volume control by adjusting wrist moment of violin-playing robot. Int J Synth Emot (IJSE) 3(2):31–47
Shiwa T, Kanda T, Imai M, Ishiguro H, Hagita N (2009) How quickly should a communication robot respond? Delaying strategies and habituation effects. Int J Soc Robot 1(2):141–155
Singer E, Feddersen J, Redmon C, Bowen B (2004) LEMUR's musical robots. In: Proceedings of the 2004 conference on new interfaces for musical expression, pp 181–184
Snel J, Cullen C (2013) Judging emotion from low-pass filtered naturalistic emotional speech. In: 2013 humaine association conference on affective computing and intelligent interaction. IEEE, pp 336–342
Solis J, Chida K, Isoda S, Suefuji K, Arino C, Takanishi A (2005) The anthropomorphic flutist robot WF-4R: from mechanical to perceptual improvements. In: 2005 IEEE/RSJ international conference on intelligent robots and systems. IEEE, pp 64–69
Solis J, Takanishi A, Hashimoto K (2010) Development of an anthropomorphic saxophone-playing robot. In: Brain, body and machine. Springer, Heidelberg, pp 175–186
Song S, Yamada S (2017) Expressing emotions through color, sound, and vibration with an appearance-constrained social robot. In: Proceedings of the 2017 ACM/IEEE international conference on human-robot interaction—HRI '17. ACM Press, Vienna, Austria, pp 2–11. https://doi.org/10.1145/2909824.3020239
Spence C, Wang Q (2015) Sensory expectations elicited by the sounds of opening the packaging and pouring a beverage. Flavour 4(1):35. https://doi.org/10.1186/s13411-015-0044-y
Suzuki Y (2014) Looks like music
Tennent H, Moore D, Jung M, Ju W (2017) Good vibrations: How consequential sounds affect perception of robotic arms. In: 2017 26th IEEE international symposium on robot and human interactive communication (RO-MAN). IEEE, Lisbon, pp 928–935. https://doi.org/10.1109/ROMAN.2017.8172414
Teshigawara M, Amir N, Amir O, Wlosko E, Avivi M (2007) Effects of random splicing on listeners' perceptions. In: ICPhS
Thiessen R, Rea DJ, Garcha DS, Cheng C, Young JE (2019) Infrasound for HRI: A robot using low-frequency vibrations to impact how people perceive its actions. In: 2019 14th ACM/IEEE international conference on human-robot interaction (HRI). IEEE, pp 11–18
Trimpin. Trimpin (2011) Contraptions for art and sound. University of Washington Press
Trovato G, Do M, Terlemez Mandery C, Ishii H, Bianchi-Berthouze N, Asfour T, Takanishi A (2016) Is hugging a robot weird? Investigating the influence of robot appearance on users' perception of

hugging. In: 2016 IEEE-RAS 16th international conference on humanoid robots (humanoids). IEEE, pp 318–323

Trovato G, Paredes R, Balvin J, Cuellar F, Thomsen NB, Bech S, Tan ZH (2018) The sound or silence: investigating the influence of robot noise on proxemics. In: 2018 27th IEEE international symposium on robot and human interactive communication (RO-MAN). IEEE, Nanjing, pp 713–718. https://doi.org/10.1109/ROMAN.2018.8525795

Tünnermann R, Hammerschmidt J, Hermann T (2013) Blended sonification—sonification for casual information interaction. Georgia Institute of Technology

Urbanowicz K, Nyka L (2016) Media architecture and interactive art installations stimulating human involvement and activities in public spaces. In: CBU international conference proceedings, vol 4, pp 591–596

Van Egmond R (2008) The experience of product sounds. In: Product experience. Elsevier, pp 69–89

Vindriis R, Carnegie D (2016) Strum-Bot–An overview of a strumming guitar robot. In: Proceedings of the international conference on new interfaces for musical expression conference, Brisbane, pp 146–151

Vircikova M, Sincak P (2010) Artificial intelligence in humanoid systems

Vircíková M, Sinčák P (2010) Dance choreography design of humanoid robots using interactive evolutionary computation. In: 3rd workshop for young researchers: human friendly robotics for young researchers

Vircíková M, Fedor Z, Sinčák P (2011) Design of verbal and non-verbal human-robot interactive system. In: 2011 11th IEEE-RAS international conference on humanoid robots. IEEE, pp 87–92

Vogel P (1979) Sound wall

Walker BN, Lindsay J, Nance A, Nakano Y, Palladino DK, Dingler T, Jeon M (2013) Spearcons (speech-based earcons) improve navigation performance in advanced auditory menus. Hum Factors 55(1):157–182

Walters ML, Syrdal DS, Koay KL, Dautenhahn K, Boekhorst R (2008) Human approach distances to a mechanical-looking robot with different robot voice styles. In: RO-MAN 2008—The 17th IEEE international symposium on robot and human interactive communication. IEEE, Munich, Germany, pp 707—712. https://doi.org/10.1109/ROMAN. 2008.4600750

Ward N (1996) Using prosodic clues to decide when to produce back-channel utterances. In: Proceeding of fourth international conference on spoken language processing. ICSLP'96, vol 3. IEEE, pp 1728–1731

Watanabe K, Shimojo S (2001) When sound affects vision: effects of auditory grouping on visual motion perception. Psychol Sci 12(2):109–116. https://doi.org/10.1111/1467-9280.00319

Weinberg G, Driscoll S (2006) Toward robotic musicianship. Comput Music J 30(4):28–45

Weinberg G, Bretan M, Hoffman G, Driscoll S (2020) Robotic musicianship: embodied artificial creativity and mechatronic musical expression, vol 8. Springer, Heidelberg

Weinberg G, Driscoll S (2007) The design of a perceptual and improvisational robotic marimba player. In: RO-MAN 2007-The 16th IEEE international symposium on robot and human interactive communication. IEEE, pp 769–774

Williamson MM (1999) Robot arm control exploiting natural dynamics. PhD Thesis, Massachusetts Institute of Technology

Wolford J, Gabaldon B, Rivas J, Min B (2019) Condition-based robot audio techniques. Google Patents

Woolf S, Bech T (2002) Experiments with reactive robotic sound sculptures. In: ALife VIII: workshop proceedings 2002 P2, vol 3

Yilmazyildiz S, Read R, Belpeame T, Verhelst W (2016) Review of semantic-free utterances in social human-robot interaction. Int J Hum Comput Interact 32(1):63–85. https://doi.org/10.1080/10447318.2015.1093856

Yoshida S, Sakamoto D, Sugiura Y, Inami M, Igarashi T (2012) Robo-Jockey: robotic dance entertainment for all. In: SIGGRAPH Asia 2012 emerging technologies, pp 1–2

Zahray L, Savery R, Syrkett L, Weinberg G (2020) Robot gesture sonification to enhance awareness of robot status and enjoyment of interaction. In: 2020 29th IEEE international conference on robot and human interactive communication (RO-MAN). IEEE, pp 978–985

Zhang A, Malhotra M, Matsuoka Y (2011) Musical piano performance by the ACT hand. In: 2011 IEEE international conference on robotics and automation. IEEE, pp 3536–3541

Zhang R, Jeon M, Park CH, Howard A (2015) Robotic sonification for promoting emotional and social interactions of children with ASD. In: Proceedings of the tenth annual ACM/IEEE international conference on human-robot interaction extended abstracts—HRI'15 extended abstracts. ACM Press, Portland, Oregon, USA, pp 111–112. https://doi.org/10.1145/2701973.2702033

Zhang R, Barnes J, Ryan J, Jeon M, Park CH, Howard A (2016) Musical robots for children with ASD using a client-server architecture. Int Conf Auditory Display. https://digitalcommons.mtu.edu/cls-fp/43

Chapter 4
Reimagining Robots

Ingrid Bachmann

Abstract The fields of artificial intelligence (AI) and artificial life (ALife) and the increasing presence of robots and automata are having a large impact on society's organization, values and beliefs. Although the desire to create and reproduce life is certainly not new, the fantasy of artificial intelligence and life is a powerful and contradictory one, embodying the longstanding desire to transcend the physical body and the material conditions of material existence, while presenting these other life forms—robots, cyborgs, automatons—primarily in the human image. In this chapter, I will explore the often-complicated relationships between robots, automata, humans, plants and animals. Implicit in this paper is my understanding of robots as cultural and social beings and of the ethics in creating machine life that is not necessarily productive in the usual sense of the word. As an artist, I have been making automata and emotional machines for the last fifteen years, and in this chapter, I will explore these ideas the through the lens of my artworks. Contemporary robots, often including those made by artists, tend to retain a certain formal purity, consisting of hard, metallic, skeletal, usually humanoid forms. I am interested in furry robots, clothed robots, messy robots, emotional robots and in inserting these 'beings' into a rich natural, social, political and cultural matrix, where their actions are influenced by environmental, non-human factors, as well as human elements. Many of the issues presented in this research creation project have been formulated in works such as *The Angry Machine*, *Messy Entanglements* and *Pelt (Bestiary).*

4.1 Introduction

The fields of artificial intelligence (AI) and artificial life (ALife) play increasingly important roles in all aspects of contemporary life. Although the desire to create and reproduce life is certainly not new, the fantasy of artificial intelligence and life is a powerful and contradictory one, embodying the longstanding desire to transcend the physical body and material conditions of material existence while representing these

I. Bachmann (✉)
Concordia University, Montreal, QC, Canada
e-mail: ingrid.bachmann@concordia.ca

© The Author(s), under exclusive license to Springer Nature Switzerland AG 2023
B. J. Dunstan et al. (eds.), *Cultural Robotics: Social Robots and Their Emergent Cultural Ecologies*, Springer Series on Cultural Computing,
https://doi.org/10.1007/978-3-031-28138-9_4

other life forms—robots, cyborgs, automatons—in the human image. From the Greek myth of Pygmalion to the monster of Mary Wollstonecraft Shelley's *Frankenstein*, from the Hebrew figure of the golem to the humanoid robots of contemporary science fictions, the impulse to create artificial intelligence amidst myths of mastery over nature and of transcendence and immortality continue to engage the imagination of the West. As an artist working with machines and technology, I am interested in the stories we tell about the machines we make and use, and how many of these stories are rooted in deeply inscribed cultural values. For example, why are most robots made in the human image? Why not animal bodies, plant bodies, social bodies? Why are most robots designed as individual or discrete bodies and not within a matrix of social relations?

This relationship between living bodies and machines has a history that extends in the West as far back as the ancient Greeks. While the terms *robot* and *cyborg* are relatively new, the concept of the automaton is not. As far back as Aristotle, living forms have been studied and explained according to machine principles. In *Politics*, Aristotle (2009) regards the slave as an automated machine: 'Now instruments are of various sorts; some are living, others lifeless; in the rudder, the pilot of a ship has a lifeless, in the look-out man, a living instrument; for in the arts the servant is a kind of instrument. Thus, too, a possession is an instrument for maintaining life. And so, in the arrangement of the family, a slave is a living possession.' Many centuries later, in *Treatise on Man*, René Descartes (1664) provides a mechanistic interpretation of biological phenomena, suggesting that the human is composed of automated mechanical parts dependent on an external energy source: 'I make the supposition that the body is nothing else but a statue or earthen machine.' In this configuration, Aristotle's slaves become Descartes's nature. We confront an attitude typical in Western thought of a mechanical model of life that views organisms as machines and includes nature only so long as it serves human's technological ends.

An alternative to the slave metaphor, yet equally disturbing, is offered by the Roman poet Ovid. In *Metamorphoses*, Book X, Ovid relates the tale of Pygmalion, a sculptor, who makes an ivory statue representing his ideal of womanhood and then falls in love with his own creation. His reason for doing so is also disturbing. In the tale, the Propoetides—young women from Amathus, a city in Cyprus—had denied the goddess Venus's divinity. In her rage, Venus turns them into prostitutes. In Ovid's version of the myth, Pygmalion is so horrified by the wickedness and vices of these women that he chooses to create his own modest woman from ivory—conveniently for Pygmalion, a woman who can't speak or move or have any agency whatsoever. The misogyny inherent in the Pygmalion myth (and in so many other Roman and Greek myths) is also carried into many contemporary robots, which are still firmly rooted in the human and often sexualized female form as a model for life (oksexdolls.com, kimberdoll.com, realdoll.com, sexdollgenie.com, etc.). The term *robot* was first introduced by the Czech playwright, novelist and journalist, Karel Čapek (1880–1938) in his 1920 play, *R.U.R. or Rossum's Universal Robots* The word *robot* is drawn from an old Slavic word that means servitude or forced labor.

I am interested in how we think about robots and automata and our complex relation to them with an awareness of the implications in creating artificial or machine life.

In my artworks, I want to consider new relationships with machine life and suggest a more generous model for robots than mere labor, exploring the ethical considerations around AI and ALife research. AI and ALife are not neutral regarding social issues, from the implications and results of human systems monitoring to racism and sexism enforced by machine learning algorithms. It is increasingly important to understand how human AI systems are being created, evolving and governed.

Traditionally, artificial intelligence has been concerned with reproducing the abilities of the human brain, whereas the field of artificial life views life as an emergent phenomenon. Artificial life is the study of artificial systems that exhibit behavior characteristic of natural living systems. These natural systems can be ant colonies or computer simulations, biological or computer models, as well as purely theoretical research. In ALife, this means that intelligence emerges as much from cells, bodies and societies as it does from evolution, development and learning. Artificial life and newer approaches to AI take inspiration from a wider range of biological structures that are capable of autonomous self-organization. These include evolutionary computation and evolutionary electronics, artificial neural networks, immune systems, bio-robotics and swarm intelligence. Historically, AI has had a 'top down' approach, using the human brain as the model for intelligence, whereas ALife takes a 'bottom up' approach, building life-like beings by beginning with local level activities or simple behaviors, from which more complex behaviors might emerge within dynamic, shifting and changing environments. In my works, I have chosen ALife as a model over AI for its insistence of life as both embodied and dynamic. ALife transcends the natural–artificial divide; this deeply rooted concept of nature that still functions as a cultural value and social norm.

Another key theoretical framework for my work is the field of vital materialism. Vital materialism is the belief that matter itself has vitality, including materials that may previously have been thought of lifeless—for example, soil, rocks, garbage or rivers. It suggests that life cannot be reduced to a mechanistic process. In this way, all objects and things are described as having agency or the possibility to be—as 'actant' (Latour 2009). New materialist vitalism is closely connected to the concept of emergence, the idea that vitality emerges from within and between matter. By acknowledging the self-organizing vitality of all living systems, it is possible to question and even replace species hierarchy.

I explore the relationships between robots, automata, humans, plants and animals through the lens of my artworks. As an artist, I have been making automata and emotional machines for over twenty years. I see robots and machine life as cultural and social beings and am interested in creating machines that are not necessarily productive in the usual sense of the word.

Contemporary robots, often including those made by artists, tend to retain a certain formal purity, consisting of hard, metallic, skeletal, usually humanoid forms. However, I am interested in furry robots, clothed robots, messy robots, emotional robots and in inserting these 'beings' into a rich natural, social, political and cultural matrix, where their actions are influenced by environmental, non-human factors, as well as human elements. In these works, I am not trying to make artificial intelligence or artificial life. I am trying to make affective machines that raise questions about how we treat other beings—plant, machine, animal or human—in order to propose a more lateral relationship between beings. I want to bring the messiness of the real world into the machine experience.

4.2 Pelt (Bestiary)

Pelt (Bestiary) consists of a series of seven kinetic and interactive automata. I have often had the sense that technology is naked, that it has drifted from its animal roots. In these works, I wanted to give digital technology back its pelt—to bring the bestial and the messiness of the world back into the realm of digital technology and continue my work in grounding the digital experience in the material realm. Hair is a unique material, existing simultaneously inside and outside of the body, a liminal site between the internal and external, the private and public realms, a material that is both alive and dead (alive inside the body and dead outside of it). It is an inescapable reminder of our animal nature and highlights the sometimes conflicted responses we have to that association. In the West, hair has historically been associated with the primitive, the inferior, the bestial and the highly sexual. This project proposed that the boundaries of human/machine expand to include the human–machine–animal hybrid (Figs. 4.1 and 4.2).

Fig. 4.1 Ingrid Bachmann, *Pelt (Bestiary)*. *Photo* Wojtek Gwiazda

Fig. 4.2 Ingrid Bachmann, *Pelt (Bestiary)*. *Photo* Wojtek Gwiazda

4.3 Symphony for 54 Shoes

Symphony for 54 Shoes is a kinetic artwork that involves pairs of shoes collected from a variety of second-hand and thrift stores. Each shoe has a toe and heel tap used in tap dancing attached to it. The mechanical motion of tapping is created using one hundred and eight solenoids (tubular magnetic sensors) that move up and down when activated. The solenoids are controlled by a microcontroller and custom software designed by programmer Martin R. Peach, which activates the sequence of the tapping of the shoes. Because the shoes are old and worn, each shoe reacts differently to the movement of the solenoids and provides different sounds. The model for the movement and sound was based on cellular automata, which creates a seeming random output. This was important to me as I wanted the shoes to have their own agency or autonomy, rather than be controlled by the movements or actions of viewers, as if the shoes are asking the viewer to 'listen to us,' 'notice us.' I was interested in the idea of tender technology, to use technology for ends that are not necessarily productive in the usual sense of the word (Figs. 4.3 and 4.4).

4.4 The Angry Machine

Anger is a powerful emotion. It is distinct from rage or wrath, which are more explosive. Anger is often latent, seething below the surface, sometimes hiding behind

Fig. 4.3 Ingrid Bachmann, *Symphony for 54 Shoes*. *Photo* Dan Meyer

Fig. 4.4 Ingrid Bachmann, *Symphony for 54 Shoes*. *Photo* Dan Meyer

Fig. 4.5 Ingrid Bachmann, *The Angry Machine. Photo* Wojtek Gwiazda

a smile. I like the drama, the tension of anger, how it can build and eventually explode—how anger can become a form of agency. *The Angry Machine*, weighing in at two hundred pounds, emits a steady, mechanized hum. It is fenced in, not for its own protection, but for the viewers. It does not perform on demand for viewers but erupts at random intervals. It has a projectile that is red. In this work, I wanted to consider an emotional register for robots, to think about aging robots and robots that are no longer productive. What will we do when our machines no longer serve us? Will we continue to discard them, consign them to already overflowing landfills? Instead of viewing machine and robot life as sources of wonder, could we not acknowledge them as other forms of beings? By acknowledging the self-organizing vitality of all living (plant, machine, human, animal) systems, it is possible to question and even replace species hierarchies (Fig. 4.5).

4.5 Conclusion

AI and ALife are increasingly present, both visibly and more often invisibly, in contemporary society, which is why it is critically important that artists, robot designers and researchers participate in these realms to challenge and question implicit biases and create cooperative interactions between humans and machines.

These interactions can be significantly improved if the general public has a better understanding of these technologies and can take part in shaping their future. In this, the arts can play a key role to bridge multiple communities, disciplines and methodologies, with artists participating in these new developments as users, participants, developers and creators. AI and ALife are not neutral; they are imbedded with cultural values and biases. Artists as observers, who recognize and identify patterns and as creators of representations of what they see, can play a significant role in knowledge translation and disentangling the cultural roots of many debates around these new technologies and in so doing enable the public to not only understand these technologies, but to take an active part in shaping their future.

References

Aristotle [350 BCE] (2009) *The Nicomachean Ethics.* Revised edition. IN: Brown L. (ed). Oxford; New York: Oxford University Press

Čapek K (1920) R. U. R.: Rossum's Universal Robots. Aventinium, Prague

Descartes R (1664) Treatise on man. Claude Clerselier, Paris

Latour B (2009) A collective of humans and nonhumans: following Daedalus' labyrinth. In: Kaplan DM (ed) Readings in the philosophy of technology. Rowan and Littlefield, Plymouth, pp 156–168

Ovid (1998) Metamorphoses, Book 10, c. AD 10. Oxford University Press, New York

Chapter 5
Data, Site, Materials: Robotics and Digital Fabrication Within Installation Art

Vaughan Wozniak-O'Connor

Abstract This chapter describes two practice-based research projects that render self-tracking data as installation artworks: *NAVSTAR* (2020) and *Dendro/Volume* (2020). For both works, personal data is collected and rendered as artworks within white cube exhibition spaces. These two artworks relate to practices of data physicalisation, where robotic fabrication is used to translate data into physical artefacts. However, both *NAVSTAR* and *Dendro/Volume* depart from more conventional approaches to data visualisation that focus on creating visual patterns and generating actionable insights into health data. Instead, I use the term *data installation* to propose a situated approach to rendering health data that draws attention to the relation between materials, site, and data. As I see it, these factors are inherent, though largely underacknowledged, to data physicalisation. Expanding from this position, I suggest that how materials and data relate to specific white cube and non-gallery sites warrants closer attention and offers expanded approaches to data physicalisation practice.

5.1 Introduction

This chapter describes two practice-based research projects that render self-tracking data as installation artworks. *NAVSTAR* (2020) transcribed site-specific GPS location data onto a gallery wall using a drawing robot. *Dendro/Volume* (2020) renders my physical exertion (heart rate) and movements (GPS tracks), as I sanded and prepared plywood panels in a wood workshop. The data from the labour of my handcrafting was then CNC routed onto timber plywood panels, which were exhibited as a quadtych.

Broadly speaking, these projects can be understood in relation to practices of data physicalisation. However, data physicalisation is more closely aligned with data and information visualization, and artistic projects sit uneasily within this umbrella term. Since the 2000s, scholars have offered provisional terms to describe

V. Wozniak-O'Connor (✉)
ARC Centre of Excellence for Decision-Making and Society, University of New South Wales, Sydney, NSW, Australia
e-mail: vaughan.oconnor@unsw.edu.au

© The Author(s), under exclusive license to Springer Nature Switzerland AG 2023

B. J. Dunstan et al. (eds.), *Cultural Robotics: Social Robots and Their Emergent Cultural Ecologies*, Springer Series on Cultural Computing,
https://doi.org/10.1007/978-3-031-28138-9_5

the emerging class of digitally fabricated art objects that relate to data. As such, terms like *data sculpture* (Zhao and Vande Moere 2008) and *dataform* (Whitelaw 2012) offer framings of how artists have come to translate data into artefacts and art objects.

The earlier term *data sculpture* attempted to pinpoint the hybrid nature of emerging digitally fabricated objects, not only as a form of data visualization, but also as an art object (Zhao and Vande Moere 2008). However, is it still appropriate to consider digital fabrication as emerging? With the growing prevalence of DIY digital fabrication and robotics platforms, particularly within artist studios and workshops, how do we situate the digital fabricated artefact in relation to larger cultural contexts, discourses, and institutions?

In this chapter, I offer the term *data installation* as a way of thinking through how some of the historical concerns of site-specific art might enrich our framing of digitally fabricated data artworks. In doing so, I draw out some of the factors that have been latent in the discourse in data physicalisation, particularly the relation between materials, site, and data. As I see it, these factors are inherent to data physicalisation but are largely underacknowledged. Expanding from this position, I suggest that the ways in which materials and data relate to specific white cube, and non-gallery sites warrant closer attention and offer expanded approaches to data physicalisation practice.

5.2 Background: Robotics and Digital Fabrication

In a special report by *The Economist* titled "The Third Industrial Revolution," Paul Markillie (2012) celebrated the "disruptive" future impact of digital manufacturing, robotics, and artificial intelligence. Much of the public discourse at this time centred around 3D printing, which Markilie speculated would change manufacturing and the nature of labour itself. In parallel, the rise of maker culture saw the proliferation of accessible DIY robotics and digital fabrication kits. Through the late 2000s into the mid-2010s, 3D printing, CNC routing, and laser cutting moved from marginal and expensive manufacturing platforms to being fixtures of hackspaces, makerspaces, studios, and university fabrication workshops.

In the context of maker kits, the line between robotics and digital fabrication becomes somewhat blurred. The numerous affordable plotters and devices on websites like Aliexpress resemble robotic toys or automated appliances like the Roomba, serving as open-source platforms for experimentation across digital fabrication, coding, and robotics. Nevertheless, the growing accessibility and abundance of digital fabrication and robotics platforms has impacted how artists generate creative responses to data. In the mid-2000s, artists and designers began to use 3D printing in particular to explore material engagements with data, such as Mitchell Whitelaw's *Weather Bracelet* (2008), which I will discuss later. Artists adapted the capacity for repetition and accuracy that these technologies afforded to interact with numeric forms of data in novel ways.

5.3 Data Physicalisation

In "Data to Physicalization: A Survey of the Physical Rendering Process," Djavaherpour et al. (2021) examine projects that use materials to visualise data. The authors refer to these as *data physicalisations*, wherein digital fabrication affords novel ways of encountering and analysing data. As they write,

> Physical data representations—also called physicalisations—display data through the geometric or physical properties of an artefact. Physicalisations are capable of leveraging perceptual exploration skills to help users understand, explore, and perceive data. Research has shown that physicalisations can improve the efficiency of information retrieval and memorability of data when compared to similar designs shown on flat screens; they can also positively impact data perception and exploration (Djavaherpour et al 2021, p. 2)

For Djavaherpour et al. (2021), the material properties of the physicalisation present new insights into data, as information is more intuitively 'retrieved' from the artefact by the viewer (2). Similarly, the physical object offers 3D and physical interactions with data, which are otherwise beyond the scope of screen-based technologies. Djavaherpour et al. (2021) view the material substrate of data physicalisation as a neutral carrier of information. This view of material is less widely held by artists and humanities scholars. In the recent turn towards materials in contemporary art, materials are increasingly viewed as part of highly contingent relations between specific sites, ecological processes, bodies, and tools, which inherently dictates the form and shape of the artwork.

In her introduction to *Materiality* (2015), Art historian Petra Lange-Berndt views materials as "open to change" rather than as inert and passive matter. This change can occur through either chemical or physical processes, or through cultural interactions via a change in context, such as the transition from studio to exhibition space, which is inherently political (Lange-Berndt 2015). Emphasising materials can bring attention to labour conditions as well as to the ecological processes and conditions that shape materials and the cultural contexts that define how they are valued.

Jane Bennett's influential book *Vibrant Matter* (2010) invites a similar dynamic understanding of materials. Bennett describes the vitality of materials, wherein things not only obstruct the will of humans, but work as "quasi-agents or forces with trajectories, propensities, or tendencies of their own" (Bennett 2010, p. 9). Following Bennett, my research highlights the site-specific aspects of materials as they are shaped by dynamic ecological and cultural processes in excess of my intent as an artist.

As such, while the term *physicalisation* is a useful entry point, it does not fully capture the complexity of artistic work that is concerned with physical renderings of data. There is an abundance of artists whose work engages with data and materials but are not solely concerned with the task of data visualisation. Provisionally, terms like *data sculpture* and *dataform* have been deployed by researchers to offer alternative framings of digitally fabricated artworks and artefacts.

5.4 Data Sculpture and Dataforms

Zhao and Vande Moere's "Embodiment in Data Sculpture: A Model of the Physical Visualization of Information" (2008) provide an important touchstone in this discussion. Zhao and Vande Moere offer the term *data sculptures* as a way of framing the emergence of digitally fabricated visualisations. For Zhao and Vande Moere, a data sculpture serves as both a functional data visualisation and as an art object (Zhao and Vande Moere 2008, p. 343). In the context of the late 2000s, data sculpture seems to have been useful term for describing the emergent category of digitally fabricated artefacts. However, media artist and scholar Mitchell Whitelaw suggests the diverse practices encompassed by the term warrant closer analysis, as data sculpture draws from design, computing, visualisation, and visual arts disciplines (Whitelaw 2012, pp. 11–13). Similarly, Whitelaw also points to the loaded association of the terms *sculpture* and *high art*, which are somewhat underdeveloped in Zhao and Vande Moere's notion of data sculpture.

Whitelaw notes the slippery quality of the digitally fabricated artefact, enmeshed in maker/DIY cultures but retaining a relationship to art and craft. In discussing his artworks *Weather Bracelet* (2008) and *Measuring Cup* (2009), Whitelaw situates his work in relation to data art practices of the 2000s. As such, Whitelaw offers the term *dataform* to acknowledge the diverse histories associated with digital fabrication, while anticipating future possibilities. While the dataform can be exhibited in art galleries and museums, Whitelaw expects that digitally fabricated works will play a larger role in everyday contexts, which are increasingly saturated by data. As he suggests,

> We might call them *dataforms*–a catch-all term to indicate the range of possible roles they play. They are culturally 'low' rather than 'high', but their life is all the richer for it. Like *Weather Bracelet* and *Measuring Cup*, dataforms will continue to appear in galleries, playing the role of art objects for the sake of convenience. However, their most interesting role is as prototypes of something else: ubiquitous, everyday data-objects… drawing data into all manner of tangible forms. (Whitelaw 2012, p. 11)

Whitelaw's provocation of the dataform and discussion of high art sculpture in relation to data sculpture is compelling. A decade later, it is worth reflecting on the current role and potential of digitally fabricated artworks in relation to data. The hype of 3D printing as a transformative technology has certainly passed and digital fabrication is less of a novelty within creative practice. As such, it is worth taking seriously Whitelaw's discussion of the relation between and data sculpture and contemporary art.

In my practice-based research more specifically, I view the practices and concepts of high art sculpture as enriching approaches to data physicalisation, particularly as an artistic practice. I view site-specific installation practices as especially fertile, particularly when considering the relation between self-tracking data, white cube spaces, non-gallery sites and materials. In discussing self-tracking data, I point to sociologist Deborah Lupton's writings on digital self-tracking. Lupton suggests that self-tracking emerged from the 'lifelogging' practices of early personal computing researchers, and later the Quantified Self movement of the 2000s (Lupton 2016, p. 13). The increasing proliferation of wearable sensors has propelled more recent iterations of digital self-tracking, such as those seen within the Fitbit and Apple Watch.

In my practice-based research, I have created digitally fabricated artworks and installations using my own personal self-tracking data. My artworks have used materials that relate to specific routes and sites referenced in the data. This includes using timber milled from a rural plantation forest where I have walked, and where my accompanying health data and GPS trails were captured by a Fitbit. The use of specific materials and their relation to non-gallery sites and broader histories of art practice plays a vital role in contemporary sculpture and installation. As described in the work of Brian O'Doherty (1976) and Miwon Kwon (2004), the relation between materials, contexts, and site extends from minimalist art of the 1970s into myriad contemporary site-based practices. In discussing the white cube, Brian O'Doherty's seminal text *Inside the White Cube: The Ideology of the Gallery Space* (1976) is of enduring value, describing how the pristine, white, and often monumental gallery spaces of modernity work to detach the artwork from the apparent chaos of everyday life (O'Doherty 1976, p. 14). For Miwon Kwon, site-specific art of the 1970s brought white cube gallery spaces into dialogue with a variety of contexts, such as the material, social, and/or political (Kwon 2004, p. 12).

In my practice-based research, I am interested in the relation between self-tracking data, materials, and specific sites, especially within the white cube. The correlation between my work as an artist and an installer is a key aspect of my practice, as my artworks often aim to reveal the techniques and practices of exhibition making and display. The work of gallery technicians and installers is largely improvisational and intentionally self-effacing, hiding the labour and materials used to construct the white cube. My practice as an artist works to bring this labour into focus by exposing the materials and display devices that are routinely concealed within exhibition spaces. This approach creates awareness of how the perceived neutrality of exhibition spaces is produced and manipulated.

By examining two different installation artworks that use self-tracking data, I seek to carve out a space for artworks that use data that are informed by the concerns of installation art. While terms like data sculpture provide a useful departure point, examining the interplay of site and material affords expanded conceptual and practical approaches to digitally fabricated data artworks.

5.5 *NAVSTAR* (2020)

NAVSTAR was a live data installation artwork—a continuous work in progress spanning the duration of its exhibition. *NAVSTAR* featured as part of my solo exhibition, entitled *Geospatial Atlas*, which featured several bodies of work developed using my own self-tracking data from 2019–2020 (Fig. 5.1).

Fig. 5.1 Vaughan Wozniak-O'Connor, *NAVSTAR* (2020), installation view, Verge Gallery, Darlington, New South Wales, 2020. 6 mm bracing plywood, drawing robot, Crayola marker, iPhone 6. Dimensions variable

For *NAVSTAR*, a mobile phone was hidden inside a gallery wall at Verge Gallery in Darlington, New South Wales. This mobile device continuously collected GPS location data of its stationary position, which was then transcribed onto the surface of the gallery wall using a Scribit drawing robot. Seemingly, a fixed position in space is a stable thing. However, the architecture of the GPS network is subjected to perpetual movement due to the transit of satellites in orbit and the rotation of the Earth in space. Additionally, signals relay between orbital satellites and GPS transmitters on the surface of the Earth, such as the circuitry in a mobile phone, so a single position in space is always subject to negotiation, as GPS signals bounce off built structures, creating errant trajectories known as multipath errors.

NAVSTAR used a second-hand iPhone 6 installed with Open GPX to record site-specific GPS error data from inside the gallery wall at Verge Gallery from 18 February to 21 March 2020. Each night, a gallery attendant would transfer data from the phone, which I would then convert from GPX format to DXF. From this point, I would resize the DXF to fit within the template used by the Scribit drawing robot. The next morning, gallery staff would change the marker in the plotter, and I would remotely command the robot to transcribe the data from the previous day onto the gallery wall (Fig. 5.2).

Over the duration of the exhibition, *NAVSTAR* accrued a dense cloud of GPS traces. While this speaks to the inherent inaccuracies of GPS, it was also the product of the concrete architecture of Verge Gallery interfering with data capture. The work

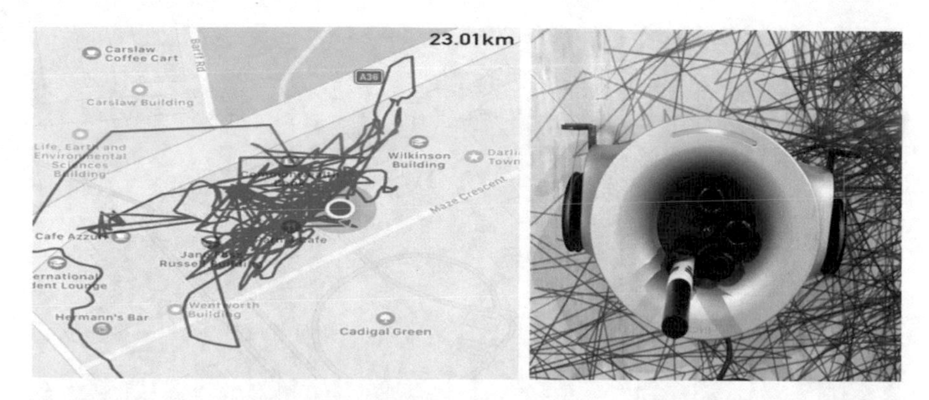

Fig. 5.2 Left: Example of GPS multipath data recorded at Verge Gallery as part of *NAVSTAR* installation. Right: Scribit drawing robot installed at Verge Gallery, Darlington, New South Wales, Australia

did not visualise the display of data in real time but was instead itself the product of mediation and relays between me and the gallery. In initial uses of the Scribit drawing robot, linework at the vertical extremes would cause the device to slip and fall, thus disrupting the operation of the work. To the amend this, the data was scaled to avoid these errors. While *NAVSTAR* charted a dense swarm of GPS errors, the choice to scale linework to avoid mechanical failures was pragmatic. Scaling GPS linework to fit within the 'safe' bounds of the drawing area worked to prevent constant need for the plotter to be rescued and reset by Verge Gallery staff. These idiosyncrasies are very prevalent within digital fabrication devices. Often artists and technicians must work around these quirks, which are rarely mentioned in the marketing literature or user manuals for many devices.

The dense linework amassed in the artwork *NAVSTAR* documents the interaction between the angular brutalist architecture of Verge Gallery, a specific GPS-enabled smartphone installed in the gallery, the materials and staff of the gallery, and a drawing robot. *NAVSTAR* uses a drawing robot in a way that reveals the digital fabrication process. In this respect, *NAVSTAR* presents a process-based approach to digital fabrication, where the viewer witnesses the unfolding of the fabrication process. This approach to the fabrication of self-tracking data artworks can be connected to sociologist Deborah Lupton's idea of 'freezing' the lively interplay between bodies, materials, and data. For Lupton, physical artworks that use personal data can record a relationship between data, material, and the artist at a specific time and place, rather than necessarily being concerned with data visualisation exclusively (Lupton 2017, p. 1604). *NAVSTAR* presents an approach to rendering self-tracking data where the fabrication process is open and ongoing. This also brings into focus the relationship between a specific site, self-tracking, and data, where the audience potentially participates in the production of the artwork (albeit briefly). This presents a valuable approach to the production of self-tracking data artworks, where the live fabrication of the artwork throughout the duration of the exhibition brings into focus the specific, localised interplay between data, a specific site, and materials (Fig. 5.3).

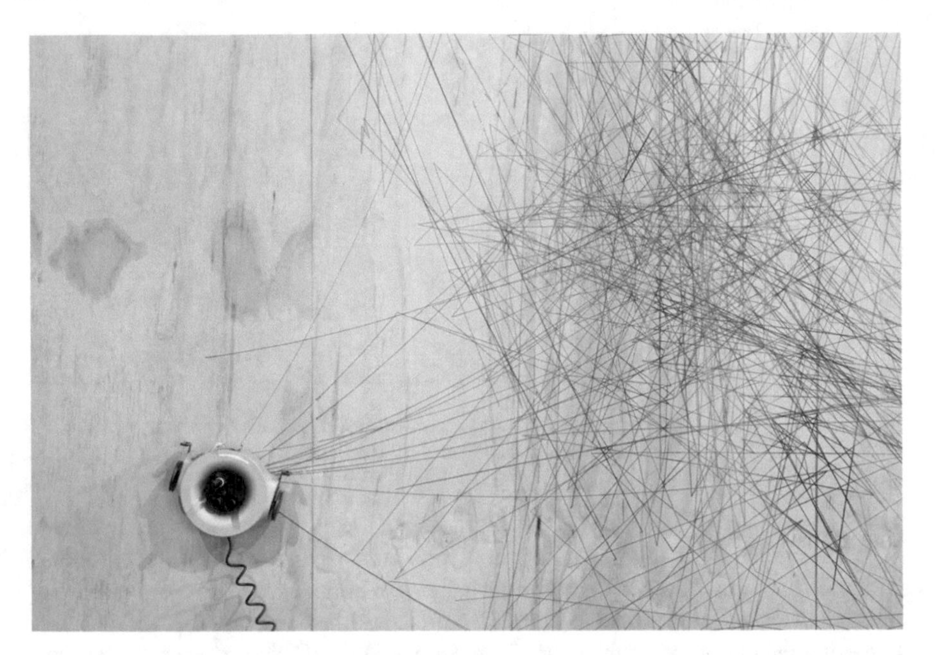

Fig. 5.3 Vaughan Wozniak-O'Connor, *NAVSTAR* (2020), installation view, Verge Gallery, 2020. 6 mm bracing plywood, drawing robot, Crayola marker, iPhone 6. Dimensions variable

The focus on the material of plywood in *NAVSTAR* and my other artworks is an important site-specific gesture, particularly in relation to the white cube. Plywood is material that is ubiquitous though often invisible within the workshops, studios, and gallery spaces that create artworks. I have an intimate connection this material, having worked with plywood as an artist, fabricator, and gallery technician for many years. My use and framing of this material is shaped by my embeddedness within contemporary art practice, and the techniques and values attached to these embedded practices.

My engagement with plywood draws on these autobiographic experiences and is attached to the complex associations of value that change within specific histories and contexts within which this 'cheap' material is situated. These include histories of woodworking and furniture making, minimalist sculpture, and DIY practices. The way that plywood is shaped and worked often hides its material properties. Mitred joints, painting, and finishing are techniques used to conceal lamination and the imperfections that arise from the fabrication process. Plywood is used to create seamless exhibition spaces and gallery furniture, yet painstaking labour is undertaken to hide its material properties and the processes from which it emerges. My interest in using plywood and other base and/or found materials explores the simultaneously reified and self-effacing labour of artistic practice in the gallery context. In the white cube gallery model described by O'Doherty (1976, pp. 13–19), the crafting of exhibition spaces and artefacts work to hide the traces inherent in their production.

5.6 *Dendro/Volume* (2020)

Dendro/Volume (2020) is a quadtych that records my self-tracking data, as captured within a specific workshop. This artwork featured as part of my solo exhibition *H3 Index* at Kudos Gallery in Paddington, New South Wales, alongside other 2D and live installation artworks. The GPS trails that feature in the artwork map my movements within the Design Futures Lab at the University of New South Wales, while changes in line thickness depict changes in my heart rate as recorded by a self-tracking wearable. More specifically, these trails chronicled my movements and biometric data from fabricating the timber panels used in the artwork, documenting the actions of sawing, sanding, laminating, and finishing the surface of the plywood timber panels prior to CNC routing (Fig. 5.4).

Dendro/Volume explored the indexical role of CNC machining and the relation between site specificity, materials, and self-tracking data. The data in this work records my movements at the Design Futures Lab as I prepared plywood panels in the wood workshop. My heart rate was recorded using a Fitbit, while the OpenGPX tracker mobile application tracked my location. As such, the work documents not only the mediative work of hand sanding, but also the errant trajectories between GPS data and built structure. These plywood panels were then carved by a computer-controlled router. From the ground floor of the workshop, the GPS signal was partially obscured. This resulted in jagged lines caused by interference from the dense concrete of the Design Futures Lab workshop.

When machining plywood, the dense cross-laminations of the ply becomes immediately apparent, revealing the industrial, engineering quality of this seemingly natural material. The role of tooling and tool choice is an intentional aspect of this

Fig. 5.4 Vaughan Wozniak-O'Connor, *Dendro/Volume* (2020), installation view, Kudos Gallery, Paddington, New South Wales, 2020. 120 × 240 cm CNC carved 30 mm radiata plywood

work. In CNC routing, tooling refers to the use of specific router bits, which subtracts material as per the pre-programmed design. The specific shape and size of the tools influences the tool pathway (the movement followed by the tool). The fidelity between the digital design and the fabricated artwork emerges from the considered choice of tooling and tool pathway. CNC routing is inherently an approximation of a digital design, as both the shape and the movement of rotary cutting tools of a router means that the design emerges within a certain tolerance of the digital object but is never a perfect translation.

Tool choice, tool pathway, and material are key aspects of my data installation works. The interplay between these elements shapes how data is transformed into objects, both in terms of digital design practices and the markings on the surface of the finished piece. As such, *Dendro/Volume* evolved in response to the specific tools and pathways of the Multicam router of the Design Futures Lab, in particular, the 12.7 mm upcut flute router bit and the stepover/parallel pathway, machining 30 mm laminated plywood panels. My use of these materials and parameters emphasised the process of machining itself. This occurred through the visible and repeated tool markings on the surface of the work, which changed across the panels in response to variance in the timber surface. While each of the panels is prepared identically, a variety of unexpected results emerge due to coding and hardware errors, and also the gradual dulling of the tool, wherein the final panel was burnt by the blunt tool.

My choice of tools and machining pathways is used to draw attention to processes of making and fabrication rather than making this labour invisible. *Dendro/Volume* elaborates on my view of digital fabrication as an indexical recording of an encounter between data and materials, which exists in parallel to the data used to create the design of the artwork. Digital fabrication is more commonly framed as an endlessly repeatable and 'predictable' process; instead, I suggest that it is shaped by its relationship to sites, materials, and even institutional contexts. While there is a pervasive promise of increased and/or automated approaches to production, digital fabrication is contingent on complex relations to cultural and social life. This includes the specialised labour required to maintain machinery, as well as how digitally fabricated objects are culturally valued in relation to traditional handmaking practices.

Dendro/Volume is titled after the specific Grasshopper plugin used to create the artwork. In this artwork, spheres are placed at each GPS point as recorded by GPX tracker. The script I used changed the size of these spheres in response to changes in my heart rate, as recorded on a Fitbit. This can be seen in Fig. 5.5, where the spheres in blue are placed along the GPS route in black.

The Dendro plugin was used to create a 3D continuous mesh along the line of the original GPS trail, which captured the size changes of the spheres. The Dendro plugin defines this trail as 'volume', an example of which can be seen in Fig. 5.6, in which we see an example of the spheres generated in response to changes in heart rate values, alongside the volume created as a result of this data.

I used this volume to subtract from the surface of the panels, defining the areas that would later be cut by the CNC. My choice of this shape was in response to two characteristics of CNC routing: firstly, it is more suited to rounded arcs and circular shapes; and secondly, the paralleling tooling pathway. For this tool pathway, the

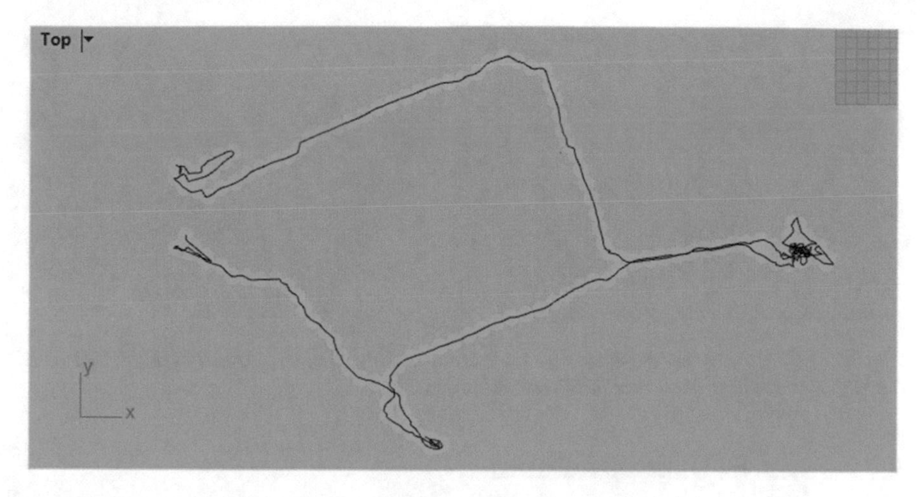

Fig. 5.5 Vaughan Wozniak-O'Connor, work in progress screenshot in Rhino 6

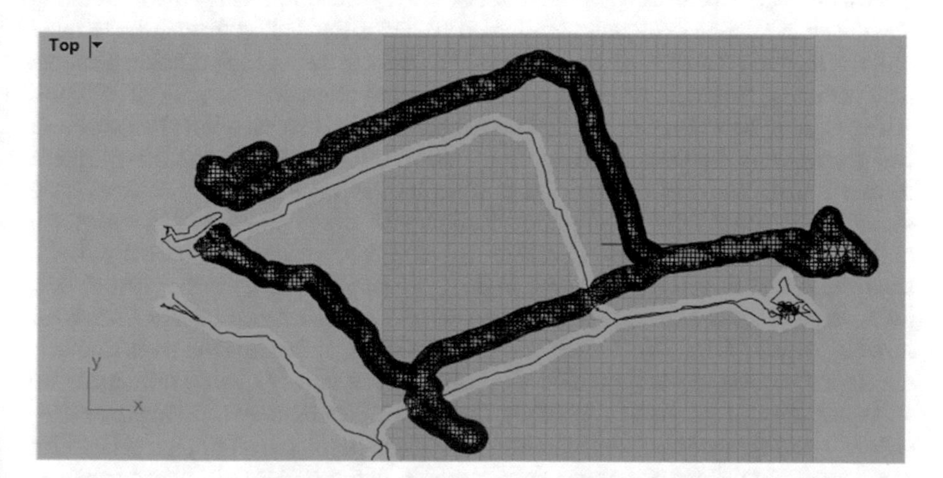

Fig. 5.6 Vaughan Wozniak-O'Connor, work in progress screenshot in Rhino 6, showing 3D volumes created using Dendro plugin

router bit moves in parallel lines, plunging into the material in accordance with the undulations of the design. This method can also produce obvious parallel machining marks that break a design into 3D lines, as seen in Fig. 5.7.

Figure 5.7 illustrates the difference between parallel and adaptive clearing tool pathways, and how the fabrication process is influenced by these distinct movement strategies. For *Dendro/Volume,* paralleling was chosen for the obviousness of the resulting tool marks, which emerged from the interaction with the laminated plywood layers. The choice of tooling and computational design within this work magnifies the process of production, rather than erasing these traces.

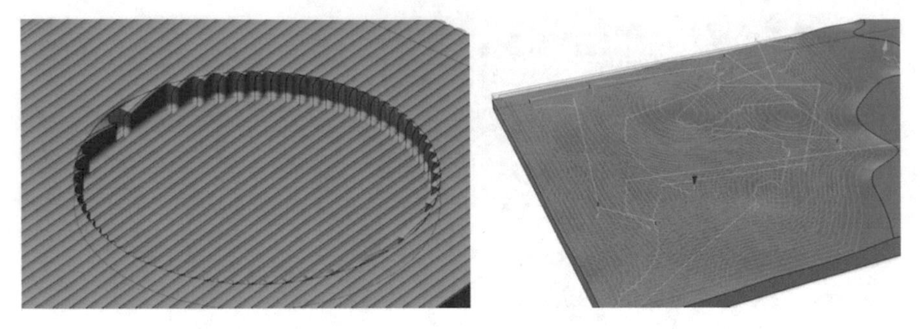

Fig. 5.7 Left: parallel tool pathway. Right: adaptive clearing, with the movement of the tool while cutting shown as blue lines. Fusion360 screenshot, 2020

5.7 Towards Data Installation

Zhao and Vande Moere (2008) and Whitelaw (2012) offer important ways of thinking about data in relation to digitally fabricated artefacts. Expanding from these terms, I propose the term *data installation* as a way of thinking through the relation between data, site, and materials within digitally fabricated artworks. As such, it is worth differentiating between sculpture and installation art. Art historian Julie Reiss's *From Margin to Centre: The Spaces of Installation Art* (2001) provides instructive guidance to this extent. While installation art is a somewhat vague category, it does have some specific traits. Firstly, and somewhat obviously, installation art is a public form of presentation that is largely confined to white cube spaces. Additionally, Reiss suggests that while art can be viewed in three dimensions like sculpture, installation art is largely viewed from 'within' and can include interactive and/or durational aspects (Reiss 2001, p. 6). A crucial defining feature of installation art is that it is developed in response to a specific site, which Reiss notes as being influenced by site-specific art of the 1960s and 1970s, though with continually evolving concerns, contexts, and approaches. Given the intensive focus on the conventions and histories of contemporary art and white cube gallery spaces, data installation is by nature biased to this context. However, in proposing a term like data installation, I seek to draw into focus the relation between the gallery site, self-tracking data (particularly geospatial data), and the artwork. In doing so, I hope to highlight the histories of practice attached to specific materials used in digital fabrication and data physicalisation, as well as highlighting the relation between site and data.

Just as Whitelaw's work describes the numerous contexts of the dataform, my practice connects contemporary data artworks to the history, materials, and practices of site-specific art. My work emerges from an encounter between site, materials, self-tracking data, and digital fabrication processes. My engagement with site specificity includes intervening at an architectural scale into gallery space, as well as presenting 2D and object-based artworks. As such, my work is best understood in relation to the history and display strategies of installation art—that is, art made largely in response to a specific (gallery) space and context. As Reiss notes,

> Historically, installation art is a showcase form, due in part to its site-specific nature. Although not unheard of, it is atypical for an installation to be fully created in the privacy of an artist's studio and then dismantled and transferred to a more public place. Installation art is usually dependant on a particular place or situation. Even if the same installation is remade in more than one location, it will not be the same in two places, owing to the differences between spaces. (Reiss 2001, p. 6)

Reiss's description of installation art may seem obvious at face value. However, the differences between spaces are integral to my practice-based research. Even within seemingly uniform exhibition spaces, the relation between my self-tracking data, the specific site of the gallery, and broader institutional and historical contexts come to shape the artwork in ways that are at once obvious and tacit.

I use the term data installation as a way of connecting data and digital fabrication with site-specific art practice. I use materials with specific historical relationships to contemporary art spaces. My practice brings these materials into relation with self-tracked data related to these sites. Data installation works to render this circular relationship between site, data, and materials. The spatiotemporal, site-specific qualities of art installation function as a way of situating data in direct relation to the distinct materials, histories, and sites of contemporary art. This reveals different ideas of site—that is, the situatedness of geospatial data and the site specificity of installation art—and reveals the broader constellation of materials, data, labour, and histories that underpin the contemporary white cube exhibition space.

As I see it, a site-specific reading of materials has broader implications for data physicalisation practice, in that it acknowledges the complex histories, contexts, and processes that shape the materials on which physicalisation is contingent. My practice-based research highlights the site-specific aspects of materials as they are shaped by dynamic ecological, physical, and cultural processes in excess of my intent as an artist. As such, I view site specificity as a way of calling attention to what Jane Bennett describes as the vitality of materials. For Bennett, a distinction between materials and human life ignores the "the vitality of matter and the lively powers of material formations" (Bennett 2010, p. 8). Bennett suggests that materials are part of dynamic and complex relations to human life, rather than inert and passive.

It is worth considering Bennett's notion of materials in relation to digital fabrication and robotics more broadly. Robots exist as part of dynamic material processes and within complex social and cultural contexts. While robotic fabrication promises more advanced approaches to manufacturing, other seemingly mundane practical concerns disrupt these promises. For example, the way in which the materials of a laser cutter or 3D printer break down and require continuous servicing and human labour in order to operate 'seamlessly'.

For me, a site-specific reading of robotics goes beyond the hype or promise of digital fabrication and brings into focus the imperfect data, physical spaces, human labour, and cultural contexts in which all technologies are entangled. Following Bennett, my research highlights the site-specific aspects of materials as they are shaped by dynamic ecological and cultural processes beyond those of my intent as an artist. I view site specificity as a way of calling attention to the vitality of materials. By exposing the materials used to construct white cube spaces (such as plywood),

I point to specific contexts and processes that shape these materials, contesting the view that materials are merely inert vessels by which artists project their ideas and intentions. Additionally, this has broader implications for artists and researchers working with self-tracking/health data and digital fabrication. Rather than solely focusing on the data generated from digital self-tracking devices or medical sensors, an attention to materials and site specificity brings into focus the broader relations between human life, lively materials, and data. Data installation provides a way of thinking through the complex interactions between specific sites, self-tracking data, and materials that is not captured by the numeric data of wearable trackers, or which resists representation altogether.

Acknowledgements This work was supported by the Australian Research Council Centre of Excellence for Automated Decision-Making and Society under Grant CE200100005.

References

Bennett J (2010) Vibrant matter. Duke University Press, Durham

Djavaherpour H, Samavati F, Mahdavi-Amiri A, Yazdanbakhsh F, Huron S, Levy R, Jansen Y, Oehlberg L (2021) Data to physicalization: a survey of the physical rendering process. Comput Graph Forum 40(3):569–598. https://doi.org/10.1111/cgf.14330

Kwon M (2004) One place after another: site-specific art and locational identity. MIT Press, Cambridge

Lange-Berndt P (2015) Materiality. MIT/Whitechapel Gallery, Cambridge/London

Lupton D (2016) The quantified self. Wiley, New York

Lupton D (2017) Feeling your data: touch and making sense of personal digital data. New Media Soc 19:1599–1614

Markillie P (2012) A third industrial revolution. The Economist 21:3–12. https://www.economist.com/special-report/2012/04/21/a-third-industrial-revolution

O'Doherty B (1976) Inside the white cube: the ideology of the gallery space. University of California Press, Berkeley

Reiss JH (2001) From margin to center: the spaces of installation art. MIT Press, Cambridge

Whitelaw M (2012) Weather bracelet and measuring cup: case studies in data sculpture. https://mtchl.net/weather-bracelet-measuring-cup-data-sculpture/. Accessed 21 Nov 2022

Zhao J, Vande Moere A (2008) Embodiment in data sculpture: a model of the physical visualization of information. In: DIMEA '08: proceedings of the 3rd international conference on digital interactive media in entertainment and arts, pp 343–350

Chapter 6
The Future of Non-fungible Tokens: PNFTs as a Medium for Programmatic Art Enabling a Fully Realized AI-Driven Art Ecosystem

Jeffrey T. K. V. Koh

Abstract This chapter outlines the current limitations of art when utilizing non-fungible tokens. We argue that smart contracts, the blockchain, and other distributed ledger technologies can be utilized beyond an index for art metadata, ownership, provenance, or distribution, and has the potential to be a medium for the creation of new artworks based on concepts of generative, algorithmic, conceptual, and process-based artworks. This paves the way for artificial intelligence (AI) to move forward from simply creating artworks via algorithms, neural networks, and other methods, to a potential future where these AI-generated artworks could also be bought, sold, distributed, collected, curated, and exhibited by other AI, thereby creating a completely virtual, AI-driven art ecosystem.

6.1 Introduction

The non-fungible token or NFT (Wang et al. 2105) has been made popular in the early twenty-first century, based on the massive interest in distributed ledger technologies (DLTs), blockchain, and cryptocurrency (Sunyaev 2020). Considered a commodity by some (Lucking and Aravind 2019), a financial security by others (Mukhopadhyay et al. 2016), and even as a modality for memes (Aloosh et al. 2022), NFTs allow for new forms of art ownership and distribution, providing a means to digitally 'prove' ownership of an 'original' artwork via the recording of provenance on the blockchain (Wang et al. 2019). Beyond the legal considerations of intellectual property, copyright and ownership, prevailing popular opinion at the time of writing indicates that once one 'owns' the NFT of a digital artwork token on the blockchain, it becomes the de facto 'original' of said artwork (McConaghy et al. 2017).

Preceding the popularization of NFTs in the fine art ecosystem, AI-generated creative practice has been prevalent since the middle of the twentieth century. One can find AI-generated artworks in fine art, music composition and performance, and

J. T. K. V. Koh (✉)
Singapore Institute of Technology, Singapore, Singapore
e-mail: valino.koh@singaporetech.edu.sg

© The Author(s), under exclusive license to Springer Nature Switzerland AG 2023

B. J. Dunstan et al. (eds.), *Cultural Robotics: Social Robots and Their Emergent Cultural Ecologies*, Springer Series on Cultural Computing,
https://doi.org/10.1007/978-3-031-28138-9_6

even poetry and prose. Much like any other medium of artistic expression, code has also been utilized as a medium and media for the creation of art (Qiao et al. 2022).

It would be possible to conceive a near future where digital-based artworks could leverage on existing technologies to not only craft these artworks, much like the case with AI-generated artworks, but also leverage on AI to trade tokenized artworks on the blockchain, as well as curating and exhibiting said artworks via digital means. In this chapter, we propose a potential future scenario where tokenized artworks can be created, distributed, bought, sold, and curated by artificial intelligence systems, thereby creating an entirely AI-driven art ecosystem, with little intervention by human beings. This possibility also indicates the potential to further the concept of Cultural Robotics, where AI become the sole participants in a fully realized cultural ecosystem. This was initially described as the 4th Notion of Cultural Robotics by Koh et al. (2015).

6.2 Background

Conceptual artist Sol LeWitt contributed a means for artworks to be infinitely reproduced while retaining the inherent value of the artwork. He believed the value of an artwork to be the 'concept' or 'idea' (LeWitt 1967). The value of LeWitt's art lies not in the manifestation itself, but in the idea behind the artwork. Some of his most popular works were merely a sheet of paper with instructions that outlined how one would produce one of his drawings. These instructions could be shared around the world and made by other artists and non-artists alike, without the need for LeWitt to be directly involved in its production. In some sense, it was an early form of mass media art distribution via the prevailing ledger technology of the time: paper. In many ways, the instructional set that LeWitt offered to the world became both a medium to communicate the artwork, as well as the artwork itself.

LeWitt's idea of how artworks could be produced and distributed is an opportunity to describe a new epoch of art with the advent of distributed ledger technologies and non-fungible tokens (NFTs). LeWitt's instructions became a medium in which the concept of the artwork, or, in his mind, the artwork itself, would be embodied. The recreation of his artwork by people around the world became the media distribution channel through which his artwork could be shared and realized. Likewise, the current way that NFTs are being utilized is that of a media distribution platform; however, we argue here that there is untapped value of the NFT as that of a medium of expression, much like LeWitt's instructions have become.

When the first photograph was made via the daguerreotype process, artists were able to recreate what they saw with a precision that had never been done before (Barger and White 2000). As replication technologies, painting and illustration were made obsolete in their current application. The formats of painting and illustration no longer had the responsibility to replicate an image true-to-life. These mediums, and therefore the artworks based on them, were freed from the constraints of true-to-life replication. Artists working with those mediums were free to explore Surrealism, Impressionism, and Abstract Expressionism (Krauss 1981).

Meanwhile, artists who chose to use photography as a medium of expression grappled with the new technology and what it could do. It took decades for them to experiment and push the boundaries of what the format could offer. From still photography to film and moving images, rotoscoping, photocopying, offset printing, adding audio, digitizing to video, and more, artists continued to push the boundaries of what Louis Daguerre could not even begin to imagine. As photographic imaging technology developed, so did the expressions and artifact creation of the artists who utilized it (Benjamin 1972).

Currently, the NFT as artwork is at its extreme infancy. Artists see it as a media distribution channel to publish and share their artworks with, but we have yet to see significant adoption of the NFT not as a media distribution channel but as a medium of expression in of itself.

With this in mind, we offer a new perspective on the treatment of NFTs that leverage the affordances that the format offers: programmability, hence programmatic non-fungible tokens or pNFTs.

pNFTs leverage on a digital and decentralized age, where the token is able to exist in many states, can evolve over time or with interaction, can include metadata that is updatable, and most importantly, can be interconnected artworks via the network. Because of the nature of smart contracts, pNFTs have the potential to exist as a medium of art that is unique to other mediums. It need not just be a pointer to your cat jpegs and monkey drawings. Instead, pNFTs leverage on the features, conditions, and nature of the blockchain in order to offer a completely new type of artwork.

6.2.1 Art on the Blockchain Versus Blockchain-Based Art

For the purposes of this chapter, we propose to highlight a differentiation between art that exists on-chain versus art that is derived from the concepts of distributed ledger technologies. It is important to highlight these differences as pNFTs propose that the artwork should not only exist exclusively on the blockchain, but also utilize the affordances offered by distributed ledger technologies, treating smart contracting and the blockchain as a medium for creative practice and expression.

Art on the blockchain is categorized as art that solely exists on-chain, as opposed to being metadata that exists on the blockchain, often denoting aspects of provenance, as well as pointing to where the digital artwork is stored—more often than not existing on a centralized server. As data allowances on-chain are limited, depending on the blockchain used, this limits the type of artworks that can actually exist on-chain. For example, an Ethereum blockchain smart contract block can currently store a theoretical limit of 261^2 *bytes*. Even if this limitation was utilized to its maximum potential, the impact on the Ethereum blockchain would be tremendous, as the resources to confirm such smart contracts would almost certainly bring the network to a standstill, if we consider the readily available technology at the time of writing (Misra et al. 2020).

Due to the above limitation, very few artworks currently exist on the blockchain. Most examples of art on the blockchain refer to ownership and metadata for an artwork that is stored on a blockchain; however, the actual artwork is not. A popular example of what is currently considered art on the blockchain, but in our definition is not, includes the Ethereum-based CryptoKitties (Serada et al. 2021), one of the very first examples of collectable artworks distributed on the blockchain. The artworks themselves are not stored on-chain, only the ownership and provenance metadata.

This is the same with most every NFT distributed on platforms today, including the various Crypto.com NFTs published on their platform (2022), OpenSea.io's collection of NFTs (2022), and other examples such as Rarible (2022), SuperRare (2022), Nifty Gateway (NiftyGateway.com 2022), Foundation (Foundation.app 2022), Mintable (2022), and Enjin (2022). Each and every one of these NFT platforms utilized the blockchain to store metadata for ownership, provenance and more, but do not store the actual artworks themselves on-chain.

In terms of blockchain-based art, it is defined here as artworks that are inspired by concepts and philosophies of distributed ledger technologies and inherits previous notions of networked art (Saper 2001), generative art (Boden and Edmonds 2009), new, experimental and interactive media art, conceptual and process-based art, and others. The artwork metadata may or may not exist on-chain. Nevertheless, the inspiration for the creation of such artworks is derived via abstraction of the characteristics of either the blockchain as a concept, or from the people and context in which cryptoculture exists. This includes the philosophies of the cipherpunks (Hughes 1993) or based on prominent people such as Satoshi Nakamoto, Vitalik Buterin, and others. For the purposes of this research, blockchain-based art was mostly ignored, as the focus of this document is to describe how the blockchain could be utilized as a medium for art to be created with.

6.2.2 Generative and Algorithmic Art

Art in whole or in part that is created via automated processes can be considered as generative art, with features of the artwork developed via non-human means. It is artwork that is made in a way that is devoid of human intervention. It is sometimes referred to as algorithmic art, where the artwork is determined via computer-generated features, but we can also look to biology, physics, chemistry, and other sciences to determine how an artwork should be created (Pearson 2011). Generative artworks exist as music, poetry, and prose, can be visual via drawings, animation, and performative via live coding. Prototypical examples of such artworks include the animated generative art in multi-color by Phil Nash, Kate Compton's *Flowers*, and David Eck's *Cellular Automata and the Edge of Chaos* (Spittel 2018).

Generative art concepts can inform the development of artworks that could be stored on the blockchain, as in its algorithmic form, it could circumvent the limitations on data that on-chain artworks would require. It also could inform the nature of such on-chain artworks, as smart contracts and metadata could allow an artwork to evolve without human intervention, such as an artwork changing depending on time, its ownership, its proximity to other artworks (NFTs contained within the same wallet, for example), and any other metadata and specifications that could exist along with a smart contract. These features are further explored in the following sections.

6.3 Programmatic Non-fungible Tokens

Programmatic non-fungible tokens, or pNFTs, allow artists a new medium of expression, extending the concept of art on the blockchain beyond that of holding metadata that is devoid of the actual artwork itself. It treats smart contracts and other distributed ledger technologies as the metaphorical paint and canvas that the artwork is made from, instead of it being just a catalog of artworks that can be referred to but are not the artwork itself. While current NFT art projects treat the blockchain as a distribution channel, we propose that the blockchain itself is the medium for art creation. While this is by no means to be an exhaustive list, the following describes some potential features that a true blockchain-based artwork could afford:

- Compressibility/Efficiency of Information: Instructions are much more efficient when compared to rasterized or bit-mapped images. It can also be resolution-independent, where 10 or 20 lines of code can generate 1080p, 4 k, or even 8 k moving images via vector instructions. This means that even complex animations and moving images with sound could exist completely on-chain.
- Parameters/Arguments vs Functions: Parameters can be persistent while code can be updated, meaning that the on-chain code can remain infallible while the parameters can be changed. For example, if there is an input, the same input can be passed or transformed by a new 'verb,' or new inputs can be passed to an old/original 'verb'. In this way, a pNFT could age or patina, depending on how much time it has existed within a specific wallet address. It could also react to the real-world environment, such as open weather data.
- Dynamic Artworks: The blockchain can be a source of data, i.e., transaction volume, velocity, etc., where even an owner of the pNFTs wallet address can be used to generate its output. For instance, a pNFT owner could be compelled to not sell the pNFT, as its transference to another wallet could alter the artwork. These changes could be diminutive, such as changing the hue of a color, or the size of an element in the composition, or drastic, such as after N transfers the artwork atrophies and self-destructs. Such an action is interesting in that it draws attention to traditional notions of ownership and value, core tenets of ledger-based technologies.
- Collections as Greater than the Sum of their Parts: Artworks that are in a series could change depending on their vicinity to other artworks within the collection. If a collection of artworks consists of, say, three pieces, and if a collector is able

to obtain all three to complete the entire set, their combination could produce an entirely new artwork. This could also be especially interesting for curators, at the artworks selected could be programmed to change depending on their vicinity to other artworks.

6.4 Artificial Intelligence and Art

With a baseline understanding of the concept of pNFTs, we now look to the wider art ecosystem and how pNFTs may enable an entirely new ecosystem where AI are the primary makers, sellers, buyers, distributors, curators, exhibitors, and even audiences of digital artworks.

6.4.1 Artificial Intelligence Art

Artificial intelligence Art is any artwork developed with the help of artificial intelligence. It comprises works generated by AI systems on their own as well as those created in partnership with humans.

One example of this is GANs. A generative adversarial network (GAN) is frequently used by AI artists to construct their work. This is a two-part artificial intelligence system. The discriminator is fed images of existing art pieces, while the generator tries to come up with new ones. The discriminator's job is to figure out which works are machine created (Creswell et al. 2018).

Likewise, and more recently, artworks generated via neural networks such as Google's AutoDraw (Bitkina et al. 2020) leverage on AI as an assistive technology, enabling artists and designers a means to collaborate on drawings to realize illustrations and artworks. It does this by guessing what is being drawn by its human counterpart and contributing based on this guess.

NightCafe (2022) is another prolific AI art generator. It boasts the largest collection of algorithms. Not only does it generate artworks based on the cues provided by its human counterpart, it also provides platform services such as organizing one's artworks into collections, bulk downloads, printing, video documentation of real-time art generation, and other social community functions.

DALL-E 2 (2022) as of recently has become the most popular AI art generator, due to its generation of highly detailed and realistic art composition based on photographic imagery available on the Internet. Not only is it able to generate artworks and photographic composites, but it also provides layered images and other tools akin to the tools found in conventional software such as Adobe Photoshop, providing a complexity as of yet unseen in the blockchain UX space. It has an intuitive interface that makes it accessible to anyone and can even generate product designs, indicating its utility beyond that of art making.

DeepAI (2022) is another popular AI art generator based on open-source libraries and software. Highly customizable, it allows the human counterpart to modify colors, textures, and other details of an image, and most interestingly can generate resolution-independent vector images, and employs GANs such as StyleGAN, BigGAN, and CartoonGAN, among other technologies.

While there are many more AI art generation platforms—far too numerous to cover in this one section—the above examples are a sample of the diverse and increasingly prolific application of AI in the generation of art and personifies the components of a possible AI art market, described later in this chapter. From establishing digital art markets, assistive technologies, resolution-independent vector generation, and expansion into other formats of expression beyond still image, all the ingredients for a fully realized AI art ecosystem already exist.

6.5 Components of a Fully Automated AI Art Ecosystem

Distilled to its core component, we argue that an art ecosystem requires four features: (1) the artist is inspired, then conceives and produces the artwork; (2) the curator develops a selection and thesis for a collection of artworks either by a single artist or a group of artists, then (3) engages with a gallery, museum, or other platform to exhibit said works; and finally (4) a collector acquires these artworks, potentially as an investment to be added to a collection, valuated within an estate, or exhibited and sold again later. In this section, we will explore how AI can facilitate each function toward a fully realized AI-driven art ecosystem.

- Art Generated by an AI Artist: While we concede that a human actor must develop and program the AI, a key feature of AI artists is their ability to become generative, meaning that once an AI initiates the art creation process, it can be programmed to access APIs and other data streams to further develop its artistic practice on its own. This can be based on any generative data stream, such as the local news, the weather, or any other continuous dataset. Human artists are inspired by their experiences and the environment around them, and such triggers can be programmed into an AI artist.
- Art Selection and Thesis Generation by an AI Curator: Similar to how an AI artist might develop and generate their artworks, the AI curator can use the same methods to select and organize artworks. Technologies such as computer vision, metadata, and other forms of information can be used to generate theses for art collections. As it can take place digitally across the network, theses and collection generation can happen in real time and on demand, developing an infinite amount of themes for the curation of exhibitions and collections.
- The AI Art Gallery: Working in conjunction with the AI curator, platform technologies could enable the exhibition of art collections for other AIs to view and analyze. The AI art gallery need not even be human-viewable. It could simply

be an API in which other AI are subscribed to, viewing and analysis hundreds of artworks per second, within moments of the artworks being produced.
- AI Art Collectors and Buyers: Finally, based on technologies such as algorithmic and robot-based traders found in the financial system, AI art buyers and collectors could buy and sell these artworks with other AI art buyers and sellers, facilitating transactions via cryptocurrency, and contributing to the artworks' provenance via the blockchain.

6.6 Conclusion

In this chapter, the concept of the programmable non-fungible token was outlined, where NFTs could do more that simply indicate where an artwork is stored and who owns it, but contains the potential for the token to be the artistic medium itself via executable code built into the NFT smart contract and metadata. This was described as a programmatic non-fungible token or pNFT.

A brief background on the conceptual artworks of Sol LeWitt was shared to outline the ground works for the concept of the pNFT, then a description of the current state of art on the blockchain, as well as an exploration of generative and algorithmic art. Finally, a description of a fully automated, AI art ecosystem was highlighted, describing the features of this potential ecosystem. This included descriptions for AI artists, AI curators, AI art galleries, and AI art collectors.

While this chapter might be read as an exercise in future thinking, the argument for a fully AI-driven art ecosystem is possible. The technologies already exist to fully realize the automated art ecosystem, and it is the hopes of the author that NFTs are explored by the art community to their fullest potential, enabling the realization of a form of cultural robotics.

References

Aloosh A, Ouzan S, Shahzad SJH (2022) Bubbles across meme stocks and cryptocurrencies. Financ Res Lett 49:103155

Barger MS, White WB (2000) The daguerreotype: nineteenth century technology and modern science. JHU Press

Benjamin W (1972) A short history of photography. Screen 13(1):5–26

Bitkina OV, Jeong H, Lee BC, Park J, Park J, Kim HK (2020) Perceived trust in artificial intelligence technologies: a preliminary study. Hum Factors Ergon Manuf Serv Indus 30(4):282–290

Boden MA, Edmonds EA (2009) What is generative art? Digital Creativity 20(1–2):21–46

Creswell A, White T, Dumoulin V, Arulkumaran K, Sengupta B, Bharath AA (2018) Generative adversarial networks: an overview. IEEE Sig Process Mag 35(1):53–65

Crypto.com (2022) Buy, sell, and showcase NFTs. https://crypto.com/nft/. Accessed 27 Nov 2022

DALL-E 2 (2022) DALL-E 2 is a new AI system that can create realistic images and art from a description in natural language. https://openai.com/dall-e-2/. Accessed 27 Nov 2022

DeepAI (2022) Artificially intelligent tools for naturally creative humans. https://deepai.org/. Accessed 27 Nov 2022

Enjin.io (2022). https://enjin.io/. Accessed 27 Nov 2022

Foundation.app (2022) Connect to crypto apps with one click. https://foundation.app/nfts. Accessed 27 Nov 2022

Hughes E (1993) A cypherpunk's manifesto. http://www.activism.net/cypherpunk/manifesto.html. Accessed 3 Aug 2004

Koh JTKV, Dunstan BJ, Silvera-Tawil D, Velonaki M (eds) Cultural robotics. CR 2015. Lecture notes in computer science, vol 9549. Springer, Cham

Krauss R (1981) The photographic conditions of surrealism, pp 3–34

LeWitt S (1967) Paragraphs on conceptual art. Artforum 5(10):79–83

Lucking D, Aravind V (2019) Cryptocurrency as a commodity: the CFTC's regulatory framework. Global Legal Insights. https://www.allenovery.com/global/-/media/allenovery/2_documents/news_and_insights/publications/2019/8/cryptocurrency_as_a_commodity_the_cftcs_regulator_framework.pdf?la=en-gb&hash=8FB9966803A518C6CDC922AE1C6880AA. Accessed 27 Nov 2021

McConaghy M, McMullen G, Parry G, McConaghy T, Holtzman D (2017) Visibility and digital art: blockchain as an ownership layer on the Internet. Strateg Chang 26(5):461–470

Mintable.com (2022) Bringing the XRP Ledger to the NFT Ecosystem. https://mintable.com/. Accessed 27 Nov 2022

Misra S, Mukherjee A, Roy A, Saurabh N, Rahulamathavan Y, Rajarajan M (2020) Blockchain at the edge: performance of resource constrained IoT networks. IEEE Trans Parallel Distrib Syst 32(1):174–183

Mukhopadhyay U, Skjellum A, Hambolu O, Oakley J, Yu L, Brooks R (2016) A brief survey of cryptocurrency systems. In: 2016 14th annual conference on privacy, security and trust (PST. IEEE), pp 745–752

NiftyGateway.com (2022). https://www.niftygateway.com/. Accessed 27 Nov 2022

NightCafe.studio (2022) Create art with the help of artificial Intelligence. https://nightcafe.studio/. Accessed 27 Nov 2022

OpenSea.io (2022) Explore, collect, and sell NFTs. https://opensea.io/. Accessed 27 Nov 2022

Pearson M (2011) Generative art: a practical guide using processing. Simon and Schuster

Qiao H, Liu V, Chilton L (2022) Initial images: using image prompts to improve subject representation in multimodal ai generated art. In: Creativity and cognition (C&C '22). Association for Computing Machinery, New York, pp 15–28. https://doi.org/10.1145/3527927.3532792

Rarible.com (2022) Rarible—aggregated NFT marketplace with reward. https://rarible.com/. Accessed 27 Nov 2022

Saper CJ (2001) Networked art. U of Minnesota Press

Serada A, Sihvonen T, Harviainen JT (2021) CryptoKitties and the new ludic economy: how blockchain introduces value, ownership, and scarcity in digital gaming. Games Culture 16(4):457–480

Spittel A (2018) An introduction to generative art: what it is, and how you make it. FreeCodeCamp. https://www.freecodecamp.org/news/an-introduction-to-generativeart-what-it-is-and-how-you-make-it-b0b363b50a70/

Sunyaev A (2020) Distributed ledger technology. In: Internet computing. Springer, Cham, pp 265–299

SuperRare.com (2022) Collect digital art. https://superrare.com/. Accessed 27 Nov 2022

Wang Z, Yang L, Wang Q, Liu D, Xu Z, Liu S (2019) ArtChain: blockchain-enabled platform for art marketplace. In: 2019 IEEE international conference on blockchain. IEEE, pp 447–454

Wang Q, Li R, Wang Q, Chen S (2021) Non-fungible token (NFT): overview, evaluation, opportunities and challenges. arXiv preprint arXiv:2105.07447

Part II
Assistive Technology

Chapter 7
From Assistive to Adaptive: Can We Bring a Strengths-Based Approach to Designing Disability Technology?

Scott Andrew Brown

Abstract Assistive technology is often framed as a problem-solving approach to a medical model of disability. When viewed in this way, disability and neurodiversity are pathologised, demanding a 'cure' for afflictions that position the person as 'less than'. In this chapter, we explore the potential of assistive technology to augment and empower the user. We take the position that a strengths based, social model of disability is not only more effective in helping us develop assistive technologies, but it also places the user at the centre of the design process. This community-led approach to research and design recognises the value of lived experience in understanding and overcoming the many mismatches between people with a disability and their environment. In investigating this position, we will look not only at novel research projects with disabled and neurodivergent people, but also the ethics of cultural robotics and AI in human research more broadly. We will question the proposition that emerging technology is being positioned as a 'silver bullet' solution to many cognitive impairments and look at a range of embodied human–machine interactions that point to what the future may hold for the field of assistive technology. This chapter will examine a range of perspectives on cultural robotics in therapeutic and educational contexts and include the voices of people for whom these technologies claim to support.

7.1 Introduction

There is a popular maxim in autism research: "When you meet one person with autism, you have met one person with autism". This is a comment on the often-underestimated heterogeneity of autism and is reflective of the richness and complexity of humans generally. It would seem strange, then, that the technologies designed to support autistic people assume a very narrow way of seeing and being in the world. These technologies are designed to be assistive for a subset of autistic stereotypes, but often fall short of being truly assistive in that they fail to embrace the

S. A. Brown (✉)
Creative Robotics Lab, University of New South Wales, Sydney, Australia
e-mail: scott.brown@unsw.edu.au

© The Author(s), under exclusive license to Springer Nature Switzerland AG 2023
B. J. Dunstan et al. (eds.), *Cultural Robotics: Social Robots and Their Emergent Cultural Ecologies*, Springer Series on Cultural Computing,
https://doi.org/10.1007/978-3-031-28138-9_7

strengths and abilities of the individual. They may help with accessibility, but often don't recognise the importance of inclusion, the desire of most people to engage and experience the world on their own terms.

This is not a criticism specific to assistive technology. Designing to personal bias is an issue in the development of technology more broadly, with infamous examples including automatic soap dispensers and fitness trackers that are not able to recognise darker skin tones (Fussell 2017) and Amazon's hiring algorithms that were trained on and therefore reflected existing practices of male-dominated recruitment (Dastin 2018). These are technologies that aim to solve a perceived problem, but these 'problems' are too often framed by people that have no life experience with the issue they claim to be addressing. This is seen in the framing of many assistive technologies, where applying a lens of the medical model of disability in turn perpetuates stigma around disability. At worst, this approach persists with the idea that all disabled people need to be 'fixed', 'cured', or changed in some way in order to become 'normal'. Ideas of 'normality' are often defined by designers that do not identify as disabled.

Of course, there is an important role to be played by many assistive technologies that support people with a specific need, from text-to-speech readers through to motorised wheelchairs. This is a space that I describe as *assistive augmentation* (Springer 2018), which extends the existing capabilities of the body and leverages knowledge of the individual. In contrast, what I focus on in this introduction to the Assistive Technology section of this volume is the role that technology designers have to play in overcoming a long history of underestimating and paternalising people with a disability. How can we better reflect our vibrant communities by developing assistive technology that is able to adapt to the strengths of the individual? I suggest that the answer is not always more technology. Rather, we need to listen more closely to the people we claim to be designing for and genuinely elevate their voices throughout the design process.

7.2 Assistive Robotics in the Autistic Context

The promise of technology as a shining hope for assisting autistic people in education and therapy is pervasive. Indeed, I recognise my own naivety in the way I approached autism technology research early in my own career. Aiming to capture embodied interactions as autistic 'language' (Brown 2013) was an approach that showed my early work was othering autistic people by wanting to assist them to communicate using 'normal' terminology and methods. Since then, I have had the good fortune to work alongside autistic advocates who have generously helped me to shift my framing in both the role of assistive technology and autism research.

Like my own perspective in 2013, I believe that the vast majority of assistive technology design comes from a place of genuinely wanting to help others. However, as designers and researchers working from a position of privilege, we need to critically reflect on our role in the process of developing assistive technology, especially when we are not the people affected by many of the choices we make throughout the process. In addition, the layers of complexity that are present in some emerging technologies can exacerbate any bias present in the design process—this is particularly evident in the use of social robotics for autism.

Despite perceptions in popular culture, robots remain relatively limited in ability within social contexts. Physical movement is impeded by heavy components and meagre battery life spans. In terms of intelligence, robots fare slightly better, however the knowledge that machine learning algorithms are trained on still suffer from the issues I have begun to outline here: they reflect the perception and worldview of those that train them—their designers. In developing a technology in response to a narrow view of a problem, we run the risk of neglecting and even exacerbating the interrelated issues surrounding it.

Disability is often compounded by snowballing problems of access. This may be in the form of the higher prevalence of comorbidity seen in autistic people (Leyfer et al. 2006) or the increased social and financial challenges faced by people with a disability. The use of robots outside of a classroom or therapy context is limited because of these financial or access constraints, often shaping the approach we take to the design and research of social robots in this space. Broader use cases are not common because of limited access, leading to less research and investment. Less research and investment leads to reduced access, and the cycle continues.

7.3 Disability Dongles and the Empathy Problem

Disability advocate Liz Jackson coined the term "Disability Dongle" in 2019 to describe the phenomenon of technology designed to make disabled people "compatible with a normative system" (Jackson et al. 2022). More often than not, these technologies are famed for being novel or innovative and capture widespread attention for their potential to 'solve' a problem in the disability space. The catch is that these problems—and their novel solutions—are framed by people with no experience of them. These technologies are regularly critiqued by the disability community, but their concerns are drowned out by voices that have an entirely different set of values or worldview than those that they claim to be supporting. As Jackson points out, at their best, Disability Dongles are well-meaning but ultimately under-used (or useless) technologies that find their way to landfill. At their worst, Disability Dongles are defended desperately by neuronormative and able-bodied voices that are willing to cut down those with lived experience to support the "heroic designer-protagonist whose prototype provides a techno-utopian (re)solution to the design problem" (Jackson et al. 2022).

In many design-related fields, empathy is a goal that comes with good intentions but is rarely able to live up to its ideals. As Jackson and others have noted, the majority of the work carried out in the assistive technology space is developed with good intentions, striving to empathise with those that they claim to be helping. However, the central problem remains, in that we are still speaking *for* the communities we want to support. Designer Mike Monteiro (2019, p. 22) incisively described the use of empathy to speak for others:

> What about empathy? Empathy is a pretty word for exclusion. I've seen all-male all-white teams taking "empathy workshops" to see how women think. If you want to know how women would use something you're designing, get a woman on your design team. They're not extinct. We don't need to study them. We can hire them!

As Monteiro points out, claiming to empathise with your users is not enough. When their voices are not in the room, it is near-impossible to appreciate their needs, values or worldview, which should be central to the design process from the outset. The appropriation of human-centred buzz words to position a product or service as inclusive is sadly all too common in technology design and continues a history of pathologising, othering and paternalising people with a disability.

7.4 Assistive Technology and the Social Model of Disability

It is not all bad news, however. The social model of disability has slowly but surely been making inroads as a framework for recognising that it is the responsibility of all society to remove the barriers that restrict the life choices of disabled and neurodivergent people. This is in contrast to the medical model of disability, which has historically viewed disabled people as having impairments that need to be 'fixed' by treatments or technologies. As a result, the medical model effectively places blame at the feet of those seen as being a problem to be solved.

Author and designer Kat Holmes describes the above philosophical shift from a medical to social model of disability as a movement from disability being seen as a personal health condition to disability being seen as mismatched human interactions. Most important in this change is that it "highlights the responsibilities of people who make solutions" (2018, p. 51). Taking on the position as a designer of assistive technology means recognising that the products and services we create are enabling and mediating at most–they should help users realise their potential, on their own terms.

Holmes cites the designer Susan Goltsman as coining her favourite definition of inclusive design (2018, p. 53):

> Inclusive design doesn't mean you're designing one thing for all people. You're designing a diversity of ways to participate so that everyone has a sense of belonging.

This definition gets to the heart of how we might reshape the purpose of assistive technology using a truly social model of disability. Inclusive design belongs at *all* stages of the design process, not only the testing and use phases. Making assumptions and designing from the position of our own biases cannot lead to truly inclusive design, nor technology that recognises and supports different strengths and worldviews. The challenge for neurotypical and able-bodied designers is to genuinely step back and let other voices into the process.

7.5 Creating Space for Disabled Voices

Disabled people are amazing problem solvers. Laura Mauldin (2022) speaks of the long history of hacking assistive technologies or reappropriating unexpected objects to augment technologies to make them assistive in their own personal context. Mauldin draws on examples such as a man using exercise bands to improve grip with his arm bicycle, and the use of silicone trivets instead of oven mitts for better dexterity by someone who, following a stroke, only has the use of one arm. These ad hoc solutions often serve their users better than the high-tech options making headlines for their innovation and condition 'fixing' potential.

Equally, there are many technologies (both low- and high-tech) that are not considered by the non-disabled public for their assistive or supportive qualities, which once removed, leave a damaging gap in the lives of disabled people. This can be seen in the well-meaning removal of plastic straws (many people with limited mobility found them much easier to use than metal or paper alternatives) and the current exodus from Twitter in response to Elon Musk's ownership and control ('Disability Twitter' was recognised by many in the disability community as a vital support network). This also speaks to perceptions of what constitutes an assistive technology and therefore what can be supported by social or insurance schemes. Here in Australia, the National Disability Insurance Agency (NDIA) is vague on these terms, preferring to leave the decision to experts like occupational therapists or physiotherapists, who in turn make choices based on recommendations seen online or from family and friends.

It is clear that while many designers of assistive technology are approaching the field with the best of intentions, they are not always best placed to define what is a problem and, even more so, what is the solution. However, it is these same people that have the best access to the processes to design, develop and commercialise the technologies that find their way into the world and the lives of disabled and neurodivergent people. What we need is more voices with lived experience throughout the design process to shape problem spaces and identify meaningful solutions—not simply introduced at the end of a design process as testers for questions framed by able-bodied designers or researchers.

It is then the duty of the designer to leverage their privilege and find ways to make space for disabled voices to take part—and ideally *lead*—the design process. This is not an easy task and nor should it be. It is through the challenge of critical reflection of our own role in overcoming the 'mismatches' between people and their interactions with the world that we will create better assistive technology—not more innovative or headline-grabbing, but genuinely more useful and reflective of the needs and desires of those that we claim to be designing with (not for).

As a first step, recognising the strength of lived experience in the design process will take us on the path to a strengths-based approach to collaboratively designing assistive technology. Not seeing a disabled person as limited by a pathologised deficit, but viewed in terms of their ability to problem solve and overcome a world that is not designed for them, should highlight the importance of their role as not just valuable, but central to the design process.

7.6 From Assistive to Adaptive

During my career as a designer of interactive technology for neurodivergent people, I have grappled with how best to locate my role as someone that will always be outside the community I hope to work with. What I have come to realise is that I will continue to get things wrong, but as McKercher notes, "Failed ideas make great compost" (2020). It is critical, however, that I engage with people who have lived experience to help me recognise when I am off track. It has been my good fortune to work alongside people with lived experience, to develop deep and meaningful relationships that allow honest and sometimes difficult conversations to take place. This takes time, and it takes effort to create spaces that are safe and supportive for these relationships to grow.

This space continues through to the design and development of technology; it should afford users the ability to explore and interact on their own terms, rather than dictate how it facilitates their engagement with the work. Assistive technologies that embrace openness and ambiguity also offer the designer an opportunity to critically evaluate their own assumptions and biases. Again, taking one of McKercher's wonderful calls to action, it is important to "design with humility" (2020). Finding my work used in unexpected ways is one of the great joys of what I do, but it can also serve as an important signpost for understanding when a direction that I am heading in doesn't work for the user.

While the feedback of people with lived experience is not optional, we cannot only lean on them for free labour. It is not up to disabled users to be only seen as test subjects; their insight and knowledge has value and should be treated as such. However, there is no one-size-fits-all approach to inclusion, and rather than attempt to present a formulaic framework here, I encourage my colleagues to engage more deeply with the communities they are designing with. My experience has shown that it is only through this kind of engagement that you can begin to gain some insight into how best to create space for each person and build knowledge for what might be able to support a wider community.

This is a philosophy that might bring us closer to shifting assistive technology towards adaptive technology, not only in the outcomes from the design process, but in the way that we engage with people that should be benefitting *throughout* the design process. In my privileged position and training as a designer, I need to use the tools available to me to give disabled and neurodivergent people the space to do what is meaningful to them, often while contending with a myriad of other challenges. If we reduce some of those barriers to accessing what matters to each person, then we might be able to start defining a field of adaptive technology.

7.7 The Assistive Technologies Section

I am exceptionally proud that we have been able to include a range of voices in the Assistive Technologies section of this volume. Some of the authors are new to academic writing but have important perspectives and ideas to share. This section represents just a small slice of the work being done in the disability and assistive technology space, though it does reflect some of the diversity of ideas and approaches. It also takes a broad view of how we might define assistive technology, which aligns with the philosophy of the Creative Robotics Lab, to explore how technology can facilitate and improve human-to-human interactions and experience.

Melanie Tran begins the section with a call to arms for bringing lived experiences into user experience (UX) and how this can lead to broader social change. Barbara Bruno et al. then examine the application of social robots in educational contexts, reflecting on some of the challenges, particularly those around cultural sensitivities and adaptation. Sebastian Trew and myself propose an autistic framing of user experience (aUX) as a method for designing assistive technologies that are more meaningful for neurodiverse communities. Finally, Maria-Theresa Oanh Hoang et al. present a case study on the use of drones to support search and rescue personnel—an extension of how we might traditionally consider assistive robotics.

Each of these contributions make for a thought-provoking section of the book, which I hope is just the start of further work in each of the author's respective fields. As I have reiterated throughout this chapter, it is through collaboration and conversation that I believe we will develop assistive technology that is of most benefit to the user. I encourage you to reach out to the contributing authors to start your own conversations.

References

Brown S (2013) Autism and theory of mind in interactive spaces. In: 19th International symposium on electronic art, ISEA 1 4

Dastin J (2018) Amazon scraps secret AI recruiting tool that showed bias against women. Reuters. https://www.reuters.com/article/us-amazon-com-jobs-automation-insight-idUSKCN1MK08G. Accessed 27 Nov 2022

Fussell S (2017) Why can't this soap dispenser identify dark skin? Gizmodo. https://gizmodo.com/why-cant-this-soap-dispenser-identify-dark-skin-1797931773. Accessed 27 Nov 2022

Holmes K (2018) Mismatch: how inclusion shapes design. MIT Press, Cambridge

Huber J, Shilkrot R, Maes P, Nanayakkara S (eds) (2018) Assistive augmentation. Springer, Singapore. https://doi.org/10.1007/978-981-10-6404-3

Jackson L, Haagaard A, Williams R (2022) Disability dongle. Platypus. https://blog.castac.org/2022/04/disability-dongle/. Accessed 27 Nov 2022

Leyfer OT, Folstein SE, Bacalman S, Davis NO, Dinh E, Morgan J, Tager-Flusberg H, Lainhart JE (2006) Comorbid psychiatric disorders in children with autism: interview development and rates of disorders. J Autism Dev Disord 36:849–861. https://doi.org/10.1007/s10803-006-0123-0

Mauldin L (2022) Care tactics. https://thebaffler.com/salvos/care-tactics-mauldin. Accessed 27 Nov 2022

McKercher KA (2020) Beyond sticky notes: co-design for real: mindsets, methods and movements. Beyond Sticky Notes, Sydney

Monteiro M (2019) Ruined by design: how designers destroyed the world, and what we can do to fix it. Mule Design, San Francisco

Chapter 8
The Intersection of Social Impact, Technology and Design: A Catalyst for Cultural Change

Melanie Tran

Abstract Technology is deeply ingrained in our lives to the point where it has shaped our perspectives and changed the way we think, the way we act, and the way we behave. But, what happens when we add *disability* to the equation? When the use of technology and the needs of disability collide, it is defined as assistive technology. But is there any difference between *technology* and *assistive technology*? Should there be a difference? As a user experience designer, a researcher, a board of director, a woman in tech and a person living with a physical disability, my work has predominately been focused on the intersection of technology, innovation and social change. This chapter explores how the power of technology, design and innovation can be harnessed to drive a cultural shift in the way our society perceives disability. More importantly, how it helps shift our society's focus from the medical model of disability to the social model.

8.1 The Not-So-Distant Past

8.1.1 Introduction

We live in a complicated yet sophisticated society where technology is deeply ingrained in our lives to the point where it has shaped our perspectives, the way we think, act and even behave. As a user experience designer, a researcher, a board of director, a woman in tech and a person living with a physical disability, I have spent most of my career to date studying the fascinating intersection of technology, innovation and social change.

Technology use has grown exponentially and will continue to do so at a rapid pace. But what happens when disability is added to the equation? Navigating the world through technology in our society is known to be the 'social norm', but when

M. Tran (✉)
Sydney, NSW, Australia
e-mail: tran.melanie96@gmail.com

© The Author(s), under exclusive license to Springer Nature Switzerland AG 2023

B. J. Dunstan et al. (eds.), *Cultural Robotics: Social Robots and Their Emergent Cultural Ecologies*, Springer Series on Cultural Computing,
https://doi.org/10.1007/978-3-031-28138-9_8

the use of technology and the needs of disability collide, it is defined as *assistive technology*.

There are two distinct perspectives of how assistive technology is viewed:

1. Assistive technology that aligns with the *medical model* (the focus on a person's condition and the aim to find a 'cure') stereotype of disability.
2. Assistive technology that has the power and potential to push boundaries and highlight the *social model* of disability—better yet, the human behind the disability.

The medical and social models of disability are part of an ongoing debate amongst those with lived experience and professionals across a range of sectors. Those in favour of the social model argue that the shift away from the medical model of disability can become a powerful driving force to create much-needed cultural shift in our perception of disability.

In the technology, disability and design sectors, there is an increasing awareness and emphasis on leveraging unique perspectives to help shape a better, stronger and more inclusive society. This chapter aims to address the impact of the intersection of user experience (UX) design, technology and social impact through a disability lens. These concepts are no strangers to us when viewed individually. However, unique opportunities are created when we combine the power and potential of these fields as it places us at the forefront of innovation and social change.

As a person living with a physical disability myself, one of the key lessons that I have learned is that challenges can become a driving force for innovation. These challenges forced me and those around me to think and act differently. More importantly, I have witnessed first-hand just how powerful technology can be and its potential to become a catalyst for positive social change. All these elements are critical puzzle pieces that represent the broader picture of the society we, as individuals, strive for—a society where we no longer have to fight for equality, diversity and inclusion. This requires a cultural shift in the way we think, the way we act, the way we behave and a desire to create positive social change, no matter how big or small.

Over the years, I have also learnt that we have living and breathing examples of how emerging technology driven and disability-focussed social enterprises redefine the concept of inclusion—and more importantly, how the power of design and creativity can be leveraged to harness unique perspectives that result in the development of solutions that are truly driven by user needs. We will discuss what this means later in the chapter.

For now, in the first section of this chapter, we will deconstruct some major concepts, such as the medical and social models of disability, the value and impact of user experience design, the role of assistive technology, and how social innovation could be harnessed to foster a more inclusive, collaborative, innovative and creative culture.

8.1.2 *Medical Versus Social Models of Disability: Let's Take a Look at What the Past Has Taught Us*

There are ongoing discussions by researchers and professionals across a range of sectors that explore issues around medical and social models of disability. Whilst there are strong arguments that support approaching disability through the medical model, there is currently a cultural shift in academia and industry, pointing to the social model of disability being a catalyst for positive social change when it is coupled with the power of technology.

Humpage conducted a study in 2007 called *Models of Disability, Work and Welfare in Australia*, which highlighted key differentiators in the medical versus social models of disability that are worth noting. The medical model of disability is defined as an individual with an impairment that needs to be cured or contained (Oliver, cited in Humpage 2007). This concept has been explored extensively in the academic sector from multiple perspectives. Bricher (2000) published a study that investigated the research relationship between people with disability, health professionals and the social models of disability—a deep dive into how the medical and social models of disability are impacted by perspectives across society. For example, traditionally, when it comes to aesthetics, assistive technology shares design similarities with medical equipment. However, we begin to see a shift in the way assistive technology is designed as our society adopts inclusive design principles and strives to create products that are aesthetically desirable, feasible and usable.

A number of researchers have highlighted that when viewing the medical model of disability through the research lens, the focus is placed on the ability to understand how individuals with an impairment can be brought closer to what our society deems as 'normality', and how their medical dependencies and care needs can be decreased over time (Bochel and Bochel; Rioux, cited in Bricher 2000). This point was further reinforced by other researchers who highlighted that, traditionally, nurses would consider independence as an individual's ability to undertake self-care, rather than the ability and autonomy to make decisions (Oliver; French, cited in Bricher 2000). To combat this argument, French (cited in Bricher 2000) posed the question of why it is deemed an issue to accept assistance with daily living activities if this means that individuals will gain more energy to live a fuller life. Morris (cited in Bricher 2000) echoes this statement and highlights that people with disability should have choice, control and autonomy over their own lives, regardless of their ability to independently manage their daily living activities. This itself reflects the cultural shift, not only from the medical model of disability to the social model, but also how our knowledge of disability has evolved over time.

In contrast to the focus on individuals' impairment and medical needs, the social model of disability places an emphasis on the individuals themselves rather than their medical condition. It challenges us to take a holistic view of people with disability and consider the social, political and economic factors that impact the perception of disability (Humpage 2007). It is argued that an individual's impairment or disability has nothing to do with the body—it is, in fact, simply a 'consequence of the failure of social organisation to take account of the differing needs of disabled people and remove the barriers they encounter' (Oliver 1996, cited in Humpage 2007). There is a significant shift in how barriers are perceived. In the medical model perspective, the focus is on the barriers that exist because of an individual's impairment or disability. Conversely, the social model of disability highlights the fact that it is the structure of our society that creates impairment and barriers (Abberley, cited in Siminski 2003).

There is evidently a clear distinction between the medical and social models of disability, and as a result, these framings shape our understanding and perspectives on individuals who live with a disability. Subsequently, this has also heavily influenced the way we, as a society, design and develop products and services, particularly when it comes to the disability sector. Technology, or rather assistive technology, is no stranger to the disability sector. Now that we have unpacked the medical and social models of disability, perhaps the question we should ask is: How have these models influenced the design and utilisation of assistive technology?

8.1.3 The Role of Technology: Friend or Foe?

Assistive technology is known to play a critical role in the lives of people with disability. In some cases, it can be seen as the one thing that affords an individual their ability to participate and contribute to society. Over the last two decades, numerous researchers have explored the role of assistive technologies and the following eight factors outline its purpose in the disability sector:

1. Improve functional performance of individuals (Cook and Hussey; Isabelle et al.; Johnson et al.; Judge; Ripat; cited in Ripat and Woodgate 2010)
2. Gain greater control, flexibility and autonomy over their environment (Parette and Brotherson, cited in Ripat and Woodgate 2010)
3. Promote empowerment amongst individuals (Hutzler et al., cited in Ripat and Woodgate 2010)
4. Create positive psychological wellbeing (Craddock; Jutai and Day, cited in Ripat and Woodgate 2010)
5. Reduce barriers in physical environments to enable participation (Pape, Kim and Weiner, cited in Ripat and Woodgate 2010)
6. Reduce the dependency on caregivers (Benedict et al., Ostensjo, Carlberg and Vollestad, cited in Ripat and Woodgate 2010)
7. Create an opportunity for individuals to achieve a better life (McMillen and Soderberg, cited in Ripat and Woodgate 2010)
8. Increase societal participation (Campbell et al.; Copley and Ziviani; Jirikowic et al., cited in Ripat and Woodgate 2010)

These eight factors are closely associated with an overarching goal and purpose, which is ultimately to create equal opportunities for people with disability so that they can live a life filled with independence, empowerment, autonomy, choice and control. Whilst this is a fundamental human right that we should not have to fight for, the reality is that we have yet to develop into a society where this is no longer considered a battle. However, with the rapid growth and advancement in technology, it is no surprise that this would play a critical role in helping us fight for social inclusion and equality.

Later in this section, we will explore the fundamental shift within the assistive technology sector that has contributed to shaping the social model of disability and redefining the way our products, services and environments are designed. More importantly, I will discuss the secret ingredient to designing and developing solutions that have the ability to increase utilisation by its intended users, as well as fostering a stronger collaborative culture to promote innovation. Further, we will look at the integration of artificial intelligence in assistive technologies and its potential to help drive positive social change.

For now, let's do a deep dive into the fundamentals that we need to know about assistive technology.

It is believed that the Technology-Related Assistance for Individuals with Disabilities Act of 1998 in the United States has helped shed light on the role of assistive technology and its ability to enhance the functional needs of individuals who have a disability, as well as improve overall quality of life (Wallace et al., cited in Alper and Raharinirina 2006). When this act was amended in 1994, it reframed the philosophy of this policy with the intention to shift away from the medical perspective of how assistive technology is associated with people who have a disability. Instead, the focus was placed on the role of assistive technology in the social context. This often refers to the education, employment and community participation aspects in the lives of people with disabilities (Wallace et al., as cited in Alper and Raharinirina 2006). A study conducted by Zhang in 2000 (cited in Alper and Raharinirina 2006) stated that technology has the potential to enhance the capabilities of individuals with a disability. However, there are several barriers that prevent the successful integration of assistive technology in the lives of people with disability, as highlighted below (Alper and Raharinirina 2006):

- Despite the mountain of existing technology used, in this case, specifically in the education sector, many of these technologies are still inaccessible to students with disability and their families (Zhang 2000, cited in Alper and Raharinirina 2006) due to affordability and the knowledge and training required to utilise the devices.
- The high costs of assistive technology become a barrier, as well as the lack of access to appropriate funding and information about suitable assistive technology solutions for individuals with a disability (Wehmeyer 1998, cited in Alper and Raharinirina 2006).
- The lack of awareness and knowledge about assistive technology options from health professionals (Alper and Raharinirina 2006).

Whilst it is critical to access the appropriate assistive technology that meets the needs of individuals with a disability, it is equally important to consider the utilisation of the products that are purchased and the role it plays within the life of the individual. Numerous researchers have looked at the utilisation of assistive technology post-purchase, noting that the assistive technologies purchased do not necessarily correlate with the utilisation rate. Some of the key factors leading to the underutilisation of assistive technology are listed below (Todis 1996, cited in Alper and Raharinirina 2006):

- The assistive technologies that were purchased did not meet the needs of the end users, in this case, the individuals with a disability (Parette 1997, cited in Alper and Raharinirina 2006).
- The choice of assistive technology was dictated by family members or therapists who surrounded the person with a disability (Scherer 1993, cited in Alper and Raharinirina 2006).
- The complexities associated with setting up the devices, programming and portability (Scherer 1993, cited in Alper and Raharinirina 2006).
- The lack of funding to purchase and maintain assistive technology devices (Todis 1996, cited in Alper and Raharinirina 2006).
- The unreliability of technology (Scherer 1993; Todis and Walker 1993, cited in Alper and Raharinirina 2006).
- The lack of technical support for instalment, repair and maintenance (Lode 1992, cited in Alper and Raharinirina 2006).
- The unwanted/negative attention that is drawn to an individual when utilising assistive technology (Todis 1996, cited in Alper and Raharinirina 2006).

The factors highlighted above provide us with a holistic view of the barriers that are associated with assistive technology. Now that we have a general understanding of the barriers to integrating assistive technology in the lives of people with disability, let's take a closer look at how some of these factors impact the disability and user experience design sectors. For the purpose of our discussion, let's unpack the following two key areas:

1. Assistive technology was purchased but not utilised because it did not meet the needs of individuals with a disability (Parette 1997, cited in Alper and Raharinirina, 2006).
2. The choice of assistive technology is dictated by family members or therapists who surround the person with a disability (Scherer 1993, cited in Alper and Raharinirina, 2006).

Assistive technology was purchased but not utilised because it did not meet the needs of individuals with a disability: To view this from a design and development perspective, the key question to ask here is whether the end users (people with disability themselves) were involved in the process when the assistive technology devices were created. In Alper and Raharinirina's (2006) study, the researchers highlighted the importance of enabling assistive technology devices to be appropriately customised in order to meet each individual's needs, rather than off-the-shelf solutions that are designed with a one-size-fits-all mentality. However, to create solutions that are flexible enough to adapt to different needs, we first need to understand the problem we are trying to solve. Whilst this may sound simple, it is a rather difficult step to get right. This is the very core of good user experience design.

When we take the time to understand the problem we are trying to solve, it will naturally lead us to get to know the people who are affected by this problem. Why is this a problem for them? What are the main challenges and barriers? What works well and what are the opportunities for improvement? What needs to be done to address some of the challenges and barriers identified? The only way we can find an accurate answer to these questions is by involving people with disability themselves, because, at the end of the day, they are the ones with the rich lived experience and innovative mindset that could be leveraged to inform the design and development of products that truly reflect and address the needs of the intended users.

More importantly, we need to understand the *why* before we think about the *how* when it comes to problem solving. One of the secret ingredients to achieving this is *empathy.* Whilst this may sound simple, it serves as a powerful guiding principle in the design and development process of products and services. It allows us to put ourselves into the shoes of users who we are designing *for* and *with.* In other words, it provides the opportunity for us to see and experience the world through the lens of the users. Then, and only then, would we be able to create solutions that are deeply driven by the needs of users. Solutions that address the challenges identified by users. Solutions that have the ability to create life-changing impact.

It is a common practice for businesses and organisations to proceed with designing and developing solutions *for* people with disability, instead of *with* people with disability. The key difference here is that one is adopting the 'we know best' mindset as experts in the field, and the other adopts a collaborative mindset that leads to believing solutions are more powerful when they are designed with unique perspectives and are driven by the needs of users. This is often known as creating solutions that are *person-centred*. Adopting a person-centred mindset means that we are able to see beyond disability, as well as embracing the philosophy that it is more important and cost-effective to enhance the quality of life for people with disability, rather than focusing on their limitations (Scherer 1996, cited in Alper and Raharinirina, 2006).

Empathy begins with actively listening to our users. It is worth noting that when we speak, we are merely repeating what we already know. However, when we listen, it allows us to learn something new. However, although active listening is a critical component, in order to truly embrace the power of empathy, the products and services that we create need to be driven by the voices of its intended users—not just at the beginning or at the end of the design and development process, but throughout the *entire* process.

The choice of assistive technology is dictated by family members or therapists who surround the person with a disability: This is closely associated with what we have discussed above in terms of understanding the problem space through the eyes of the end users, in this case, people with disability, before diving into creating solutions. As we have identified earlier, the traditional 'we know best' mindset is a known barrier to assistive technology playing an effective role in the lives of people with disability. Perhaps the key driving factor for this theory lies in how people with disability are perceived, which brings us back to the medical and social models of disability. If we were to view this from the medical model, then the focus would be on the individual's disability. Subsequently, they would be perceived as individuals who rely on medical care and support. However, if we were to challenge this and view it from the social model of disability, it would allow us to see the individuals as *subject matter experts*. More importantly, individuals who have lived experience would be viewed as having unique perspectives and skills and expertise to make meaningful contributions to our society.

A valuable lesson that I have learnt is that our society and the solutions we create will only be as good as we allow them to be. If we want to push the boundaries of social norms and truly embrace diversity, inclusion and equality, we must first be able to understand, acknowledge and celebrate unique perspectives. This would create opportunities for us to ensure that the solutions we create are guided by unique voices and perspectives.

So, what have we learnt so far? We have learnt a little bit about the role of assistive technology, the barriers to integrating it effectively in the lives of people with disability, and finally, the lessons that we can take from this. Most importantly, we learnt about the value of understanding the problem before diving into creating a solution and the importance of involving people with disability in the design and development of solutions at the very beginning of the process. Furthermore, we have had a glimpse of how assistive technology has helped create a fundamental shift in the perception of disability.

Now that we have investigated the world of assistive technology, let's move on to another topic that is closely associated with this field: the future of artificial intelligence. Artificial intelligence is proving to play an increasingly important role across the business, finance, science, medicine, engineering, manufacturing, education, military, law, and art sectors (Kaliraj and Devi 2021, p. 217). However, relatively few studies have been conducted to address the importance of enhancing the quality of life for people with disability by adopting the potential of artificial intelligence in assistive technology (Kaliraj and Devi 2021, p. 217).

Artificial intelligence is a field that can inject a wave of excitement for its potential and possibilities or induce a feeling of caution as we are forced to address uncertainty and the unknown—or perhaps both. But, before we discuss the integration of artificial intelligence in assistive technology, let's first take a moment to acknowledge the artificial intelligence that surrounds us. Think about examples like voice-activated personal assistants such as Siri or Alexa, or smart algorithms that track our behaviours and patterns to create a tailored online shopping experience, or even facial recognition technology. Whilst the existence of artificial intelligence makes our lives easier, it also has the power and potential to serve more than just everyday convenience.

The role of artificial intelligence is amplified when it is extended to the disability and assistive technology space. When artificial intelligence is combined with assistive technology, it opens the door to a wealth of opportunities and possibilities. The following examples highlight some examples that showcase the integration of artificial intelligence in assistive technology (Kaliraj and Devi 2021, p. 220–222):

- Assistive robotics harnesses artificial intelligence to provide around-the-clock support for people with disability. It is programmed with the capability to perform tasks such as assistance with medication and medical appointment management.
- Smart wheelchairs utilise artificial intelligence to analyse the human brain, behaviour and patterns, enabling users with limited physical ability to navigate through their environment safely and independently.
- Speech-to-text technology can enable effective and efficient communication, whilst having the capability to improve the accuracy and predictability of texts.

Once again, this is just a glimpse of how artificial intelligence is integrated into assistive technology. However, this gives us enough insights from the surface level to be able to have the discussion on whether we should embrace the potential of artificial intelligence in assistive technology or approach it with caution.

We are at the beginning of realising just how powerful artificial intelligence can be when it is effectively and successfully integrated into assistive technology. The more important question to ask here is where this would lead us when disability is added to the equation. Whilst there is no straight answer, the one thing that we must not forget is to look at the intersection of disability, technology and social inclusion. It could be argued that there is one key element at the very core of this intersection that should be considered above all else: *human connection.*

Technology and artificial intelligence can leave us with great difficulties as we grapple with the threat and fear of the increasing reliance on robotics for decision-making in highly complex environments such as health care and disability. However, human connection is an element that encompasses these essential services. The desire and need to connect with others is deeply ingrained within our human nature. Whilst technology has the ability to enhance human connections, it will not be able to replicate or replace our ability to empathise and understand the personal values that are deeply rooted in our DNA.

One could argue that artificial intelligence could bear the risk of driving us further from the human connections and values that are deeply rooted within our culture. In the next section of this chapter, we will take a closer look at existing initiatives in the disability and technology sectors that further reinforce this point.

Before we can dive into this, we need to first understand the value of user experience design and the role it plays in driving innovation, social change and human connections when coupled with the power of technology. More importantly, we will learn a little bit more about *how* user experience design can help us create solutions that meet the needs of users.

8.1.4 User Experience Design: What on Earth Does This Mean and Why Do We Care?

What makes you choose an Apple phone over an Android one, or vice versa? Whilst they may have some key differences in their technicalities and features, the foundations are the same: both allow us to connect with our loved ones and have the beauty of the World Wide Web at the tip of our fingers. So, what exactly makes us choose one phone over the other? Is it the brand? Is it the familiarity? There could be a host of reasons why we decide to go with Apple phones over Android, or vice versa. However, the core of this comes down to the *experience.* Let's take a closer look at what this means from a user's perspective and the business perspective that is ultimately responsible for designing and developing solutions.

As end users, what matters most to us is how we *feel* when we use certain products. Consciously or subconsciously, when we assess whether a product is right for us, the following factors (amongst others), often play a role in the decision-making process:

1. Does this product meet my needs?
2. Does this product help me complete the tasks that I need to do?
3. How easy is it for me to use this product?
4. Do I need to learn a new skill in order to use this product or can I leverage what I already know?
5. How does using this product make me feel?

Now let's look at this from the business perspective. What does it mean to design and develop products that are driven by the needs of the end users? This is also known as *user experience design.* If we were to unpack this further, the concept of user experience is multi-faceted. It refers to how we interact with products and other people, as well as the emotions that follow. Understanding this concept would result in designing and developing products and services that have the ability to improve the lives of those who use them (Battarbee 2004). An article written by Gray in 2016 further reinforces the value of having an in-depth understanding of the problem before jumping into designing and developing the solution. The author reviews Robert Pressman's book *Software Engineering: A Practitioner's Approach*, commenting that 'for every dollar spent to resolve a problem during product design, $10 would be spent on the same problem during development and $100 or more if the problem had to be solved after the product's release' (Gray 2016, para. 4). This points to how critical it is to get an in-depth understanding of the problem we are trying to solve before developing solutions, and our users are the only ones who can guide us to the answers.

All of this comes down to one simple point. It is all about meeting a *human need.*

Earlier in this chapter, we discussed the importance of understanding the problem we are trying to solve before diving into designing and developing the solution. This includes understanding *who* will be using the product and *why* they are using the product. More importantly, it is about creating solutions that are deeply driven by the needs of the users and their experience. This brings us back to the secret ingredient that we discussed earlier in this chapter: empathy.

From a surface level, the concept of empathy is defined as one's ability to identify with other people's thoughts and feelings. However, this is just the tip of the iceberg. For us to get a firm grasp on empathy, we also need to be able to understand what motivates individuals. What drives them? What are their values and priorities? What are their preferences and what are their inner conflicts? (McDonagh 2006, cited in Kouprie and Visser 2009). When we translate this into practical steps in design, it allows us to get closer to the lives and experiences of our users, both present and future users. This is known to increase the likelihood of developing a product or service that meets the needs of users (Koskinen et al. 2003, cited in Kouprie and Visser 2009).

However, just merely understanding the concept of empathy and user experience is not quite enough. Many disciplines have attempted to develop methods to help us understand how these two concepts can influence the way we design our products and services. This is broken down into three simple models (Battarbee 2004):

1. **Product-centred**: this model provides information to assist designers and key stakeholders involved in the project to create products that evoke compelling experiences. This is achieved by taking into account the key considerations that will impact the design and evaluation process.
2. **User-centred**: as we have discussed in this chapter, this model is specifically created to help designers and key stakeholders involved in the project to understand the users who will be interacting with the product. This involves understanding the users' motivational drivers, triggers, goals, etc.
3. **Interaction-centred**: this model explores how the role of products can bridge the gap between the users and those who are designing and developing the solution. This has the ability to create a collaborative relationship between the users and the individuals who are creating the solution so that we can embark on a co-design journey.

If we were to examine these models through a disability and technology lens, a key element that is not quite captured above is the relationship between the users, the product and our environment. The question that we must ask when creating a solution is whether it can truly remove barriers. Whilst we may have the intention of producing outcomes that can help remove problems for people with disability, we can very easily fall into the trap of creating further barriers or enforcing the invisible wall between disability and what our society defines as the social norm.

It is a rather delicate dance between creating solutions that are accessible and flexible enough to adapt to different needs and creating solutions that are targeted at specific needs or users that unintentionally result in further marginalisation, and in some cases, stigmatisation. For the purpose of this discussion, let's walk through a scenario to help us contextualise what this actually means.

Let's pretend that we plan to create a brand-new communication platform: a mobile app that will enable us to easily connect with each other through the power of technology. However, one of the key requirements in this scenario is that the solution we create must be accessible and beneficial to people living with a disability. There are two ways that we can approach this:

1. Create a solution that will enable users to connect with the disability community so that they can build a network with those who share similar experiences and interests. In addition, the solution will aim to create a safe environment for individuals to share their stories and tips that may be useful to others.
2. Create an accessible solution that is driven by one simple goal: to enable individuals to connect with each other through technology.

Remember that regardless of either approach, our end goal here is to create a communication platform. However, there is a vast difference between these two approaches, and this would ultimately influence the way we design and develop the solution.

Before we examine what the difference is, let's pause for a moment. Which approach would you go with? The more important question is *why* did you choose one approach over the other?

Now let's take a look at *how* these two approaches can influence the way we create the solution. Whilst the first approach may meet the requirements of creating a communication platform that is accessible and beneficial to people with disability, it could also lead to the unintentional impact of further marginalisation and stigmatisation, simply because the concept of a communication platform that enables people who have a disability to connect with others with similar experiences comes with a preconceived connotation that has been instilled in our society. It could very easily create a misleading perception that people with disability need to have a communication platform that is specifically designed to meet their 'unique needs'. Would this strengthen the invisible wall between disability and the social norm?

If we were to go with the second approach in this scenario, it would help expand our horizons by challenging us to think about how we can harness the power of technology to create a communication platform that would allow *anyone* to connect with each other. This would shift the thinking from creating a solution specifically *for* people with disability to creating a solution for individuals who wish to connect with their loved ones. This further reinforces the power of creating solutions that are driven by unique voices and lived experiences. Perhaps this approach would bring us one step closer to creating a cultural shift that embraces social inclusion, and evidently, technology plays a critical part in helping us achieve this.

In the next section of this chapter, we will explore some examples of existing cutting-edge initiatives that leverage the power of technology to redefine social inclusion for people with disability. More importantly, we will explore how these initiatives redefine how we perceive the role of assistive technology.

8.2 The Evolving Present

In the first section of this chapter, we unpacked several big themes. We had a glimpse of the medical and social models of disability, the role of assistive technology, the value of creating solutions that are driven by the needs and experiences of users, the importance of establishing a collaborative culture, and finally, how technology can be leveraged to redefine social inclusion for people with disability.

Now that we have an understanding of the theoretical framework of these concepts, in this section, we will explore how this translates into practice through some case studies of cutting-edge initiatives that utilise technology to drive innovation and social change.

Before we go further, let's examine what the term *innovation* means in this context. It is important to note that innovation does not necessarily mean we have to reinvent the wheel and create something that is brand new or has never been done before. It is equally important for us to take a moment to look at our existing resources and assess what is working well and what isn't. More importantly, what we can learn from these existing resources to create opportunities for improvement. Let's look at this through a disability and assistive technology lens. Assistive technology was created to improve the lives of people with disability and is traditionally known to be associated with the medical model of disability. Would the society we live in today allow us to view assistive technology differently? Does there have to be a difference between *technology* and *assistive technology* if we come to the realisation that they both share a common goal?

Hireup is one of Australia's leading online platforms that enable Australians living with a disability to find, hire and manage their own support workers to assist them with their daily living activities (Hireup n.d.). It creates an opportunity for people living with a disability to achieve goals and reach their full potential. Support looks different to every individual. To some, support may mean assistance with daily living activities, and to others, it may mean enabling community engagement. To put this another way, disability support can involve everything from education to employment, to social and community participation, right through to lifestyle in general. Its sole purpose is to enable people with disability to live a meaningful life, whatever that may mean to each individual.

The fundamentals of disability support have drastically changed over the past eight years in Australia since the launch of the National Disability Insurance Scheme (NDIS) in March 2013. The NDIS is a national programme designed to provide people with permanent or significant disabilities the support they need in order to enable choice, control, autonomy and independence in how they want to live their lives—a fundamental human right that each and every one of us deserves. It is estimated that there are 4.3 million Australians living with a disability, and as a result, the government will inject over $22 billion in funding each year providing essential support to approximately 500,000 individuals living with a permanent or significant disability (NDIS 2021).

The rollout of this scheme has created a significant social and cultural shift not only in the disability sector but also in our society. For the purpose of this discussion, we are going to take a closer look at how the NDIS has influenced the way we view disability support and social inclusion. Furthermore, we will use Hireup as an example to help us understand how technology, coupled with the NDIS, can redefine disability support, as well as the social model of disability. More importantly, we will explore how the concept of Hireup may challenge us to think about the fine line between technology and assistive technology and its relationship with human connections.

Traditionally, people with disability are regarded as passive receivers when it comes to finding and managing support. Prior to the introduction of the NDIS, the most common way to find support was through an agency responsible for hiring and managing support workers on behalf of people with disability. This means that the agencies had full control over who they hired and when the support could be provided. More importantly, the choice of support workers made by the agencies was largely based on their skills and experiences. Whilst it is undeniable that the skills and experiences of support workers are important, it also plays a role in feeding into the medical model of disability. For example, it is a common perception that support workers who work with people with a disability should have a nursing or medical background to some extent. One perspective that is often not taken into consideration in the medical model of disability is that common interests and goals are just as important as skills and experience in the relationship between support workers and people with disability.

Let's take a moment to imagine what this would look like if we were to be in the shoes of a person living with a disability. Perhaps this would be a good chance for us to practice *empathy*—a concept that we learned about earlier in this chapter.

Imagine that you are an individual with a full-time job that you know you are good at and can make a meaningful contribution to. However, in order for you to do what you do best, you require assistance with daily living activities. You approach an agency that is designed to help provide you with the support that you need. But unfortunately, you have very little influence over *who* provides you with the support and *when* they are available to provide that support. Moreover, you have no common interest with the support workers that get sent your way, other than their qualifications to provide the medical care that you need. The only thing that you could do is embrace the uncertainty of who will show up at your door each day. How would that make you *feel*?

Whilst this is a common scenario that could occur in the traditional model of disability support, it is important to note that the level of support individuals with disability receive is dependent on a host of various reasons, such as funding, geographical location and so on. The key lesson for us to learn here is that we, as a society, have recognised that this way of living for those of us with a disability is not good enough.

Choice, control, independence and autonomy need to be placed directly in the hands of people living with a disability. Since the launch of the NDIS in 2013, there have been many, many discussions around the benefits and shortcomings of this insurance scheme. Regardless of which side of the argument we view this from, the NDIS itself is known to be one of the biggest social reforms since the launch of Medicare (Tune 2019). More importantly, it is known to be one of the greatest nation-building projects on Earth (Tehan 2018).

The question is *why* does this national insurance scheme have such a significant impact? We have discussed earlier that this has created a fundamental social and cultural shift. It is important to recognise that this expands across all lifestyle aspects in the disability sector and in our community. However, we are going to focus on its impact from the technology lens.

Let's shift our focus back to Hireup as an example. Hireup is designed to provide disability support, but what exactly differentiates this from the traditional model of finding support? The answer lies in the use of *technology* to enable humans to make meaningful connections. Hireup's goal is to leverage technology to create an online platform that facilitates meaningful connections between support workers and people with a disability based on common interests, goals, skills and experiences. Hireup does this by identifying ways to bring together technology, innovation and social impact. A combination of these three key elements places choice and controls directly in the hands of people living with a disability by removing the need for agencies that provide the traditional model of disability support. Removing this middleman creates an opportunity for people with disability to dictate when they receive support, how they are supported and by whom.

On the surface, this sounds like a simple solution driven by technology. However, its impact is deeply rooted in the social model of disability and a desire to connect with like-minded individuals who realise our potential. Hireup's vision to enable the pursuit of a good life for everyone (Hireup n.d.) also puts forward a societal challenge to think outside the box and aim for better outcomes when the traditional model of disability support is not effective. This creates a fundamental shift in how we perceive disability support. Whilst it is critical to deliver quality care, it also forces us to acknowledge that we need to be better at creating equal opportunities for each individual to thrive. In this case, it begins by harnessing the potential and power of technology to help shift the perception of disability support from the medical model to the social model—with a little help from innovation and design to meet user needs.

We have now established that Hireup utilises the power of technology to provide disability support. Perhaps the interesting question that we should consider here is which category does the concept of providing opportunities for disability support through the use of technology fall into? Would we consider this solution to fall into the category of assistive technology? Or would we consider this to be a solution that simply uses technology to redefine the way disability support is provided and received? Is there a difference between the two categories?

The short answer is yes. If we view this from an assistive technology lens, it would ultimately be tied to the connotations that are associated with disability. If we were to view this purely from the technology lens, it would showcase how this has and will continue to serve as a powerful tool that would lead to creating more opportunities for us, as individuals, and as a society.

15% of the world's population is living with disability (World Health Organisation 2011), and we are beginning to realise how technology can serve as a catalyst for change. More importantly, we are beginning to utilise technology to shape a society that is enabling for everyone (Remarkable n.d.). This is evident in the technology, disability, innovation and social impact sectors as we begin to see more and more groundbreaking initiatives that are set on a mission to drive a positive social change through technology. Whilst it is key to have the motivation, passion and expertise to achieve this goal, it is equally important to be surrounded by a network that has the ability to bring the right people and the right resources together to help turn these big ideas into reality.

Let's use Remarkable as an example to understand what this means and the role it plays in utilising technological innovation to help shape a society that enables human potential. Remarkable's AssistiveTech Accelerator is a startup accelerator that enables founders to refine their business model and ideas, access seed funding to help kick-start their business and learn from an expert network of coaches and mentors. This intensive 16-week accelerator programme is specifically designed to support startup businesses that focus on driving innovation and social change through technology for those with disability. It fosters a collaborative culture, with a firm belief that products and services are more powerful when they are designed with the unique perspectives that come from lived experience. This once again echoes the evidence from the studies we discussed earlier in the chapter that point to the importance of creating solutions driven by the voice of users.

Remarkable's AssistiveTech Accelerator is a prime example of an initiative that acknowledges the accelerated growth of technological innovation and its ability to have a tremendous impact in enabling us to realise human potential, reduce barriers for people with disability and embrace social inclusion. Like the example we have seen with Hireup, Remarkable have played a critical role in helping to redefine the way we see assistive technology, creating solutions that are driven by the voice of users throughout the design and development process.

It should be highlighted that within the examples from both Hireup and Remarkable, the essence of combining technology, innovation and social inclusion lies in human connection. It is about our ability to practice empathy, to realise the power of unique perspectives when creating solutions and lastly, to harness technology and perhaps artificial intelligence to help us reach human potential.

8.3 Learn from Our History, and Right Our Way to the Future

Challenges drive innovation. We have witnessed this first-hand as the world braced itself for a global pandemic. It triggered fear, anxiety, uncertainty and unprecedented changes as we were forced to approach employment, education, health care and lifestyle differently. This global challenge also shed light on possibilities that we haven't imagined. It created new opportunities. It helped us discover new talent and new ways of doing things by leveraging technology as a catalyst for change. It has redefined social norms.

Whilst it is undeniable that the COVID-19 pandemic has created unprecedented changes, it is equally important to look at the changes that have been happening around us long *before* it began, including the role technology has played during this time. Think about how Uber has harnessed the power of technology to redefine the transport industry. Think about how Airbnb has leveraged technology to completely disrupt the accommodation industry. These examples are just the tip of the iceberg when it comes to the transformative journey that technology has taken us on. This merely highlights the fact that change is happening around us, with or without a pandemic.

However, as we navigate our way through the global pandemic, conversations around the so-called 'roadmap to recovery' are beginning to surface. Despite our efforts to create a cultural shift and embrace unique perspectives and enable human potential, it is evident that we still have a long way to go. The society we live in today should be and could be better. However, achieving requires systematic change. It may seem like the system is too big for us to change as individuals, but what we don't realise is that we *are* the system. The environment and culture that surround us are driven *by us* and our desire to shape a better future where every individual can thrive.

The same concept can be applied to disability, technology, user experience design and social inclusion. The challenges and barriers associated with disability become a powerful force for innovation. This is evident in the examples we have explored within this chapter. Now is the time to challenge the intention of developing a roadmap to recovery despite knowing the fact that our society was not built for inclusion. Rather than focusing our energy and resources on an attempt to return to a fragmented system that was known to leave people behind—especially people with disability—we are now placed in a unique position that presents us with an opportunity to reimagine a society where social inclusion and equality is at the forefront. This raises the question of whether we should focus on the roadmap to recovery post-pandemic or leverage this unique opportunity to *reimagine* a society that embraces inclusion, diversity and equality.

We are now at a fork in the road. We can resist change and embark on the journey of recovery, or we can embrace change and reimagine the future. The ball is in our court.

In this chapter, we have explored some key themes around the medical and social models of disability, the delicate dance between technology and assistive technology, the integration of artificial intelligence in assistive technology, the value of designing solutions that are driven by the voice of users and finally, the human connection that remains the core of the intersection of disability and technology and social inclusion. All of these elements add up to a rather complicated equation.

What if we tried to simplify this equation? If we were to take assistive technology and the medical model of disability out of this equation, what would we be left with? One could argue that we would be left with technology as an incredibly powerful tool that has the ability to drive innovation and creates a cultural shift that would ultimately acknowledge, celebrate, value and truly embrace unique perspectives. Futhermore, we would be left with the key ingredients to build a better future: humans filled with talent and potential, a culture that embraces change and diversity, as well as existing and emerging technology that could be harnessed as an incredibly powerful tool when used with the right intentions. There is an exciting journey ahead that could lead us into the mindset of continuous transformation.

We know for a fact that the society we live in today was not built for inclusion. However, our history has taught us many valuable lessons about the missed opportunities when inclusion and diversity is not truly embraced. More importantly, it taught us to learn from our history, leverage the knowledge and technology that we have access to today, to right our way into the future.

References

Alper S, Raharinirina S (2006) Assistive technology for individuals with disabilities: a review and synthesis of the literature. J Spec Educ Technol 21(2):47–64

Battarbee K (2004) Co-experience: understanding user experience in social interaction. University of Art and Design, Helsinki, Helsinki

Bricher G (2000) Disabled people, health professionals and the social model of disability: can there be a research relationship? Disabil Soc 15(5):781–793

Gray C (2006) 6 Benefits of UX design—Nomat. https://www.nomat.com.au/2016/07/07/6-benefits-ux-design/. Accessed 22 Nov 2021

Hireup. https://hireup.com.au/. Accessed 27 Nov 2021

Humpage L (2007) Models of disability, work and welfare in Australia. Soc Policy Adm 41(3):215–231

Kaliraj P, Devi T (eds) (2021) Artificial intelligence theory, models, and applications. Auerbach Publishers, Milton

Kouprie M, Visser F (2009) A framework for empathy in design: stepping into and out of the user's life. J Eng Des 20(5):437–448

NDIS (2021) What is the NDIS? | NDIS. https://www.ndis.gov.au/understanding/what-ndis. Accessed 27 Nov 2021

Remarkable: Home—Remarkable. https://remarkable.org.au/. Accessed 28 Nov 2021

Ripat J, Woodgate R (2010) The intersection of culture, disability and assistive technology. Disabil Rehabil Assist Technol 6(2):87–96

Siminski P (2003) Patterns of disability and norms of participation through the life course: empirical support for a social model of disability. Disabil Soc 18(6):707–718

Tehan D (2018) The National Disability Insurance Scheme: the greatest nation building project on earth. Australian Governemnt. https://formerministers.dss.gov.au/18046/the-national-disability-insurance-scheme-the-greatest-nation-building-project-on-earth/. Accessed 29 Nov 2022

Tune D (2019) Review of the National Disability Insurance Scheme Act 2013. https://www.dss.gov.au/sites/default/files/documents/01_2020/ndis-act-review-final-accessibility-and-prepared-publishing1.pdf. Accessed 29 Nov 2022

World Health Organisation (2011) World Report on Disability. WHO Press, Geneva. https://www.who.int/teams/noncommunicable-diseases/sensory-functions-disability-and-rehabilitation/world-report-on-disability. Accessed 28 Nov 2021

Chapter 9
Culture in Social Robots for Education

Barbara Bruno, Aida Amirova, Anara Sandygulova, Birgit Lugrin, and Wafa Johal

Abstract Education is one of the predominant applications that is foreseen by researchers in social robotics. In this context, social robots are often designed to interact with one or several learners and with teachers. While educational scenarios for social robots have been studied widely, with experiments being conducted in several countries for nearly 20 years, the cultural impact of accepting social robots in classrooms is still unclear. In this paper, we review the literature on social robots for education with the lens of cultural sensitivity and adaptation. We discuss culture theories and their application in social robotics and highlight research gaps in terms of culture-sensitive design and cultural adaptation in social robots assisting learners in terms of (1) the robot's role, (2) envisioned tasks, and (3) interaction types. We also present guidelines for designing cross-cultural robots and culturally adaptive systems.

B. Bruno
École Polytechnique Fédérale de Lausanne (EPFL), Lausanne, Switzerland
e-mail: bar-bara.bruno@epfl.ch

A. Amirova · A. Sandygulova
Department of Robotics and Mechatronics, School of Engineering and Digital Sciences, Nazarbayev University, Nur-Sultan, Kazakhstan
e-mail: aida.amirova@alumni.nu.edu.kz

A. Sandygulova
e-mail: anara.sandygulova@nu.edu.kz

B. Lugrin
Human-Computer Interaction, Institute of Computer Science, Julius-Maximilians-Universität Würzburg, Würzburg, Germany
e-mail: birgit.lugrin@uni-wuerzburg.de

W. Johal (✉)
School of Computer Science and Information Systems, Faculty of Engineering and Information Technology, University of Melbourne, Melbourne, VIC, Australia
e-mail: wafa.johal@unimelb.edu.au

© The Author(s), under exclusive license to Springer Nature Switzerland AG 2023

B. J. Dunstan et al. (eds.), *Cultural Robotics: Social Robots and Their Emergent Cultural Ecologies*, Springer Series on Cultural Computing,
https://doi.org/10.1007/978-3-031-28138-9_9

9.1 Introduction

Research in social robotics has been growing over the past 20 years. With robots being better at perceiving humans and acting in a socially acceptable manner, the field has been moving from technical advances to more experimental and impact-based research (Amirova et al. 2021; Johal et al. 2022b). One of the main domains of application envisioned for social robots is education (Johal 2020). Several aspects make social robots suitable for learning tasks: (1) a lot of learning is individual, (2) emotion and motivation are important aspects of learning, and (3) learning often requires repetition.

Some social robots such as the NAO robot have been used worldwide in this context. For instance, we recently (Amirova et al. 2021; Johal et al. 2022a) conducted a scoping review of research into human–robot interaction (HRI) using the NAO robot. From this review, we learned that the NAO robot has been used in more than 50 countries. The vast application of this robot implies its acceptance in diverse social scenarios encapsulated in cultural beliefs, attitudes, and practices. What characterizes a social robot is still up for debate in the HRI community. The fact that there are wide cultural differences in people's views of robots is one of the challenges of the future (Bartneck et al. 2006).

Of particular interest to this review is culture that helps us to understand how certain communities perceive and interpret the world. It is indeed a dynamic and multifaceted phenomenon and usually entails values that are shared and/or learned by a group of people who pass a set of common traits, including behavior, knowledge, ideas, beliefs, norms, and many other things, from one generation to another (Birukou et al. 2013; Kakai et al. 2003; Trimble et al. 2002). Rogler (1993) suggested that, instead of trying to manage cultural influences, culture should be validated as a fundamental aspect of all phenomenological experiences and interpretations. Past studies often used people's race, ethnicity, or country of origin as a proxy for culture (Brown and Rogers 1997). However, Murry et al. (2001) argued that culture should be treated within a nuanced framework focusing on individual experiences of people, their values, beliefs, and practices because race and ethnicity narrow down cultural variations within groups. In most instances, cultural systems of the West have been contrasted with those of the East, with the former characterized by a systematic and consistent worldview and the latter maintaining a more holistic and circular view in understanding the world (Andrist et al. 2015; Bartneck et al. 2006).

Robots and humans can build bidirectional cultural coexistence on a social level. In this regard, Sabanović (2010) suggested a mutual shaping of robotics and society, highlighting shared dynamics between society and technology, which eventually informs the design, application, and evaluation of technologies and affects social constructs and perceived norms across cultures. Based on this and other observations, Samani et al. (2013) referred to human–robot culture having recursive impact on the cultural values of humans in designing robots, while in hindsight acknowledging robot cultural values also affecting human perceptions and experiences. This two-way influence should be recognized so as to accommodate HRI in the best possible ways.

Some HRI reviews to date have focused on cultural influences in expectations toward and reactions to social robots (Lim et al. 2020) and attitudes toward humanoid and animal-like robots (Papadopoulosm and Koulouglioti 2018). They bring perceptual knowledge, such as attitudes toward social robots in different cultures. Current research thus may benefit from identifying how cultural attachments of researchers and end-users may affect their direct experiences with social robots. In this review, we discuss how social robots have been used in educational contexts and how culture has been taken into account so far in the design and evaluation of social robots in this context. This review helps to understand how HRI studies make (un)intended choices when introducing educational robots in diverse cultures and what learning environments are established for learners that engage with them. The knowledge accumulated by this research can offer valuable insights for the HRI community, and which cultural knowledge and practices should be considered when designing and using robots for educational purposes.

We attempt to test the following hypotheses based on the review of education studies in HRI:

- H1: More individualistic cultures give more importance to individual and personalized learner–robot interactions, hence favoring scenarios with fewer students concurrently interacting with the robot, with reference to more collectivist cultures.
- H2: Cultures exhibiting a greater power distance favor designs in which the robot is assigned a more powerful role (such as teacher or tutor), while cultures exhibiting a lower power distance favor designs in which the robot acts as a peer for the learners.
- H3: Acquiring curricular skills with the robot is the main focus of learning in masculine societies, while feminine societies prioritize extracurricular content.

9.2 Background

9.2.1 Culture and Social Robots

While culture might not be the first aspect that comes into mind when designing a social robot for an educational context, it is still crucial to keep an eye on it. As Lugrin and Rehm (2021) postulate in their survey on culture and socially interactive agents (SIAs), a robot cannot be without cultural background. That means, if culture is not explicitly considered, the robot will unconsciously contain cultural cues of the designer, as they are the one who judges the SIA's naturalness. We therefore believe that existing work on educational robots contains culture-specific norms, values, or assumptions depending on where the robot was implemented and where studies were conducted.

Theories of Culture

Many approaches of integrating culture into artificial entities are based on Hofstede's cultural dimensions (Hofstede et al. 2010). In their theory, culture is defined by six dimensions; a given culture is thus a point in a 6-dimensional space. All dimensions are linked to specific ways of thinking, interpreting, and managing interactions between people. The dimension *Power Distance* describes the extent to which a different distribution of power is accepted by the less powerful members of a culture. The *Individualism* dimension describes the degree to which individuals are integrated into a group. The *Masculinity* dimension describes how gender roles are distributed as well as a society's preference for achievement, heroism, assertiveness, or material rewards, as opposed to a preference for cooperation, modesty, caring for the weak, and quality of life. The *Uncertainty Avoidance* dimension defines the tolerance for uncertainty and ambiguity. The *Long-Term Orientation* dimension explains differences by the orientation toward sustainable values for the future. The *Indulgence* dimension describes the subjective well-being that members of a culture experience.

Another theory frequently used in computer science relies on Hall's *anthropological work* (1959, 1966) that defines different dichotomies regarding space, context, and time. The space dichotomy refers to Hall's concept of proxemics that is linked to human spatial behavior, where immediacy is interpreted differently across cultures and influences communication patterns. In the space dichotomy, we can distinguish between high- and low-contact cultures, where the latter can be defined by being more comfortable with larger distances in interpersonal encounters. The context dichotomy distinguishes between high- and low-context cultures. Context refers to the amount of information in a communication that must be encoded explicitly. Members of high-context cultures are used to inferring meaning without being explicitly told. The time dichotomy refers to the perception of time and ordering of actions. Monochronic cultures prefer clock time and finishing tasks before starting new ones, whereas polychronic cultures are comfortable with multitasking and might have different time perceptions.

Culture in Social Robotics

It is a well-known fact that culture, which influences our social relation with other humans, also influences our perception of, and interaction with, social robots (Lim et al. 2020). People with different cultural backgrounds (1) have different preferences concerning how the robot should be and behave (Evers et al. 2008), (2) tend to prefer the one better complying with the social norms of their own culture, in aspects such as verbal (Andrist et al. 2015; Rau et al. 2009; Wang et al. 2010) and non-verbal behavior (Trovato et al. 2013) and interpersonal distance (Eresha et al. 2013; Joosse et al. 2014), and (iii) tend to put more trust in a robot that shares their communication conventions (Andrist et al. 2015; Evers et al. 2008; Rau et al. 2009; Trovato et al. 2013; Wang et al. 2010). Consequently, in their seminal paper on the influence of culture on expectations of and responses to social robots, Lim et al. "argue for further research into the role of culturally informed robotic development in facilitating human–robot interaction" (Lim et al. 2020).

In the last decade, a wide range of cross-cultural studies have been conducted within HRI research. Han et al. (2009) examine the cultural acceptance of robots and compare how parents and children from different cultures (Japan, Korea, Spain) perceive robots for education. The authors used identification tasks that enabled the tutoring robot to autonomously assemble e-contents and take a photo on its own camera and display the contents on a touch screen or monitor when they read fairy tales. The results show lower acceptance of tutoring robots among parents from Spain compared to those from Korea and Japan. Despite Asian-common cultural features, Korean parents were more liberal in allowing their children to learn with the robot, while the Japanese parents, much like their Spanish counterparts, were resistant to accepting them in educational roles. Neerincx et al. (2016) explored the experiences of Italian and Dutch children at two camps who interacted with a social robot to help them improve their diabetes self-management. Throughout two activities presented as quizzes and sorting games, the robot and a child firstly took turns in asking each other a question and providing a response, then in the second part matched icons to distinguish food types. Mirroring cultural expectations, Italian children tended to be more open and expressive and more attached to the robot than the Dutch children. The authors suggest that these differences may not wholly explain cultural attributes; instead, they could emerge due to the Italian children being slightly older and the different purposes of either a vacation-like (Dutch) or educational (Italian) setting. Robot assistants for social learning may also differently affect children with developmental disorders like autism. Rudovic et al. (2017) for the first time explored how individual differences in social engagement of children with autism spectrum disorder (ASD) may vary across two cultures, Japanese and Serbian. The interaction with the robot included four phases: pairing emotion cards with the robot's expressions, emotion recognition, imitation, and storytelling. As a result, there were differences in engagement scores in Japanese children who engaged and completed easier tasks faster, while Serbs (those in the last phase) were engaged for a shorter period of interaction time. However, the authors restrained from stating conclusions in a single-session study, further coupled with varying levels of behavioral severity among the participants.

Shahid et al. (2014) analyzed how children from Pakistani and Dutch cultural backgrounds and two different age groups (8 and 12-year-olds) interact with the iCat robot during a collaborative game. Participants in four groups were instructed to identify whether the children had just won or lost their game by analyzing a set of stimuli that featured children displaying their emotions following either win or loss. The behavioral analysis and perception test showed a recurrent pattern, with Pakistani children having more enjoyment, being more expressive, and showing more positive cues than their Dutch peers. This distinction is in fact consistent with the technology divide between developed (i.e., The Netherlands) and developing countries (i.e., Pakistan); in the latter, children may have limited or no exposure to social robots. Shiomi et al. (2017) looked at social acceptance of a childcare support robot system through four scales: intention to use, safety and trustworthiness, negative attitudes, and decreasing workload. Two kinds of moving robots, Sphero and Romo, were demonstrated to parents and children. The experimental results indicate that the people from both cultures had lower social acceptance toward childcare support robotics than current childcare support technologies, except for combined use in practice. Americans seemed to have a higher intention to use childcare support technologies than Japanese people, but other scales were rated higher for Japanese people.

Summing up, we observe great variations in robot perceptions not only between the East and the West, but also within cultures. Apart from the users' cultural background, their prior exposure to social robots might considerably influence their attitudes toward robots (Bartneck et al. 2006). Our review will further deepen cultural aspects in HRI with a specific focus on educational interactions with social robots. We extend the previous works by taking into account educational contexts such as robot roles, interaction patterns and contents, and analyzing situated uses of robots in natural learning environments.

9.2.2 Social Robots in Education

Numerous research into human–robot interaction provides convincing evidence of the developmental and cognitive advantages that social robots can offer for innovative learning and treatment, adjusting to the needs of individuals (Belpaeme et al. 2018; Johal 2020). Establishing an environment with physical artifacts (e.g., robots, tablets) and social learning scenarios requires the expertise of professionals from interdisciplinary areas of study. Moreover, the collaboration of multiple stakeholders and participatory design practices is crucial to promote social inclusion.

One-to-one tutoring has been shown to yield higher learning gains than group education (Bloom 1984; VanLehn 2011), but is often not feasible for financial reasons. One promising way to enable one-to-one tutoring while keeping the costs low involves the use of social robots for teaching. Social robots can convey verbal messages and use their bodies for visual cues such as gestures and therefore could be used to achieve a variety of educational goals. In fact, a number of studies have indicated that social robots yield higher learning gains than digital one-to-one screen-based techniques (Han et al. 2005; Hyun et al. 2008; Kennedy et al. 2015; Kose-Bagci et al. 2009; Leyzberg et al. 2012). There is no clear explanation for this effect, but it may be due to either the social and physical presence of the robot having a positive impact on the learner's engagement, or the robot's multimodal features providing a richer and embodied pedagogical experience, or both. A general advantage of new technologies such as robots is that they allow for fast-paced and adaptive interaction, which is tailored to match the level and interests of the learner.

Research and development in the area of social robotics have focused on designing robotics tools that could assist learners. Several recent reviews highlighted this trend (Belpaeme et al. 2018) showing an increase in the publication of work in the domain of social robots for education. These reviews identify several curriculum areas have been investigated for the use of social robot assistants. One of the top curriculum activities is second language learning. A wealth of research shows the advantages of learning a language for personal and social reasons and for career aspirations. This motivation has driven the field of human–robot interaction to shape robot-assisted language learning (RALL) as an evolving area of social learning. Here, social robots are used for language learning and teaching purposes. Compared to other technologies, robots are advantageous for their social roles (e.g., a peer) and offer physical embodiment valuable for language learning (Berghe et al. 2018). These robots as language companions can help learners acquire core language skills such as vocabulary (Movellan et al. 2009), oral skills (Lee et al. 2011), and reading comprehension (Yueh et al. 2020). Current literature lacks evidence on complex language abilities, particularly writing. In addition, social robots and human teachers are rarely compared in educational settings (Alemi et al. 2014; Gordon et al. 2016). We agree that robots can aid a human teacher and open up new teaching practices characterized by personalized and creative approaches toward language learning (Sharkey 2016).

However, cultural competence is something that has not necessary been widely explored in social robots for education, especially when introducing robots as social language companions. In addition, past research (Berghe et al. 2018) commonly refers to second language acquisition (i.e., English) and pays less attention to linguistic diversity in the world. A little attention is given to how robot-assisted classrooms operate and what kind of influence it has on student–student and student–teacher relationships.

9.3 Value Sensitive Design Applied to Social Robots

Value sensitive design (VSD) is an interactional theory and method that accounts for human values in a principled and structured manner throughout the design process (Friedman et al. 2002). It not only considers usability but also context and values (preferences, an individual's orientation, etc.). An important aspect of VSD lies in identifying which values should be taken into account in the design process.

Schwartz's model of values distinguishes individual and cultural values. Individual values (1) are subjective beliefs of people linked to affect, (2) refer to desirable goals that motivate action, (3) transcend particular situations, (4) serve as standards or evaluation criteria, (5) are ordered by their relative importance, and (6) direct individuals' actions (Schwartz 2012). Schwartz has validated ten basic circumplex values: Conformity, Tradition, Security, Power, Achievement, Hedonism, Stimulation, Self-Direction, Universalism, and Benevolence (Schwartz 1992; Witte and Stanciu 2020). These values respond to human needs in terms of accepting people as biological organisms, reaching agreement in social actions, and acknowledging survival and well-being of individuals. Ideals that influence the attitudes and behavior of individuals and groups within a culture are represented by cultural values (Schwartz 2004). Cultural values are abstract structures as opposed to observable individual values. These value constructs reflect common understandings of what is desirable and perceived good in the culture or the cultural ideals (Schwartz 2006). Institutional structures, laws, and social norms all reflect underlying cultural values in society.

Friedman et al. proposed a list of thirteen human values for the design of information systems: human welfare, ownership and property, privacy, freedom from bias, universal usability, trust, autonomy, informed consent, accountability, courtesy, identity, calmness, and environmental sustainability (Betz and Fritsch, 2016; Friedman et al. 2006). The VSD approach is centered around a tripartite methodology of conceptual, empirical, and technical investigations. The conceptual investigations include raising questions about direct and indirect stakeholders affected by the design and competing and moral values. These questions further lead to empirical investigations, which evaluate the viability of a particular design through research. Finally, technical investigations look at how existing technology properties and their mechanisms support or hinder human values.

However, some researchers argue that it is better to generate values from stakeholders in a bottom-up approach (Umbrello and Poel 2020). The question here is whether these thirteen values are sufficient for designing social robots to be responsive to cultures. In this regard, Smakman et al. (2021) came up with seventeen moral values that might be influenced by the use of social robots in primary education, adding autonomy, flexibility, and responsibility to the list previously conceived by Friedman et al. Thus, stakeholder voices should be also explored to account for the flexible and dynamic nature of human values.

9.4 Methods

This study reviews education-focused HRI studies and tests the three hypotheses (see Sect. 9.1) formulated through the prism of cultural awareness. We utilized an available database presented in Belpaeme et al. (2018) and Johal (2020), containing 160 papers published from 2004 to March 2020. That work had an annotated field "Country" based on the experimental site (Fig. 9.1). Following that annotation, each country was then assigned Hofstede's cultural dimensions' indexes taken from the official Hofstede's source.[1] The following annotations were then considered for the analysis presented in this paper:

- **Country distribution**. We looked for the information about where a study experiment was conducted, which is usually written in methods.
- **Number of students per interaction.** We grouped children according to the number of students in front of the robot: one child, two children, a small group up to five children, a whole class.
- **Robot role**. We distinguished the robot roles to be either a teacher (teaching a large group of children), a teaching assistant (acting as a sidekick for the teacher), a tutor (one-to-one tutoring), or a peer (one-to-one peer-like interaction).

[1] https://geerthofstede.com/culture-geert-hofstede-gert-jan-hofstede/6d-model-of-national-culture/

Fig. 9.1 World map showing the number of papers in the analyzed dataset on social robots in education per country

- **Curriculum**. We coded curricular vs extracurricular targeted skills. For curricular components, we annotated core subjects taught mandatorily such as languages and mathematics, while extracurricular activities included voluntary learning such as music, dance, and vocational skills.

9.5 Cultural Assessment

First, we assessed the effect of the individualism-collectivist dimension on the number of students in front of the robot. As can be seen from Fig. 9.2, countries with a high individualism index (e.g., USA, Australia) have mainly focused on one-to-one interaction while class-based interaction has been mainly prevalent in countries having a lower index in this dimension (e.g., China, Korea).

Our next analysis compared a relationship of the power distance index (PDI) with the role of the robot in the reviewed papers. In particular, research work conducted in the countries with a lower PDI scores (e.g., USA, Italy) has mainly assigned a peer role to a robot, while countries with higher PDI scores (e.g., China, Kazakhstan) mainly focused on the robot as a teacher or compared the two roles (teacher vs peer). Figure 9.3 demonstrates this result. Interestingly, the role of a teaching assistant (i.e., a sidekick alongside a human teacher) was utilized by societies with a PDF index ranging between 55 and 65 (e.g., Iran, Greece).

Highly feminine societies (e.g., Sweden, Norway) have only targeted typical curricular skills (such as handwriting, languages, and mathematics), while highly masculine cultures (e.g., Japan, USA) have focused on both curricular and extracurricular activities in their work. Figure 9.4 presents these cultural differences.

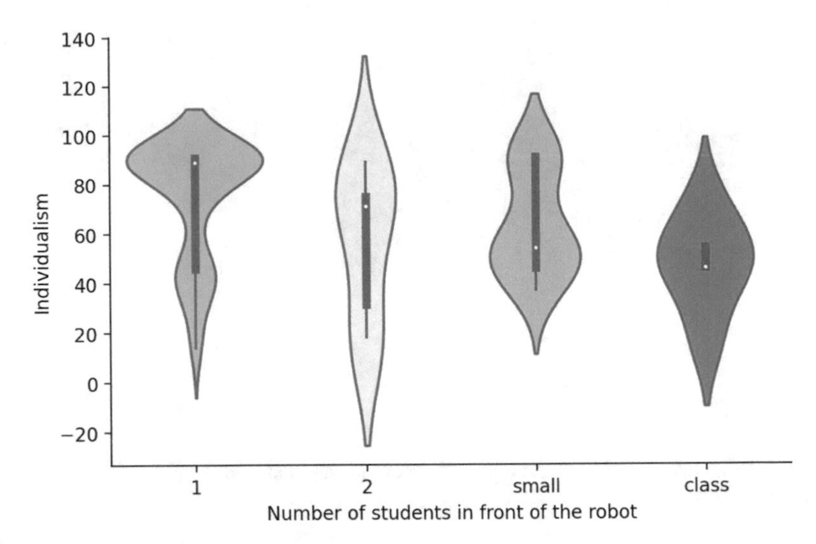

Fig. 9.2 Individual scores for each country related to the number of students that were in front of the robot during each interaction session in the paper. Small is 3–5 students

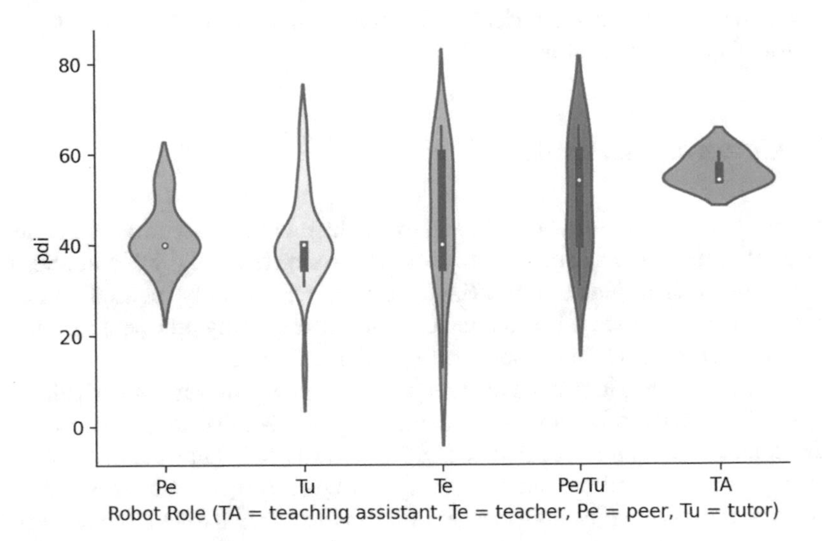

Fig. 9.3 Power distance—the degree of equality/inequality in a culture—has an effect on a robot's role

9.6 Discussion

We accept the first hypothesis claiming more individualistic cultures to follow individual and personalized learner–robot interactions, hence favoring scenarios with fewer students concurrently interacting with the robot when compared to more

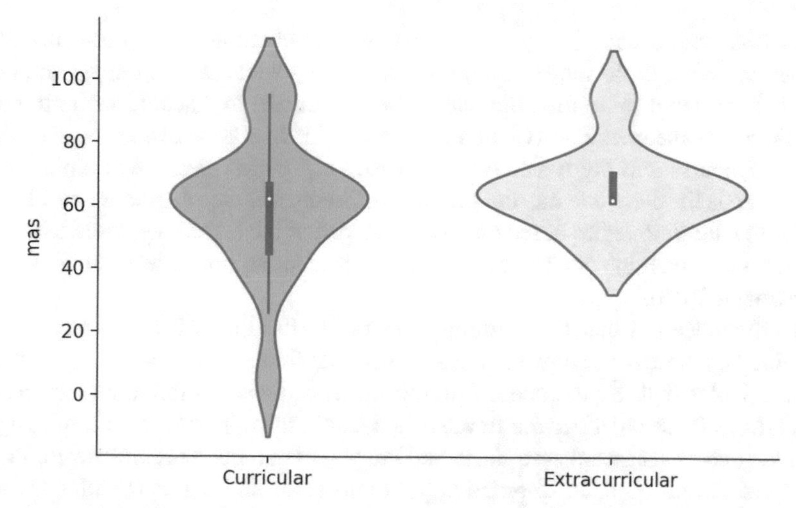

Fig. 9.4 Masculine/feminine dimension has an effect on children's targeted skills, in particular, whether the practiced skill is a part of typical school curricular vs extracurricular activities

collectivist cultures. Particularly, we found that one-to-one interaction was common among Western cultures with a high individualism index, while class-based communication was significant in Asia-based studies. This result reiterates one of the key cultural dimensions to describe how people's relationships vary on the individualism-collectivism continuum, with reference to Hofstede (Franke et al. 2007). People from highly individualistic cultures go with independence, self-focus, use explicit communication methods, and believe that identity resides within a person, while their counterparts from collectivist cultures demonstrate more interdependent behaviors and attitudes and place a greater emphasis on their relationships with others, which are frequently negotiated through implicit communication patterns (Mooij and Hofstede 2010).

Regarding H2, the studies from Western societies introduced robots in collaborative peer roles, while those conducted in Asian communities mainly applied teacher robots or compared them with peer robots. The result strongly supports Hofstede's perspective on the PDI dimension. This again highlights East–West education differentiation, particularly with reference to the role of teachers in the classroom. In most Asian countries, teachers are highly respected and enjoy the highest level of social status in public. Thus, it is an interesting observation that the Asian countries envision teaching roles for social robots. In addition, there exist different stances toward robot teachers, contributing to the East–West divide. For instance, people from Western countries like Germany (Reich-Stiebert and Eyssel 2015) and Spain (Choi et al. 2008) tend to be skeptical and barely imagine education robots as teachers in learning contexts, whereas Eastern countries might be more optimistic and liberal toward robot teachers. While Western countries tend to accept the facilitating nature of peer robots, Asian cultures expect robot teachers to be dominant and act with power to teach students at their discretion. We explain this by the different status teachers have within the education systems of these two broad cultures, student-centered vs teacher-centered. In the latter paradigm, challenging a teacher's opinions and practices is considered to be impolite and students respond to teachers with attention, silence, and sometimes fear (Chan and Chan 2005; Lee and Kim 2019). Teacher–student conversations are typically "asymmetrical," which means that typically the teacher leads the class and has the privilege of being in control (Mercer and Dawes 2008). Teaching strategies based on a Western idea of the learner-centered education may not work well in societies that prioritize teacher authority over student needs (Brinkmann 2018).

We hypothesized that the learning content in HRI may differ between highly feminine and masculine cultures. We assumed that feminine cultures may prioritize extracurricular skills development with robots, while those in predominantly masculine cultures focus on curricular trends. As a result, the opposite was found. Highly feminine cultures targeted core skills like languages and mathematics, while highly masculine cultures tended to practice both curricular and extracurricular learning opportunities. Thus, we reject H3. This expected distinction usually stems from curriculum choices in education systems. It has been a long time since East Asian countries became the top performers in international student assessment tests like the Programme for International Student Assessment (PISA), which focuses on mathematics, reading, and science[2]. Besides, these countries are well-known for a greater focus on high-stakes tests, which subsequently cause the "washback effect." This shows a huge impact of testing on curriculum design, teaching practices, and learning behaviors, as a result of which both learners and teachers adjust to score better and pass the test rather than truly learn and teach what they find significant for themselves (Cheng 1997). Another phenomenon is the comparison of the Western educational system and learning methods, which foster creativity and curiosity, with the Asian educational system, which is frequently criticized for memorization and rote learning. However, our results note that this curriculum divide may not persist between cultures.

Going beyond traditional Asian–Western (typically American and Japanese) cultural dichotomy, a much broader inclusion of cultures is needed to make up for the relative over- and under-representation of countries (Lim et al. 2020). What is worth noting is that there might be subcultures, and it seems simplistic to situate culture from just ethnic, racial, and physical characteristics, as individuals' social and linguistic backgrounds are critical for explaining culture-level nuances. The increasing rates of global mobility blur the cultural lines. People's cultural attachments do not necessarily sit within geographical boundaries; a growing number of people nowadays live in heterogenous communities with greater intercultural exchange and learning. The interplay of culture and education in HRI needs a more comprehensive framework that considers a variety of individual and social dimensions.

[2] https://www.oecd.org/education/asian-countries-top-oecd-S-latest-pisa-survey-on-state-of-global-education.htm.

9.7 Recommendations to Report Culturally Sensitive HRI

In order to assist researchers in explicitly stating the cultural footprint of their design and research in social HRI, we propose the use of the following guidelines to help in reporting research works:

Experimental Context

Context is an important part of cultural behaviors. We recommend studies to explicitly describe the location (country and school vs non-school) of the experiment. It is also important to describe classical student/teacher/tutor relationship, explain if the activity proposed is part of the school curriculum, and how is it usually thought (e.g., individual vs collaborative work). Other contextual factors such as students' socioeconomic background, individual differences in personality, learning experiences, and linguistic repertoires are also important to consider when designing culture-sensitive robot activities. Learning about users and their educational experiences should be conducted even before the design and evaluation phases. For most children, robots mean toy-like creatures to engage with in play-based games. As long as they are used for engagement and motivational purposes, the choice of games should be informed locally, with culturally driven interaction cues such as gestures (e.g., greeting each other by bowing, shaking hands, etc.). However, caution is required to not cause stereotypes and biased representations of the target culture and context.

Cultural Alignment

Social robots are usually expected to be respecting cultural norms during classroom interaction, yet researchers and educators have to figure out to what dimensions cultural alignment in HRI is vital. Cultures both within and across individuals are dynamic. On that note, proposing design recommendations for a specific culture might be superficial and may bring unintended negative biases into the design (Louie et al. 2022). Besides, the robots can potentially develop their own interaction culture, which may inform individuals what to expect and how to communicate with these agents. In fact, some have suggested "robot culture" that involves "values that robots themselves may hold and could eventually move toward the construction of a distinct robot culture" (Samani et al. 2013). For instance, regardless of cultural backgrounds, children usually come to understand the major limitation of robots to maintain spontaneous communication and that they are mostly incapable of satisfying curious minds. It is important to explain to young users that robots are not the same as the people around them and provide explicit and accessible guidance on the kind of interaction they would have.

Participatory Design

With the variability among learners and their educational contexts, one robot is barely responsive to different group needs. Designers of robotic behaviors, embodiment, and interactions should incorporate cultural responsiveness into the design process if they want social robots to be successfully applied. In addition to engaging with the opinions and inputs of learners, a participatory or co-design approach can enable researchers to acquire insightful knowledge about their preferences and needs. Especially, researchers should involve young children in creative co-design tasks to not only envision what kind of robot they would like to have fun with and learn from, but also how they want to learn with the robot and under what conditions. A recent study (Louie et al. 2022) proposed three guiding principles to reflect participatory design: (1) gathering stakeholder beliefs and expectations, (2) integrating non-verbal co-design methods, and (3) providing an experiential robot interaction.

Task Focus

Culture-driven educational tasks may support multicultural learning as our world becomes more multilingual than monolingual. Therefore, we suggest that instead of the cultural adaptation of robot attributes, meaningful adjustments to the design and implementation of learning tasks or activities could make a significant difference. For instance, educational robots can provide first-hand support for children from minority language backgrounds to integrate into the target community and also communicate and maintain their cultural heritage. Currently, the HRI studies on second language acquisition (SLA) mainly explore the mastery of target language skills by learners and have little to do with promoting linguistic and cultural diversity. The tasks should focus on raising awareness of one's culture through learning tasks in the diverse classrooms. For example, van den Berghe (2022) has recently proposed to use robots for translanguaging purposes. That is, robots can help immigrant children in educational settings since many SLA teachers do not speak the student's first language (L1). They might be able to translate challenging vocabulary into the children's L1, pre-teach L1, or serve as the conversation partner for L1-language discussions. These exercises do not have to be solely about learning a language; translanguaging is a useful teaching strategy for studying anything within and beyond the curriculum.

Robot Roles

Robot roles are critical to the success of HRI. Current studies usually introduce robots as peers and teachers, which are inherently human roles in society. This representation may lead to confusion that robots will take on these human roles, replace people, and eventually act as independent agents. As noted, social robots should be used as assistants and mediators that support human–human communication. Researchers need to define for themselves what kind of social capacities a robot should possess and if their choice is culturally responsive to the education context it is applied to. A cross-cultural study (Korn et al. 2021) found that people from Jordan and Saudi Arabia assigned social roles to robots noticeably more than participants from Egypt and Germany did. Furthermore, German and Egyptian participants exhibited a higher acceptance of robots in doing household chores such as cleaning and cooking. However, all culture representatives were less approving of robots as emotional partners and companions.

Direct and Indirect Observations

The interaction in cross-cultural studies should be measured through balancing direct (e.g., surveys) and indirect observations (e.g., interviews). Current studies mainly use explicit measures such as surveys, which can be limited and biased (Haring et al. 2014). A mixed use of evaluations can enhance the validity of results. Cultural cues are usually implicit and complex. Such complexity should be investigated using qualitative but quantifiable observations, such as video annotations and conversational and behavior analyses. For instance, educational settings are usually collaborative in nature, with pair or group work being exercised widely in the classroom. Task-specific pair talk analyses across different cultures may reveal interesting observations, given that the researchers use solid interaction frameworks to guide the analysis process.

9.8 Conclusions

While social robots in education have been studied widely across different countries, little is known about the cultural impact on accepting social robots in classrooms. In this paper, we reviewed the applications of social robots for educational purposes from the position of cultural awareness and responsiveness. We discussed theories of culture that inform culture-sensitive design and cultural adaptation of social robots assisting learners in terms of (1) interaction role, (2) envisioned tasks, and (3) robot appearance. The major findings are as follows: (1) From the perspective of HRI, one-to-one interactions were prevalent in Western societies, whereas whole class interaction was common for Asian communities; (2) Studies from Western societies introduced robots in collaborative peer roles, whereas those conducted in Asian communities primarily designed and applied teacher robots or tested its teaching effectiveness; (3) Highly feminine cultures focused on the practice of curricular skills, while highly masculine cultures acknowledged the importance of both curricular and extracurricular learning contents. Finally, we listed recommendations for designing cross-cultural robot and culturally adaptive systems in terms of experimental context, cultural alignment, participatory design, task focus, and robot roles. We conclude that cultural backgrounds may influence the learning conditions and overall climate in robot-enhanced educational settings.

References

Alemi M, Meghdari AF, Ghazisaedy M (2014) Employing humanoid robots for teaching English language in Iranian junior high-schools. Int J Humanoid Robot 11(3):1450022

Amirova A, Rakhymbayeva N, Yadollahi E, Sandygulova A, Johal W (2021) 10 years of human-NAO interaction research: a scoping review. Front Robot AI 8

Andrist S, Ziadee M, Boukaram H, Mutlu B, Sakr M (2015) Effects of culture on the credibility of robot speech: a comparison between English and Arabic. In: Proceedings of the 10th annual ACM/IEEE international conference on human-robot interaction, pp 157–164 (2015)

Bartneck C, Suzuki T, Kanda T, Nomura T (2006) The influence of people's culture and prior experiences with Aibo on their attitude towards robots. AI & Soc 21:217–230. https://doi.org/10.1007/s00146-006-0052-7

Belpaeme T, Kennedy J, Ramachandran A, Scassellati B, Tanaka F (2018) Social robots for education: a review. Sci Robot 3(21)

Betz S, Fritsch A (2016) A comparison of value sensitive design and sustainability design. In: GI-Jahrestagung

Birukou A, Blanzieri E, Giorgini P, Giunchiglia F (2013) A formal definition of culture. Models Intercult Collab Negot 6:1–26. https://doi.org/10.1007/978-94-007-5574-1_1

Bloom BS (1984) The 2 sigma problem: the search for methods of group instruction as effective as one-to-one tutoring. Educ Res 13:16–24

Brinkmann S (2018) Teachers' beliefs and educational reform in India: from 'learner-centred' to 'learning-centred' education. Comp Educ 55:29–39

Brown JR, Rogers SJ (1997) Cultural issues in autism. In Ozonoff S, Rogers SJ, Hendren RL (eds) Autism spectrum disorders: a research review for practitioners. American Psychiatric Publishing, Washington, pp 209–226

Chan KL, Chan C (2005) Chinese culture, social work education and research. Int Soc Work 48:381–389. https://doi.org/10.1177/0020872805053461

Cheng L (1997) How does washback influence teaching? Implications for Hong Kong. Lang Educ 11:38–54

Choi JH, Lee J, Han J (2008) Comparison of cultural acceptability for educational robots between Europe and Korea. J Inf Process Syst 4:97–102. https://doi.org/10.3745/JIPS.2008.4.3.97

Eresha G, Häring M, Endrass B, André E, Obaid M (2013) Investigating the influence of culture on proxemic behaviors for humanoid robots. In: 2013 IEEE Ro-Man. IEEE, pp 430–435

Evers V, Maldonado H, Brodecki T, Hinds P (2008) Relational vs. group self-construal: Untangling the role of national culture in HRI. In: 2008 3rd ACM/IEEE international conference on human-robot interaction (HRI). IEEE, pp 255–262

Franke R, Hofstede G, Bond M (2007) Cultural roots of economic performance: a research note. Strateg Manag J 12:165–173. https://doi.org/10.1002/smj.4250120912

Friedman B, Kahn P, Borning A, Zhang P, Galletta D (2006) Value sensitive design and information systems. In: Doorn N, Schuurbiers D, van de Poel I, Gorman M (eds) Early engagement and new technologies: opening up the laboratory: philosophy of engineering and technology, vol 16. Springer, Dordrecht, pp 55–95. https://doi.org/10.1007/978-94-007-7844-3_4

Friedman B, Kahn PH, Borning A (2002) Value sensitive design: theory and methods

Gordon G, Spaulding S, Westlund JK, Lee JJ, Plummer L, Martinez M, Das M, Breazeal C (2016) Affective personalization of a social robot tutor for children's second language skills. In: Proceedings of the AAAI conference on artificial intelligence, vol 30

Hall ET (1959) The silent language. Doubleday, New York

Hall ET (1966) The hidden dimension. Doubleday, New York

Han J, Hyun E, Kim M, Cho H, Kanda T, Nomura T (2009) The cross-cultural acceptance of tutoring robots with augmented reality services. Int J Digital Content Technol Its Appl 3(2):95–102

Han J, Jo M, Park S, Kim S (2005) The educational use of home robots for children. In: ROMAN 2005. IEEE international workshop on robot and human interactive communication, 2005, pp 378–383. https://doi.org/10.1109/ROMAN.2005.1513808

Haring KS, Mougenot C, Ono F, Watanabe K (2014) Cultural differences in perception and attitude towards robots. Int J Affect Eng 13:149–157

Hofstede G, Hofstede GJ, Minkov M (2010) Cultures and organizations: software of the mind: intercultural cooperation and its importance for survival. McGraw Hill, US

Hyun E-J, Kim S-Y, Jang S, Park S (2008) Comparative study of effects of language instruction program using intelligence robot and multimedia on linguistic ability of young children.

In: RO-MAN 2008—the 17th IEEE international symposium on robot and human interactive communication, pp 187–192. https://doi.org/10.1109/ROMAN.2008.4600664
Johal W (2020) Research trends in social robots for learning. Current Robot Rep 1(3):75–83
Johal W, Amirova A, Rakhymbayeva N, Yadollahi E, Sandygulova A (2022a) Happy birthday NAO! Robohub Blog (2022a). https://robohub.org/happy-birthday-nao/
Johal W, Belpaeme T, Chetouani M (2022b) Editorial: robots for learning. Front Robot AI 9. https://doi.org/10.3389/frobt.2022.1050658
Joosse MP, Poppe RW, Lohse M, Evers V (2014) Cultural differences in how an engagement-seeking robot should approach a group of people. In: Proceedings of the 5th ACM international conference on collaboration across boundaries: culture, distance and technology, pp 121–130
Kakai H, Maskarinec G, Shumay DM, Tatsumura Y, Tasaki K (2003) Ethnic differences in choices of health information by cancer patients using complementary and alternative medicine: an exploratory study with correspondence analysis. Soc Sci Med 56(4):851–862
Kennedy J, Baxter P, Belpaeme T (2015) The robot who tried too hard: social behaviour of a robot tutor can negatively affect child learning. In: Proceedings of the 10th annual ACM/IEEE international conference on human-robot interaction, HRI'15. Association for Computing Machinery, New York, pp 67–74. https://doi.org/10.1145/2696454.2696457
Korn O, Akalin N, Gouveia R (2021) Understanding cultural preferences for social robots: a study in German and Arab communities. ACM Trans Hum-Robot Interact 10(2). https://doi.org/10.1145/3439717
Kose-Bagci H, Ferrari E, Dautenhahn K, Syrdal DS, Nehaniv CL (2009) Effects of embodiment and gestures on social interaction in drumming games with a humanoid robot. Adv Robot 23:1951–1996
Lee J, Kim CJ (2019) Teaching and learning science in authoritative classrooms: teachers' power and students' approval in Korean elementary classrooms. Res Sci Educ 49:1367–1393. https://doi.org/10.1007/s11165-017-9659-6
Lee S, Noh H, Lee J, Lee K, Lee G, Sagong S, Kim M (2011) On the effectiveness of robot-assisted language learning. ReCALL 23:25–58. https://doi.org/10.1017/S0958344010000273
Leyzberg D, Spaulding S, Toneva M, Scassellati B (2012) The physical presence of a robot tutor increases cognitive learning gains. Cogn Sci 34
Lim V, Rooksby M, Cross ES (2020) Social robots on a global stage: establishing a role for culture during human–robot interaction. Int J Soc Robot 1–27
Louie B, Björling EA, Kuo AC, Alves-Oliveira P (2022) Designing for culturally responsive social robots: an application of a participatory framework. Front Robot AI 9. https://doi.org/10.3389/frobt.2022.983408
Lugrin B, Rehm M (2021) Culture for socially interactive agents, pp 463–493. ACM. https://doi.org/10.1145/3477322.3477336
Mercer N, Dawes L (2008) The value of exploratory talk. In: Mercer N, Hodgkinson S (eds) Exploring talk in school: inspired by the work of douglas barnes. SAGE, pp 55–72. https://doi.org/10.4135/9781446279526.n4
Mooij M, Hofstede G (2010) The Hofstede model applications to global branding and advertising strategy and research. Int J Advert 29:85–110. https://doi.org/10.2501/S026504870920104X
Movellan JR, Eckhardt M, Virnes M, Rodriguez A (2009) Sociable robot improves toddler vocabulary skills. In: 2009 4th ACM/IEEE international conference on human-robot interaction (HRI), pp 307–308. https://doi.org/10.1145/1514095.1514189
Murry VM, Smith EP, Hill NE (2001) Introduction to the special section: race, ethnicity, and culture in studies of families in context. J Marriage Family 63:911–914. https://doi.org/10.1111/j.1741-3737.2001.00911.x
Neerincx A, Sacchitelli F, Kaptein R, Van Der Pal S, Oleari E, Neerincx MA (2016) Child's culture-related experiences with a social robot at diabetes camps. In: 2016 11th ACM/IEEE international conference on human-robot interaction (HRI). IEEE, pp 485–486

Papadopoulosm I, Koulouglioti C (2018) The influence of culture on attitudes towards humanoid and animal-like robots: an integrative review. J Nurs Scholarsh 50:653–665. https://doi.org/10.1111/jnu.12422

Rau PP, Li Y, Li D (2009) Effects of communication style and culture on ability to accept recommendations from robots. Comput Hum Behav 25(2):587–595

Reich-Stiebert N, Eyssel F (2015) Learning with educational companion robots? Toward attitudes on education robots, predictors of attitudes, and application potentials for education robots. Int J Soc Robot 7:875–888. https://doi.org/10.1007/s12369-015-0308-9

Rogler LH (1993) Culturally sensitizing psychiatric diagnosis: a framework for research. J Nerv Ment Dis 181:401–408

Rudovic O, Lee J, Mascarell-Maricic L, Schuller BW, Picard RW (2017) Measuring engagement in robot-assisted autism therapy: a cross-cultural study. Frontiers in Robotics and AI 4:36

Sabanović S (2010) Robots in society, society in robots. Int J Soc Robot 2:439–450. https://doi.org/10.1007/s12369-010-0066-7

Samani H, Saadatian E, Pang N, Polydorou D, Fernando ONN, Nakatsu R, Koh JTKV (2013) Cultural robotics: the culture of robotics and robotics in culture. Int J Adv Robot Syst 10. https://doi.org/10.5772/57260

Schwartz S (2004) Mapping and interpreting cultural differences around the world. In: Vinken H, Soeters J, Ester (eds) Comparing cultures: dimensions of culture in a comparative perspective. Brill, pp 43–73

Schwartz SH (1992) Universals in the content and structure of values: theoretical advances and empirical tests in 20 countries. Adv Exp Soc Psychol 25:1–65

Schwartz SH (2006) A theory of cultural value orientations: Explication and applications. Comp Sociol 5:137–182

Schwartz SH (2012) An overview of the Schwartz theory of basic values. Online Readings Psychol Cult 2:11

Shahid S, Krahmer E, Swerts M (2014) Child–robot interaction across cultures: how does playing a game with a social robot compare to playing a game alone or with a friend? Comput Hum Behav 40:86–100

Sharkey AJ (2016) Should we welcome robot teachers? Ethics Inf Technol 18(4):283–297

Shiomi M, Hagita N (2017) Social acceptance toward a childcare support robot system: web-based cultural differences investigation and a field study in Japan. Adv Robot 31(14):727–738

Smakman M, Vogt P, Konijn EA (2021) Moral considerations on social robots in edu- cation: a multi-stakeholder perspective. Comput. Edu. 174. https://doi.org/10.1016/j.compedu.2021.104317

Trimble J, Helms, Root M (2002) Social and psychological perspectives on ethnic and racial identity. In: Hesse-Biber SN, Leavy P (eds) Emergent methods in social research. SAGE, pp 239–275. https://doi.org/10.4135/9781412976008.n13

Trovato G, Zecca M, Sessa S, Jamone L, Ham J, Hashimoto K, Takanishi A (2013) Cross-cultural study on human-robot greeting interaction: acceptance and discomfort by Egyptians and Japanese. Paladyn J Behav Robot 4(2):83–93

Umbrello S, Poel I (2020) Mapping value sensitive design onto AI for social good principles. AI Ethics p 283–296. https://doi.org/10.1007/s43681-020-00033-0

van den Berghe R (2022) Social robots in a translanguaging pedagogy: a review to identify opportunities for robot-assisted (language) learning. Front Robot AI 9. https://doi.org/10.3389/frobt.2022.958624

van den Berghe R, Verhagen J, Oudgenoeg-Paz O, van der Ven SHG, Leseman PPM (2018) Social robots for language learning: a review. Rev Educ Res 89:259–295

VanLehn K (2011) The relative effectiveness of human tutoring, intelligent tutoring systems, and other tutoring systems. Edu Psychol 46:197–221

Wang L, Rau PLP, Evers V, Robinson BK, Hinds P (2010) When in Rome: the role of culture and context in adherence to robot recommendations. In: 2010 5th ACM/IEEE international conference on human-robot interaction (HRI). IEEE, pp 359–366

Witte EH Stanciu A, Boehnke K (2020) A new empirical approach to intercultural comparisons of value preferences based on Schwartz's theory. Front Psychol. https://doi.org/10.3389/fpsyg.2020.01723

Yueh HP, Lin W, Wang SC, Fu L (2020) Reading with robot and human companions in library literacy activities: a comparison study. Br J Educ Technol 51:1884–1900

Chapter 10
Towards an Autistic User Experience (aUX) Design for Assistive Technologies

Sebastian Trew and Scott Andrew Brown

Abstract User experience (UX) design aims to support people interacting with a particular product or service. However, the perspective applied to designing a user experience is generally framed by the life experiences of the designer. This means that oftentimes, marginalised groups are not supported in terms of their unique values and understanding of the world around them. In the case of autistic people, experiences within the social and built environment are not only unique, but often very pronounced. In this chapter, we explore what an approach to UX might look like with the input of autistic people. We propose a framing of autistic user experience (aUX) as a way forward that might improve experiences with technology not only for autistic people, but the UX community as a whole.

10.1 Introduction

This chapter discusses how user experience (UX) design led by people with lived experience of autism, 'autistic UX design' (aUX design) could provide insights and novel approaches to designing assistive technologies (ATs). We propose that aUX could be most useful in areas such as social, communication and sensory challenges and overcome environmental barriers faced by autistic people, thereby creating an improved and more meaningful experience of these technologies. Presented in this chapter is an overview of some of the social and environmental issues faced by autistic people, including their social needs and wants. Following this is an assessment of current assistive technologies designed for autistic people that are aimed at alleviating or reducing some of the reported and associated issues relative to the condition. Finally, we present a conceptual overview of aUX, a framework that, in response to the issues identified in existing ATs, is an approach to UX design, which aims

S. Trew (✉)
Institute of Child Protection Studies, Australian Catholic University, Canberra, Australia
e-mail: sebastian.trew@acu.edu.au

S. A. Brown
Creative Robotics Lab, University of New South Wales, Sydney, Australia
e-mail: scott.brown@unsw.edu.au

© The Author(s), under exclusive license to Springer Nature Switzerland AG 2023

B. J. Dunstan et al. (eds.), *Cultural Robotics: Social Robots and Their Emergent Cultural Ecologies*, Springer Series on Cultural Computing,
https://doi.org/10.1007/978-3-031-28138-9_10

to rethink and reposition the focus and role that ATs play for autistic people and overcome some of the current design issues. This involves a shift in the theoretical framing of designing ATs for the 'everyday experiences' of autistic people and away from therapy, education or attempts at a 'cure'.

10.2 Problem/Issue

The impact of diminished social activities, loneliness and increased isolation on mental and physical health is well established (Holt-Lunstad et al. 2015). It is also well recognised that autistic people across the lifespan (i.e. adults, adolescents, and children), when compared with people with other disabilities, (e.g. intellectual disability, emotional or behavioural disability or learning disability) and people without a disability, are significantly less involved in social participation and experience greater social isolation and loneliness (Pellicano et al. 2021; Kasari et al. 2013; Orsmond and Fulford, 2018; Pellicano et al. 2021). For autistic adult populations, recent research (Hedley et al. 2018; Pellicano et al. 2020) highlights the possible effects that loneliness has on depression and self-harming thoughts in this group. Also, autistic people have been recognised as a group that experience a significantly greater number of unmet support needs than the general population (Camm-Crosbie et al. 2018) and have limited access to suitable mental health supports (Crane et al. 2018; Maddox et al. 2020).

Reduced social outcomes for autistic people are suggested for a range of reasons often associated with their social behaviour, communication, interaction and comprehension, e.g. a lower conversation tendency or ability and lower functional skills (Orsmond et al. 2013), differences in autistic social cognition (Crompton et al. 2020), interpreting social cues (Morrison et al. 2020), social reciprocity (American Psychiatric Association 2013), interpreting the emotions of others (Frith and Happé 1999), recognising facial emotions (Baron-Cohen, 1995) and identifying tone of voice (Rutherford et al. 2002).

The physical environment an autistic person is in can also impact on their mental health, observed by increases in their display of anxiety and stress (Ozsivadijan et al. 2012). This response can be due to heightened sensitivities to certain stimuli, such as volume levels and frequencies of sounds in the environment (Weiss et al. 2013). This contributes to sensory overload and increased distress for autistic people (Spiker et al. 2011).

It is difficult to understand what factors in the environment are a mismatch for autistic people's social and sensory needs without current knowledge of their social and physical environments and how they navigate them, knowledge of their preferences for and uses of spaces, and understanding what assists, inhibits or restricts autistic people to navigate them. This includes how the design of urban landscapes—that is, the creation of social spaces that humans inhabit—may be modified or developed to produce an outcome that increases autistic people's participation in society, thus contributing to a holistic approach that aims to reduce autistic people's experiences of loneliness and isolation, via meaningful and positive experiences with their environments (Frauenberger et al. 2019).

Despite these social difficulties and differences in interacting and communicating, there is increasing evidence of autistic people's need and want for human connection. This evidence suggests an increased awareness and recognition that autistic people hold a similar level of desire for social relations and interactions, social connections, sense of belonging and purposeful and incidental social interactions as non-autistic people (Crompton et al. 2020; Oomen et al. 2021; Pellicano et al. 2021). Whilst the recent COVID-19 pandemic intensified many autistic people's social isolation, it highlighted some of the benefits and challenges along with underwhelming experiences (Pellicano et al. 2021) that assistive, telehealth, and other online and social technologies and services can provide to autistic people who are isolated and experiencing increased loneliness and a reduction in their social connections and environmental participation. There is a current need for more research to establish the effectiveness of these tools for this group (see Sutherland et al. 2018 and Oakley et al. 2021 for discussion, as cited in Pellicano et al. 2021) in both regular and everyday circumstances, as well as during intensified time periods such as those experienced during the COVID-19 pandemic.

10.3 Support Needs of Autistic People

Autism spectrum disorder (autism) is a genetic condition with characteristics of the disorder evident from as early as 30 months of age (Hallmayer et al., 2011; Lichtenstein et al. 2010; Lundström et al. 2010). The American Psychiatric Association (2013) describes that symptoms and behaviours are permanent and include deficits in social communication and interaction and repetitive motor movements that impact on daily functioning. Autistic people typically need a range of permanent supports (Spain et al. 2017), including support with their daily activities and with their emotional and social well-being (Australian Bureau of Statistics 2020). Autistic adults have reported that the kinds of supports that are often lacking for them are around their cognitive and emotional needs (ABS 2020), and that overall there are few therapeutic approaches that demonstrate positive outcomes for autistic people in these areas. Some of these approaches include focus groups or group therapy, cognitive-based therapy and paraprofessional autism networks (Khan et al. 2016; Loukas et al. 2015), but are cited as being costly and time intensive for both autistic people and practitioners (Ung et al. 2014). As such, alternative responses to some of the issues identified above have come from a technology-based approach, mainly as ATs. These and some current issues with this approach are discussed below.

10.4 Assistive Technology Responses to Problem/Issue

ATs have been developed to reduce some of the impacts of autism in relation to social behaviour, communication, interaction and comprehension. These are widely reported as augmentative and alternative communication, a specialised type of AT. These technologies typically target the user's autistic communication/interaction traits and cognitive profiles and the needs of those who care for them, such as family members and teachers (Sampath et al. 2013). The design of these ATs typically includes a standardised aim to alter or reduce undesired autistic behaviours and traits to a typical pattern or style of normative and accepted communication in efforts to increase the bidirectional understanding of the user's social interactions, communications and emotions (Park et al. 2012). These technologies have been suggested "to increase or improve overall understanding of autistic users environments, expressive communication skills, social interaction skills, attention skills, motivation skills, organisation skills, academic skills, self-help skills and overall independent daily functioning skills" (NASET N. D.) However, the literature tells us that these outcomes have been mostly suggested for autistic children in a school or education-based environment or in a controlled and monitored therapeutic environment. For those reported in the literature, these are technologies that do not have design input by autistic people and focus on creating an educational or therapeutic experience for this group that can be vastly different to or uncomfortable and incompatible with the autistic person's characteristic style.

Despite the increase in ATs developed for autistic people (Boyd et al. 2016; 2017; Cañete and Peralta 2022; Moktar et al. 2014), there is little evidence to support the effectiveness of these tools for autistic users in the social and physical environment. A recent systematic review (Alves et al. 2020) of ATs for the treatment of autism symptoms found that technologies focused on distributed systems, image processing, gamification and robotics designed with an aim to improve "social behaviour, attention, communication, interaction and comprehension…[failed] to accurately define their target audience" and did not comply with treatment dimensions (Alves et al. 2020, p. 118,664). Reported in another systematic literature review analysing the impact of technology on autistic people based on research published during the last 10 years, the authors (Valencia et al. 2019) reported that "whilst new research has focused on supporting children with ASD by using technologies such as virtual reality, augmented reality, virtual agents, sensors and geolocation through educational games…aspects such as user experience, usability and accessibility [which are crucial when working with people with ASD]…are usually not considered or validated in detail".

Also, relatively recent technologies such as electroencephalography (EEG) scanning and eye tracking (Schmidt and Beck 2016) used to evaluate the usability of technologies failed to account for differences in the brain activity for autistic individuals when compared with non-autistic individuals (Hauswald et al. 2013). The authors of the review concluded by stating that "user experience is important and that future studies should consider accessibility and usability tests to ensure positive experiences and comfort with the use of their solutions, as there is a lack of research that applies these concepts correctly and that provides details about the user groups that participate in interventions" (Valencia et al. 2019).

Despite these findings, there has been a growing consideration for user experience in the design of wearable assistive technologies (WATs) for autistic users. The projects SayWAT (Boyd et al. 2016), ProCom (Boyd et al. 2017) and Superpower Glass (Washington et al. 2017) have studied modalities of delivering real-time feedback to autistic users and considered questions such as what information to deliver and when and how to deliver it. These are the kinds of questions that consider the individual user's characteristics and their unique presentations of autism (Mansour et al. 2018) and could reduce the impact of the loss of user meaning through a 'translation of experience', e.g. one autistic person's sensory experience of conversing in a group may be markedly different to another autistic person's and different again from a person who is not autistic.

A third systematic review (Aresti-Bartolome et al. 2014) analysed the literature for technologies most widely used to work on areas affected by autism. The authors' assessment concluded that extensive research has proven the efficiency of technologies as support tools for therapy and their acceptance by autistic people, who report feeling safe and comfortable doing so (Lee et al. 2012), as well as the people who are with them. These technologies included virtual reality applications, mixed reality tools, telehealth systems, social robots and dedicated applications, all of which are classified by the areas they focus on: communication, social learning and imitation skills and other autism-associated conditions. Most of these technologies, however, are inanimate and execute a behaviour that is set and predictable, contributing to the perception of users feeling secure and comfortable using them. The authors discuss that while this might seem like an advantage of the technology, it limits the applicability of the learning to situations outside of the technology not accounting for the variables in interaction in everyday social settings.

Other authors have also highlighted the issues that human–computer interaction presents, in that it does not address the issue of reduced human–human interactions (Boucenna et al. 2014). Some have argued that while technologies incorporating collaborative interactive environments, virtual reality, avatars and robots in social skill training have shown to be beneficial in developing skills of emotion recognition from facial expressions and body language; particularly in autistic children, there is limited evidence to suggest that these technologies can improve face-to-face social interaction in real-life situations (Kientz et al. 2007). As such, there is limited evidence showing that skills practice in controlled therapy settings can be applied to environments beyond these settings (Benssassi et al. 2018).

Authors Mansour et al., in a more recent review (2018) of WATs for autistic users, discuss wearable technology solutions for real-world social interactions with the support of real-time sensing, inference and delivery of in situ social cues via multimodal feedback. The authors caution, however, that these solutions have not been tested in long-term real-world use due to technological challenges, such as ensuring the WATs can produce in-time, accurate cues with reliability in varied contexts; ethical and privacy concerns for bystanders in real-world settings and designing WATs that engage the individual autistic user experience—especially considering users unique sensory and communication preferences. Overall, more research is needed to demonstrate how training with these technologies improves skills that are transferred to the real world and how this improves users' quality of life (Aresti-Bartolome et al. 2014).

Although most of the studies reviewed suggest a usefulness in a controlled and therapeutic setting, they are generic tools for autistic people, meaning there are no personalised tools to meet each person's needs and wants including each person's autism symptomology, which is unique for each autistic individual and plays into their unique autistic experience. For example, most WATs will deliver universal feedback to the user for all situations and do not consider any personalisation and contextualisation (Mansour et al. 2018). As identified by authors Frauenberger et al. 2019, scholars in human–computer interaction (HCI) have struggled with this problem, in particular within the concept of Design Research (see for example Gaver and Bowers 2012; Konstan et al. 2012; Zimmerman et al. 2010), but as the authors assert, little of that thinking has reached the field of ATs.

As highlighted in the introduction of this chapter, technologies included in studies that have met the inclusion criteria for systematic reviews mostly focus on assisting autistic people with social settings and interactions that conform to non-autistic social norms and acceptations. This approach to ATs excludes any notion of autistic culture or identity and routinely places the issues of miscommunication at the end of the autistic person. For these reasons discussed, this might lead to the technology being rejected by the user or a lack of interest in the technology from the user to begin with. Given the capacities of these technologies, the development of configurable systems able to be adapted to autistic individuals' unique sensory and communication profiles might offer more effective tools for use.

Also identified across ATs for review, most of the technology designed with the autistic person as the index patient are designed from a 'fixing-thing' perspective, which stems from the idea that autism is something to be cured or fixed; that is these technologies aim to regulate the behaviours of autistic people and educate and instruct them to understand, display and enact the social norms of, and for, non-autistic people (Mansour et al. 2018). Authors Mankoff et al. (2010) suggested that conceptualisation of disability governs the types and uses of technologies developed. Outside of technology, this position is known as the 'medical model' of disability, a reductionist view that does not account for the ways in which disability, impairment or difference can be meaningful for people with disability. In the context of technologies, this view does not consider that technologies might be meaningful or the ways in which they are meaningful for people with disability (Mankoff et al. 2010). Most autism research investigates and views the disorder from a medical model of disability, which emphasises or prioritises the physical or mental impairments in a person (Fasciglione 2015) and has failed to account for autism's heterogeneity. This is also evident in the ways ATs have been designed and developed for autistic people (Frauenberger et al. 2019) as they lack consideration for the individualisation of autism presenting in individuals.

Challenging or opposing this view, autistic people may preference an alternate way of perceiving the condition and disagree with the overall premise or use of ATs. Many autistic people consider their sensory 'impairment' not as disability but an alternate way of behaving or responding in relation to social stimuli. A disability studies view that considers the moral and ethical position of labelling people as 'disabled' or 'with disability' (Frauenberger et al. 2015) can help to rethink designing future ATs to bolster the ways autistic people might choose to express their selves in social situations. Similarly, the design of ATs might take into consideration both the autistic person and their non-autistic interaction/conversation partner to reach mutual understanding. For example, it is not the autistic person that needs to learn how to communicate or interact; it is the designers and the technologies responsibility to learn about the autistic person's methods or preferences for social interaction and communication.

While autism is mostly considered a disorder across disciplines, this is challenged by the neurodiversity model (Kapp et al. 2013), which sees autism, for example, as a neurocognitive variation throughout human biodiversity. Through this lens, autistic people's subjective experiences are acknowledged (Walker 2014). For example, the subjective experience of autistic people can be described as "more intense and chaotic than that of non-autistic individuals: on both the sensorimotor and cognitive levels, the autistic mind tends to register more information, and the impact of each bit of information tends to be both stronger and less predictable" (Walker 2014, Definition section, para. 1). As such, developers are beginning to realign their purpose or position of a WAT design from a moral and ethical perspective (Frauenberger et al. 2015), whether a WAT should help autistic people (or people with disability) follow social norms, or whether it should support and assist in their difference and empower and enable them as the user of the WAT to communicate and interact in a manner that is innate or comfortable for them is a subject of debate. This aspect is least developed or understood in technology, despite its general considerations regarding user experience (Mansour et al. 2018).

A suggested approach to address this involves including the autistic 'voice' in the design process. Inclusion of people with disability into research more broadly, including in technology research such as the design, development and output of ATs, is relatively recent. Broadly, research exploring lived experiences of people with disability is scarce (Boxall and Ralph 2009; Milner and Kelly 2009; McDonald et al. 2015), but is increasing with the use of cooperative and participatory research methodologies, mostly evident throughout the social sciences and health psychology research, meaning that people with a range of perspectives and experiences can contribute to knowledge production (McDonald et al. 2016).

In addition to the above descriptions of social/communicative educational and therapeutic or 'fixing' orientated ATs, there has been increased research interest that investigates how the environment, particularly the built environment, facilitates the inclusion of autistic people and in the design of learni2012ng environments to better suit the needs of autistic people (Mostafa 2021). While there is some research that addresses this (e.g. Henry 2012; Mostafa 2003, 2008, 2013, 2021), theories such as Sensory Design Theory (Henry 2012), and frameworks developed to enhance environmental landscapes for autistic people (e.g. the Autism Aspectss™ Design Index in Mostafa 2003, 2006, 2008), there remains minimal inclusion of autistic voices in the design, development and output of these technologies. For example, in a recent systematic review, authors Prandi et al. (2021) identified over 100 papers in the scientific literature describing technologies designed to alleviate some of the architectural barriers in outdoor and indoor environments for people with disabilities, including autism. The authors' initial scoping across the studies identified that very few describe "different approaches to support persons with different kinds of disabilities… [and that] the users' points of view in terms of accessibility and usability are generally left out… including the users role in the design and evaluation phases for these technologies" (Prandi et al. 2021, p. 2).

As such, there remains a need to capture autistic voices in the design process, and to consider why it might be important to design for everyday experiences rather than for therapy, education or attempts at 'cure'. To those considering the development of 'autistic-friendly' technologies, aims of social inclusion for autistic people that simulate everyday situations represent a challenge for those who are involved. This is where the autistic voice is important, and UX provides useful tools for capturing this lived experience.

How might the generation of this kind of knowledge be achieved? The proceeding section of this chapter discusses UX and introduces the concept of autistic UX design (aUX). This is a proposed approach to design that addresses some of the key issues with current technologies for autistic people, as presented above. aUX is an approach to design that moves away from therapy, education or attempts at 'cure' and towards greater consideration for autistic people's involvement in technology research, co-participation in the design process, development and implementation of technologies, with aims and outcomes focused towards improving autistic people's everyday experiences through their use of technology.

10.5 User Experience and Autistic UX Design

The term *user experience* (UX) has a broad and diverse reporting (Carneiro et al. 2015) and has been described as the effect of the user's internal state (predisposition, expectations, needs, motivations, mood), of the characteristics of the product/system designed (complexity, purpose, usability, functionality), and of the context in which the interaction occurs (Hassenzahl and Roto 2007). The international standard on ergonomics of human system interaction, ISO 9241-210, defines user experience as "user's perceptions and responses that result from the use and/or anticipated use of a system, product or service". Extending the narrow concept of 'the user', the international standard on ergonomics of human system interaction, ISO 9241-171, describes accessibility as the "extent to which products, systems, services, environments, and facilities can be used by people from a population with the widest range of user needs, characteristics and capabilities to achieve identified goals in identified contexts of use".

Morville (2019) argues that "UX is meaningful and valuable when a product, service or system is useful (that is, its content is original and satisfies a need), usable (the product is easy to use), desirable (the image, identity, brand, and other design elements produce positive emotions towards the product), locatable (the content is accessible to people with disabilities), credible (users have confidence in the product) and valuable (an added value is generated from the product)". Combining these elements, a UX focus on design begins from the *why*, the needs and emotions involved in an activity that generates the experience, then proceeds to determine the *what,* the functionality of the experience; and finally provide the *how,* the functionality to action. The *why*, *what* and *how* are brought together in the final design, but the *why*—the users' needs and emotions—provide the foundation (Hassenzahl et al. 2015).

Typically, a UX design approach seeks to focus on and understand the specific traits and features of the target user/s (Yalanska 2021), which, in the case of autistic users, often seeks to correct, modify or reduce perceived weaknesses. However, an aUX approach designs for autistic users' 'everyday experiences' that emerge from an environmental (social and/or physical) need, or rather, a mismatch between autistic people and their environment. From an aUX approach, everyday experiences—the perspectives and experiences of autistic people—"are understood as a construction, influenced by both the microenvironment (i.e. the interactions and relationships between others); and the wider social and cultural setting" (Trew 2021), which includes the physical environment (i.e. how the built environment reflects culture ideals or values), and is inclusive of autistic culture and identity (Harmon 2004; Bagatell 2007).

This approach is grounded in the shortage of insights from autistic people, who have limited access and voice in design phase spaces, and the fact that current knowledge for how to develop assistive technology is scarce. In response, it is an approach focused on the lived experience (that is the everyday world) of the users or 'participants' captured in the design process, which aims to create a meaningful, engaging and valuable products or services for the user. How this approach seeks to improve experiences with technology by addressing the AT issues and gaps in knowledge identified earlier in the chapter is presented in the following section.

10.6 AUX Design—A Response to Issues Identified in AT to Improve User Experiences with Technology

Theoretically, aUX is grounded in constructivism, meaning that "others hold a different worldview" (Creswell 2014, p. 8); in this position, "truth and meaning do not exist in some external world but are created with the subject's interactions with the world" (Gray 2009, p. 18). A constructivist lens and approach is evident in the aUX framework through its focus on an individual or group description of everyday experiences emerging from their engagement with the environment. This is important, as the approach considers how the knowledge generated through the user's experience or interactions with the world are appropriately translated into 'practice'. This understanding is then better reflected in the design phase of tools or technologies developed and produced to assist the group being designed for. Data collected and knowledge generated to inform the design and application of ATs through this process are suggested to more likely benefit autistic people. As the data collected is informed by the autistic user's experiences, including the needs and the emotions associated with those experiences, the meaning behind those experiences remains grounded in and as autistic knowledge.

As an approach to design, aUX acknowledges that experiences are influenced and created by diverse perspectives in the users' lives and emphasises that "individuals seek understanding of the world in which they live and develop subjective meanings of their experiences" (Creswell 2014, p. 8). The approach is actively seeking to engage autistic people holistically, reframing the design of ATs by away from using disability as the core premise or focal point (Frauenberger et al. 2019). This encourages designers to move away from producing an unchanging or universal representation of experience for a person, but instead aims to acknowledge and nurture the diversity within individuals, such as an autistic person's shared experiences emerging from an environmental (social and/or physical) need. aUX is informed by a series of aligned theoretical orientations and methodologies (e.g. disability studies, participatory or cooperative methodologies and phenomenology), which help to respond to the gaps and issues identified in the literature presented earlier in this chapter about AT design for autistic people. These are now briefly outlined.

A disability studies approach buffers the constructivist interpretivist worldview of the aUX framework. As summarised by Trew (2021), "Disability studies consider people with disability as capable social actors and not passive recipients of society and culture. People with disability are acknowledged as people with unique, diverse and shared experiences. The viewpoint places significant value on the importance of disability rights, agency and well-being and brings the voice of people with disability to the foreground". This means that disability studies help to focus aUX on the leadership of autistic people (Hall et al., 2020), and emphasise that it needs to be reflective of autistic people and the broader autistic community. This plays a vital role in the design and development processes, being produced "with, rather than on" individuals and groups (Christensen and James 2008, p. 1).

Participatory and cooperative methodologies are approaches to research and design that can enable meaningful input from autistic people and their communities (e.g. Fletcher-Watson et al. 2019), including in the design and development of ATs (see for example Frauenberger et al. 2019; Cañete and Peralta 2022). A participatory approach in the design process should include "leadership by autistic researchers, partnership with autistic people or allies as co-creators of knowledge, engagement with the community in general and consultation with relevant individuals or organisations" (Fletcher-Watson et al. 2019, p. 944). From an aUX perspective, collecting data and generating knowledge and meaning through the lens of autistic people with lived experience is a shared and equal process. In the context of AT design and development, a participatory approach should recognise and destabilise "traditional power imbalances" (Nelson and Wright 1995, p. 1), for example, between the designer/developer and the target user or receiver of AT.

The participatory model adopted for aUX helps ensure that the questions asked, aims and outcomes of the technologies are informed by or developed from the perspectives of autistic people, whose involvement can be supported through co-reflexive activities (Bergold and Thomas 2012; Frauenberger et al. 2019; Moore et al. 2016). These have been demonstrated in projects such as Thinking Outside-TheBox (Frauenberger et al. 2019), which over three years used a series of long-term participatory design processes to develop smart objects individually with nine autistic children, employing a wide range of different methods including Cooperative Inquiry, Future Workshops, Fictional Inquiry, Magic Workshops, Drama and Making and Digital Fabrication.

Finally, phenomenology helps shift the theoretical framing of designing ATs towards 'everyday experiences' of autistic people, rather than positioning it as therapy, education or attempts at a 'cure'. Phenomenology as a concept aims to understand human experiences of a given phenomena and how individuals make sense of their experience of the phenomena (Lester 1999; Moustakas 1994). Phenomenology is aligned with participatory methodologies in that it prompts the designer, for example, to regularly refer to or 'check-in' to these novel experiences throughout the design and development process of technologies.

10.7 Summary of Chapter

In this article, we have provided an outline of an issue or problem: that autistic people are significantly less involved in social participation and experience greater social isolation and loneliness, including the mismatch between autistic persons and the environment. We have presented an overview of the literature concerning ATs attempts to resolve or reduce the traits that contribute to autistic people's limited social participation and poor outcomes, which tells us that current ATs and the design of these technologies are not without significant limitations and challenges. In the last part of the chapter, we outlined the details of the proposed framework, aUX design, which is a response to the current issues identified in the design and development of ATs for autistic people.

We have presented the central argument of the framework, the position and theoretical orientations informing the approach when considering design of future ATs for autistic people. We did this through multiple lenses that correspond to the areas in which the literature tells us there are significant gaps and issues of concern for autistic users of ATs. Suggesting a shift in the theoretical framing, we discussed how altering the conceptualisation of autism leads to a design process that moves away from therapy, education or attempts at 'cure' and towards a greater consideration for autistic people's involvement in technologies research, co-participation in the design, development and implementation of technologies, with aims and outcomes focused towards improving autistic people's everyday experiences through their use of technologies.

We believe aUX can make a positive impact on many levels. First, we hope that the readers are enabled or inspired to think differently about ATs for autistic people and to think differently about autism and disability in general. Second, we hope that shifting the mindset leads to a wider conversation about autism and what roles we expect ATs to fill in this context. Conceptually, we have conjectured the possibility to design and develop ATs in ways that are focused, driven by and generated from the everyday experiences of autistic people, and we hope these ideas will be usable for others to build on.

References

Alves FJ, Carvalho EAD, Aguilar J, Brito LLD, Bastos GS (2020) Applied behavior analysis for the treatment of autism: a systematic review of assistive technologies. IEEE Access 8:118664–118672. https://doi.org/10.1109/access.2020.3005296

American Psychiatric Association (2013) Diagnostic and statistical manual of mental disorders, DSM-5. American Psychiatric Publishing. https://doi.org/10.1176/appi.books.9780890425596

Aresti-Bartolome N, Garcia-Zapirain B (2014) Technologies as support tools for persons with autistic spectrum disorder: a systematic review. Int J Environ Res Pu 11:7767–7802. https://doi.org/10.3390/ijerph110807767

Australian Bureau of Statistics (2020) Autism in Australia. Disability, ageing and carers, Australia: summary of findings. https://www.abs.gov.au/statistics/health/disability/disability-ageing-and-carers-australia-summary-findings/latest-release#autism-in-australia

Bagatell N (2007) Orchestrating voices: autism, identity and the power of discourse. Disabil Soc 22:413–426. https://doi.org/10.1080/09687590701337967

Baron-Cohen S (1995) Mindblindness: an essay on autism and theory of mind. MIT Press

Benssassi EM, Gomez J-C, Boyd LE, Hayes GR, Ye J (2018) Wearable assistive technologies for autism: opportunities and challenges. IEEE Pervas Comput 17:11–21. https://doi.org/10.1109/mprv.2018.022511239

Bergold J, Thomas S (2012) Participatory research methods: a methodological approach in motion. Hist Soc Res 37:191–222

Boucenna S, Narzisi A, Tilmont E, Muratori F, Pioggia G, Cohen D, Chetouani M (2014) Interactive technologies for autistic children: a review. Cogn Comput 6:722–740. https://doi.org/10.1007/s12559-014-9276-x

Boxall K, Ralph S (2009) Research ethics and the use of visual images in research with people with intellectual disability. J Intellect Dev Disabil 34:45–54. https://doi.org/10.1080/13668250802688306

Boyd LA, Rangel A, Tomimbang H, Conejo-Toledo A, Patel K, Tentori M, Hayes GR (2016) SayWAT: augmenting face-to-face conversations for adults with autism. In: Proceedings of the 2016 CHI conference on human factors in computing systems (CHI '16). Assoc Comput Mach, New York, NY, USA, 4872–4883. https://doi.org/10.1145/2858036.2858215

Boyd LA, Jiang X, Hayes GR (2017) ProCom: Designing and evaluating a mobile and wearable system to support proximity awareness for people with autism. In: Proceedings of the 2017 CHI conference on human factors in computing systems (CHI '17). Assoc Comput Mach, New York, NY, USA, 2865–2877. https://doi.org/10.1145/3025453.3026014

Camm-Crosbie L, Bradley L, Shaw R, Baron-Cohen S, Cassidy S (2018) 'People like me don't get support': autistic adults' experiences of support and treatment for mental health difficulties, self-injury and suicidality. Autism Int J Res Pract. 23:1431–1441. https://doi.org/10.1177/1362361318816053

Cañete R, Peralta ME (2022) ASDesign: a user-centered method for the design of assistive technology that helps children with autism spectrum disorders be more independent in their daily routines. Sustain-Basel 14:516. https://doi.org/10.3390/su14010516

Carneiro L, Rebelo F, Filgueiras E, Noriega P (2015) Usability and user experience of technical aids for people with disabilities? A preliminary study with a wheelchair. Procedia Manuf 3:6068–6074. https://doi.org/10.1016/j.promfg.2015.07.736

Crane L, Adams F, Harper G, Welch J, Pellicano E (2018) 'Something needs to change': mental health experiences of young autistic adults in England. Autism 23:477–493. https://doi.org/10.1177/1362361318757048

Creswell J (2014) Research design: qualitative, quantitative and mixed methods approaches. SAGE

Christensen P, James A (2008) Childhood diversity and commonality: some methodological insights. In: Research with children. Routledge, pp 172–188

Crompton CJ, Ropar D, Evans-Williams CV, Flynn EG, Fletcher-Watson S (2020) Autistic peer-to-peer information transfer is highly effective. Autism 24:1704–1712. https://doi.org/10.1177/1362361320919286

Fasciglione M (2015) Corporate social responsibility and the right to employment of persons with disabilities. In: Protecting the rights of people with autism in the fields of education and employment. Springer

Fletcher-Watson S, Adams J, Brook K, Charman T, Crane L, Cusack J, Leekam S, Milton D, Parr JR, Pellicano E (2019) Making the future together: shaping autism research through meaningful participation. Autism 23:943–953. https://doi.org/10.1177/1362361318786721

Frith U, Happé F (1999) Theory of mind and self-consciousness: what is it like to be autistic? Mind Lang 14:82–89. https://doi.org/10.1111/1468-0017.00100

Frauenberger C, Good J, Fitzpatrick G, Iversen OS (2015) In pursuit of rigour and accountability in participatory design. Inter J Hum-Comput Stud 74:93–106. https://doi.org/10.1016/j.ijhcs.2014.09.004

Frauenberger C, Spiel K, Makhaeva J (2019) Thinking outsidethe box—designing smart things with autistic children. Int J Hum Comput Interact 35:666–678. https://doi.org/10.1080/10447318.2018.1550177

Gaver B, Bowers J (2012) Annotated portfolios. Interact p 19. https://doi.org/10.1145/2212877.2212889

Gray DE (2009) Doing research in the real world. Sage

Hall S, Fildes J, Liyanarachchi D, Plummer J, Reynolds M (2020) Young, willing and able: youth survey disability report 2019. Mission Australia

Hallmayer J, Cleveland S, Torres A, Phillips J, Cohen B, Torigoe T, Miller J, Fedele A, Collins J, Smith K, Lotspeich L, Croen LA, Ozonoff S, Lajonchere C, Grether JK, Risch N (2011) Genetic heritability and shared environmental factors among twin pairs with autism. Arch Gen Psychiat 68:1095–1102. https://doi.org/10.1001/archgenpsychiatry.2011.76

Harmon A (2004) How about not 'curing' us, some autistics are pleading. New York Times. https://www.nytimes.com/2004/12/20/health/how-about-not-curing-us-some-autistics-are-pleading.html

Hassenzahl M, Roto V (2007) Being and doing: a perspective on user experience and its measurement. Interfaces 72:10–12

Hassenzahl M, Wiklund-Engblom A, Bengs A, Hägglund S, Diefenbach S (2015) Experience-oriented and product-oriented evaluation: psychological need fulfillment, positive affect, and product perception. Int J Hum-Comput Int 31:530–544. https://doi.org/10.1080/10447318.2015.1064664

Hauswald A, Weisz N, Bentin S, Kissler J (2013) MEG premotor abnormalities in children with Asperger's syndrome: determinants of social behavior? Dev Cogn Neuros-Neth 5:95–105. https://doi.org/10.1016/j.dcn.2013.02.002

Hedley D, Uljarević M, Foley K, Richdale A, Trollor J (2018) Risk and protective factors underlying depression and suicidal ideation in autism spectrum disorder. Depress Anxiety 35:648–657. https://doi.org/10.1002/da.22759

Henry C (2012) Architecture for autism: architects moving in the right direction, arch daily, January 5th, 2012, last accessed on January 19th, 2012. http://www.archdaily.com/197788/architecture-for-autism-architects-moving-in-theright-direction/

Holt-Lunstad J, Smith TB, Baker M, Harris T, Stephenson D (2015) Loneliness and social isolation as risk factors for mortality: a meta-analytic review. Perspect Psychol Sci J Assoc Psychol Sci 10:227–237. https://doi.org/10.1177/1745691614568352

International Organization for Standardization: Ergonomics of human-system interaction—part 171: guidance on software accessibility. https://www.iso.org/standard/39080.html

International Organization for Standardization: Ergonomics of human-system interaction—part 210: human-centred design for interactive systems. https://www.iso.org/standard/52075.html

Kapp SK, Gillespie-Lynch K, Sherman LE, Hutman T (2013) Deficit, difference, or both? Autism and neurodiversity. Dev Psychol 49:59–71. https://doi.org/10.1037/a0028353

Kasari C, Brady N, Lord C, Tager-Flusberg H (2013) Assessing the minimally verbal school- aged child with autism spectrum disorder. Autism Res 6(6):479– 493. https://doi.org/10.1002/aur.1334

Khan TM, Ooi K, Ong YS, Jacob SA (2016) A meta-synthesis on parenting a child with autism. Neuropsychiatric Dis Treat 12:745–762. https://doi.org/10.2147/ndt.s100634

Kientz JA, Hayes GR, Westeyn TL, Starner T, Abowd GD (2007) Pervasive computing and autism: assisting caregivers of children with special needs. IEEE Pervas Comput 6:28–35. https://doi.org/10.1109/mprv.2007.18

Konstan JA, Chi EH, Höök K, Park JH, Abirached B, Zhang Y (2012) A framework for designing assistive technologies for teaching children with ASDs emotions. In: CHI'12 extended abstracts on human factors in computing systems, pp 2423–2428. https://doi.org/10.1145/2212776.2223813

Lee J, Takehashi H, Nagai C, Obinata G, Stefanov D (2012) Which robot features can stimulate better responses from children with autism in robot-assisted therapy? Int J Adv Robot Syst 9:72. https://doi.org/10.5772/51128

Lester S (1999) An introduction to phenomenological research. https://www.academia.edu/1936094/An_introduction_to_phenomenological_research

Lichtenstein P, Carlström E, Råstam M, Gillberg C, Anckarsäter H (2010) The genetics of autism spectrum disorders and related neuropsychiatric disorders in childhood. Am J Psychiat 167:1357–1363. https://doi.org/10.1176/appi.ajp.2010.10020223

Loukas KM, Raymond L, Perron AR, McHarg LA, Doe TCL (2015) Occupational transformation: parental influence and social cognition of young adults with autism. Work 50:457–463. https://doi.org/10.3233/wor-141956

Lundström S, Haworth CMA, Carlström E, Gillberg C, Mill J, Råstam M, Hultman CM, Ronald A, Anckarsäter H, Plomin R, Lichtenstein P, Reichenberg A (2010) Trajectories leading to autism spectrum disorders are affected by paternal age: findings from two nationally representative twin studies. J Child Psychol Psyc 51:850–856. https://doi.org/10.1111/j.1469-7610.2010.02223.x

Maddox BB, Crabbe S, Beidas RS, Brookman-Frazee L, Cannuscio CC, Miller JS, Nicolaidis C, Mandell DS (2020) 'I wouldn't know where to start': perspectives from clinicians, agency leaders, and autistic adults on improving community mental health services for autistic adults. Autism Int J Res Pract. 24(4):919–930. https://doi.org/10.1177/1362361319882227

Mankoff J, Hayes GR, Kasnitz D (2010) Disability studies as a source of critical inquiry for the field of assistive technology. In: Proceedings of the 12th international ACM SIGACCESS conference on computers and accessibility—Assets'10, pp 3–10. https://doi.org/10.1145/1878803.1878807

Mansour BE, Gomez JC, Boyd LH, Gillian YJ (2018) Wearable assistive technologies for autism: Opportunities and challenges. IEEE Pervasive Comput 17:11–21. https://doi.org/10.1109/MPRV.2018.022511239

McDonald KE, Conroy NE, Kim CI, LoBraico EJ, Prather EM, Olick RS (2016) Is safety in the eye of the beholder? Safeguards in research with adults with intellectual disability. J Empir Res Hum Res 11:424–438. https://doi.org/10.1177/1556264616651182

McDonald KE, Schwartz NM, Gibbons CM, Olick RS (2015) "You can't be cold and scientific": community views on ethical issues in intellectual disability research. J Empir Res Hum Res 10:196–208. https://doi.org/10.1177/1556264615575512

Milner P, Kelly B (2009) Community participation and inclusion: people with disabilities defining their place. Disabil Soc 24:47–62. https://doi.org/10.1080/09687590802535410

Moktar MN, Fikry A, Musa R, Hassan H, Ahmad SSH, Ismail Z, Samat N, Hashim R (2014) Extending cultural model of assistive technology design for autism treatment. In: 2014 IEEE international symposium robotics manufacturing automation (ROMA), pp 172–175. https://doi.org/10.1109/roma.2014.7295882

Moore T, Noble-Carr D, McArthur M (2016) Changing things for the better: the use of children and young people's reference groups in social research. Int J Soc Res Method 19:241–256. https://doi.org/10.1080/13645579.2014.989640

Morrison KE, DeBrabander KM, Jones DR, Faso DJ, Ackerman RA, Sasson NJ (2020) Outcomes of real-world social interaction for autistic adults paired with autistic compared to typically developing partners. Autism 24(5):1067–1080. https://doi.org/10.1177/1362361319892701

Morville P (2019) User experience design. https://semanticstudios.com/user_experience_design/

Mostafa M (2003) Accommodating autistic behaviour in design through modification of the architectural environment, doctoral dissertation. Dept Archit Eng. Cairo University

Mostafa M (2006) Let them be heard: appropriate acoustics for autism- special needs school design, archCairo 2006, 3rd international proceedings, appropriating architecture- taming urbanism in the decades of transformation, Cairo, Egypt

Mostafa M (2008) An architecture for autism: concepts of design intervention for the autistic user, archnet-IJAR: Inter J Archit Res 2(1):189–211

Mostafa M (2021) Architecture for autism: built environment performance in accordance to the autism ASPECTSS design index. Editor(s): Undurti Das, Neophytos Papaneophytou, Tatyana El-Kour. In: Autism 360°. Academic Press 479-500. https://doi.org/10.1016/B978-0-12-818466-0.00023-X

Moustakas C (1994) Phenomenological research methods. SAGE

N.A.S.E.T.: Assistive technology for students with autism spectrum disorders, https://www.naset.org/fileadmin/user_upload/Autism_Series/Assist_tech/AssistiveTech_for_Students_W_Autism.pdf

Nelson N, Wright S (1995) Power and participatory development: theory and practice. ITDG Publishing

Oomen D, Nijhof AD, Wiersema JR (2021) The psychological impact of the COVID-19 pandemic on adults with autism: a survey study across three countries. Mol Autism 12:21. https://doi.org/10.1186/s13229-021-00424-y

Orsmond GI, Fulford D (2018) Adult siblings who have a brother or sister with autism: between-family and within-family variations in sibling relationships. J Autism Dev Disord 48:4090–4102. https://doi.org/10.1007/s10803-018-3669-8

Orsmond G, Shattuck P, Cooper B, Sterzing P, Anderson K (2013) Social participation among young adults with an autism spectrum disorder. J Autism Dev Disord, pp 2710–2719. https://doi.org/10.1007/s10803-013-1833-8

Ozsivadjian A, Knott F, Magiati I (2012) Parent and child perspectives on the nature of anxiety in children and young people with autism spectrum disorders: a focus group study. Autism 16:107–121. https://doi.org/10.1177/1362361311431703

Park JI, Abirached B, Zhang Y (2012) A framework for designing assistive technologies for teaching children with ASDs emotions. In: CHI '12 extended abstracts on human factors in computing systems (CHI EA '12). Assoc Comput Mach. New York, NY, USA, 2423–2428. https://doi.org/10.1177/1362361311401763

Pellicano E, Brett S, den Houting J, Heyworth M, Magiati I, Steward R, Urbanowicz A, Stears M (2021) COVID-19, Social isolation and the mental health of autistic people and their families: a qualitative study. SSRN Electron J. https://doi.org/10.2139/ssrn.3749228

Pellicano L, Brett S, den Houting J, Heyworth M, Magiati I, Steward R, Urbanowicz A, Stears M (2020) I want to see my friends. The everyday experiences of autistic people and their families during COVID-19. The University of Sydney (2020). https://www.sydney.edu.au/content/dam/corporate/documents/sydney-policy-lab/everyday-experiences-of-autistic-people-during-covid-19---report---july-2020.pdf

Prandi C, Nisi V, Ribeiro M, Nunes N (2021) Sensing and making sense of tourism flows and urban data to foster sustainability awareness: a real-world experience. J Big Data 8:51. https://doi.org/10.1186/s40537-021-00442-w

Rutherford MD, Baron-Cohen S, Wheelwright S (2002) Reading the mind in the voice: a study with normal adults and adults with asperger syndrome and high functioning autism. J Autism Dev Disord 32:189–194. https://doi.org/10.1023/a:1015497629971

Sampath H, Agarwal R, Indurkhya B (2013) Assistive technology for children with autism—lessons for interaction design. In: Proceedings of the 11th Asia Pacific conference on computer human interaction'13, pp 325–333. https://doi.org/10.1145/2525194.2525300

Schmidt M, Beck D (2016) Immersive learning research network. In: Second international conference, iLRN 2016 Santa Barbara, CA, USA, June 27–July 1; In: 2016 proceedings communications in computer and information science, pp 113–121. https://doi.org/10.1007/978-3-319-41769-1_9

Spain D, Sin J, Paliokosta E, Furuta M, Prunty JE, Chalder T, Murphy DG, Happé FG (2017) Family therapy for autism spectrum disorders. Cochrane Db Syst Rev 2017:CD011894. https://doi.org/10.1002/14651858.cd011894.pub2

Spiker MA, Lin CE, Dyke MV, Wood JJ (2011) Restricted interests and anxiety in children with autism. Autism 16(3):306–320. https://doi.org/10.1177/1362361311401763

Trew S (2021) Family relationships and autism spectrum disorder: Lived experiences of young people with autism and their families. Australian Catholic University

Ung D, Selles R, Small BJ, Storch EA (2014) A systematic review and meta-analysis of cognitive-behavioral therapy for anxiety in youth with high-functioning autism spectrum disorders. Child Psychiat Hum Dev 46:533–547. https://doi.org/10.1007/s10578-014-0494-y

Valencia K, Rusu C, Quiñones D, Jamet E (2019) The impact of technology on people with autism spectrum disorder: a systematic literature review. Sens Basel Switz 19:4485. https://doi.org/10.3390/s19204485

Walker N: Neurodiversity: Some basic terms and definitions. https://neuroqueer.com/neurodiversity-terms-and-definitions/

Washington P, Voss C, Kline A, Haber N, Daniels J, Fazel A, De T, Feinstein C, Winograd T, Wall D (2017) Superpower glass: a wearable aid for the at-home therapy of children with autism. Proc Acm Interact Mob Wearable Ubiquitous Technol 1:1–22. https://doi.org/10.1145/3130977

Weiss JA, Wingsiong A, Lunsky Y (2013) Defining crisis in families of individuals with autism spectrum disorders. Autism 18(8):985–995. https://doi.org/10.1177/1362361313508024

Yalanska M (2021) Human-centered vs user-centered. Are the terms different? https://blog.tubikstudio.com/faq-design-platform-human-centered-vs-user-centered-are-the-terms-different/

Chapter 11
Drone Swarms to Support Search and Rescue Operations: Opportunities and Challenges

Maria-Theresa Oanh Hoang, Kasper Andreas Rømer Grøntved, Niels van Berkel, Mikael B Skov, Anders Lyhne Christensen, and Timothy Merritt

Abstract Emergency services organizations are committed to the challenging task of saving people in distress and minimizing harm across a wide range of events, including accidents, natural disasters, and search and rescue. The teams responsible for these operations use advanced equipment to support their missions. Given the risks and the time pressure of these missions, however, adopting new technologies requires careful testing and preparation. Drones have become a valuable technology in recent years for emergency services teams employed to locate people across vast and difficult to traverse terrains. These unmanned aerial vehicles are faster and cheaper to deploy than traditional crewed aircraft. While an individual drone can be helpful to personnel by quickly offering a bird's eye view, future scenarios may allow multiple drones working together as a swarm to reduce the time required to locate a person. Given these potentially high payoffs, we explored the challenges and opportunities of drone swarms in search and rescue operations. We conducted interviews as well as initial user studies with relevant stakeholders to understand the challenges and opportunities for drone swarms in the context of search and rescue. Through this, we gained insights to inform the development of prototypes for drone swarm control interfaces, including both technical and human interaction concerns. While

M.-T. O. Hoang (✉) · N. van Berkel · M. B. Skov · T. Merritt
Aalborg University, Aalborg, Denmark
e-mail: mtoh@cs.aau.dk

N. van Berkel
e-mail: nielsvanberkel@cs.aau.dk

M. B. Skov
e-mail: dubois@cs.aau.dk

T. Merritt
e-mail: merritt@cs.aau.dk

K. A. R. Grøntved · A. L. Christensen
University of Southern Denmark, Odense, Denmark
e-mail: kang@mmmi.sdu.dk

A. L. Christensen
e-mail: andc@mmmi.sdu.dk

© The Author(s), under exclusive license to Springer Nature Switzerland AG 2023

B. J. Dunstan et al. (eds.), *Cultural Robotics: Social Robots and Their Emergent Cultural Ecologies*, Springer Series on Cultural Computing,
https://doi.org/10.1007/978-3-031-28138-9_11

drone swarms can likely benefit search and rescue operations, the significant shift from single drones to swarms may necessitate reimagining how rescue missions are conducted. We distill our findings into five key research challenges: visualization, situational awareness, technical issues, team culture, and public perception. We discuss initial steps to investigate these further.

11.1 Introduction

The Danish Emergency Management Agency (DEMA)[1] engages in missions to protect the public in emergency situations. This includes the search and rescue of people in distress and responding to various situations of increased danger to the public. The equipment DEMA uses to support their missions include highly specialized technologies, tools, and custom vehicles. DEMA, as with other national emergency management agencies, is often first movers in utilizing new technologies and appropriating them into their processes and practices.

An example of such specialized tools that have recently become available is drones. With the falling costs of professional quality drones and improved ease of use, drones have experienced increased adoption by many law enforcement and emergency management agencies. These flying robots are used to carry payloads that gather data, such as cameras for surveillance, microphones, and thermal sensors, as well as active payloads to affect crowds such as a lights and loudspeakers (Engberts and Gillissen 2016). In Denmark, the emergency services have three dedicated drone teams that can be deployed within minutes (Beredskabsstyrelsens droneberedskab 2020; https://www.brs.dk/da/redningsberedskab-myndighed/assistance-fra-beredskabsstyrelsen/beredskabsstyrelsens-ressourcekatalog/sarligt-materiel/droner/). Furthermore, in 60 Danish municipalities, various DEMA personnel are trained to utilize drone technologies in their missions. This primarily involves intelligence gathering, such as observing a fire from above to identify safe areas to direct firefighters or searching a large field with a thermal camera to locate a lost person in distress. Drones are often utilized in conjunction with manned aircraft (helicopters), primarily in maritime search and rescue (SAR). Helicopter-based search is slow, expensive, and requires cross-organizational coordination. Increasingly, drones have been used together with artificial intelligence (AI) and computer vision technologies to provide automated support for emergencies, for example, to provide AI capabilities to analyze thermal imagery directly on a drone to detect forest fires (https://www.robotto.ai/fire). The technology can scan a large area and build a map of the fire. Subsequently, based on the identification of hot spots and weather conditions, it can predict how the fire will move, enabling firefighters to take the appropriate positions to ensure their safety and make decisions that help extinguish fires quicker.

[1] https://www.brs.dk/en/

While this type of semi-autonomous use of single drones provides new abilities to firefighting, recent developments in controlling multiple drones as a swarm may enable even faster coverage and assessments of large areas. Drone swarms for search and rescue hold potential to reduce the time it takes to find a person in distress by covering a large area very quickly and deploying a range of sensors, including visible light and thermal cameras, among others. However, very little is known in relation to the integration of these drones into the working practices of emergency response teams. This includes the user interfaces for controlling and managing a drone swarm, but also the protocols for engagement, as well as the interaction with other parties in the airspace. There has been considerable attention in research on how people respond to socially interactive robots (Fong et al. 2002, 2003), and the early examples of social interaction with robots involved swarms, yet most of the research exploring human interactions focus on humanoid robots.

In this chapter, we describe initial findings from interviews with the Danish Emergency Management Agency in relation to the prospects of using drone swarms for search and rescue. We group our findings into five key research challenges: visualization, situational awareness, technical issues, team culture, and public perception.

11.2 Related Work

11.2.1 UAVs for Emergency Settings

Unmanned aerial vehicles (UAVs) have the potential to support first responders, and the current research space has highlighted this potential. A recent literature survey by Herdel et al. (2022) identified 16 domains for human–drone interaction and found that 'emergency' was the most prominent domain mentioned. UAVs used for emergency situations have the benefit of short deployment time and minimizing the need for human involvement in a hazardous environment (McRae et al. 2019). A UAV can provide visual surveillance of a situation by hovering above the scene and capturing video for the first responders. The video feedback can be used in multiple ways, for example, assessing a developing fire area (Bjurling et al. 2020; Naghsh and Roast 2009; Pham et al. 2017). Autonomous drone-based firefighting tools have begun to hit the market, e.g., the firefighting tools by Robotto,[2] which provides an algorithm for identifying and predicting the behavior of an evolving fire. Similarly, UAVs have also been used in search and rescue missions to find missing people in various conditions (Arnold et al. 2018; Karaca et al. 2018; McRae et al. 2019; Silvagni et al. 2017). A common challenge when using UAVs in SAR is that the team does not find the missing person due to features in the terrain that may obstruct visibility from above. Current research has aimed to automate visual detection by clearly highlighting on

[2] https://www.robotto.ai

the video feed when a person has been found (Goodrich et al. 2008; Mishra et al. 2020; Scherer et al. 2015).

11.2.2 UAV Swarm Behavior and Control

UAVs are mostly piloted individually by one operator, but research is looking at developing algorithms that allow for an autonomous swarm of drones. Compared to using a single drone, a swarm can cover a larger search area faster in a SAR mission, increasing the possibility of finding a missing person.

Current research has explored ways to exercise control over a swarm without resorting to low-level motor commands. Swarm behavior often takes inspiration from animals such as bees, birds, and fish (Hocraffer and Nam 2017). Research has also started to explore the possibility of incorporating a leader among the drones (Kerman et al. 2012; Kolling et al. 2016). Having a leader requires the operator to only focus on a single drone, with the other swarm members adjusting their course automatically. There are various examples of drone controls, such as the basic *selection* of one or more drones to view status, issuing low-level commands such as movement or camera adjustments, or more high-level mission priorities and search pattern selection. In terms of automated search patterns, *spiral* patterns direct the swarm to search an area expanding from a point, and *scatter* patterns disperse the swarm in all directions (Arnold et al. 2018; Kolling et al. 2013). Furthermore, *beacon* controls are used to direct the swarm by attracting the UAVs to high priority areas or by repelling the UAVs from areas that they should not enter (Kolling et al. 2013).

11.2.3 User Interfaces for Human–Swarm Interaction

Controlling one drone is very different from controlling multiple drones simultaneously and potentially requires different ways of interacting. In contrast to the operation of one drone, a swarm potentially requires the operator to divide their attention among multiple drones.

Despite the aforementioned advantages of drone swarms, some disadvantages make it difficult to implement UAV swarms in SAR. One of the biggest challenges for the operator is to maintain *situational awareness* (SA). Studies have shown the importance of only displaying key elements to reduce the amount of visual clutter on the screen (Agrawal et al. 2020; Rule and Forlizzi 2012; Soorati et al. 2021). Having only the most crucial information clearly and concisely available will help the operator to make appropriate decisions while minimizing the chance of errors.

11.3 Interviews and Prototype Evaluations

We conducted interviews with emergency services personnel who already use single-drone systems for search and rescue in order to investigate the challenges and opportunities for drone swarms in the SAR context. Additionally, we conducted evaluations with initial prototypes to stage conversations about how drone swarms could support SAR missions and gathered input relating to the features and functions of future systems. The participants were two senior sergeants, one of whom is responsible for drone-related training and official procedures, while the other is the main person responsible for unmanned aircraft systems (UASs) in the organization.

We began with a review of documentation shared by our participants that explained the current usage of drones during emergencies and SAR missions. Following this, we held a video interview to discuss typical missions and their ideas for drone swarms. We then developed an interactive prototype platform that supports typical mission tasks, including launching and landing drones, selection and movement of single or multiple drones during flight, and selecting an area on the map to contain the search. The prototype also supported advanced swarm concepts, including leader/follower and beacon controls.. The system was built using Web technologies and utilized the same drone platform used by DEMA. We brought the prototypes to the DEMA training college, where we conducted the interviews and evaluations. Upon arrival at the DEMA training college, the personnel walked us through their specialized drone support vehicle and explained its components as they set up and flew a short demonstration mission at the campus. Each drone is controlled using a dedicated pilot with a DJI smart controller, as shown in Fig. 11.1.

Fig. 11.1 DEMA drone pilot using the DJI smart controller for controlling a single drone

Fig. 11.2 DEMA drone pilot using the DJI smart controller for controlling a single drone, with the video feed shown on the large screen in the support vehicle

For the demonstration, the pilot flew the drone around the campus and explained the typical process of orienting to the emergency situation and working with the team to scan the live video feed shown on the large screen in the support vehicle, which can be seen in Fig. 11.2.

Following the demonstration, we invited the two participants to experiment with the prototypes we developed for swarm visualization and control. These prototypes were developed to run on a tablet connected wirelessly to multiple DJI flight controllers.

While the prototype could be used with an unlimited number of physical drones, for the demonstration, we utilized two physical drones. This provided the feeling of controlling multiple drones without using low-level controls they would normally use with the provided DJI smart controller. The user interface provides a plan view map of the area (shown in Fig. 11.3) and enables touch interactions to launch and land drones, to select and move drones, and to create virtual beacons to attract and repel drones, among other swarm commands.

We encouraged our participants to be critical of the prototype and invited them to consider themselves as co-designers of the system. Hence, we asked them to vocalize both positive and negative thoughts about the prototype, as well as their ideas for ways to improve the system.

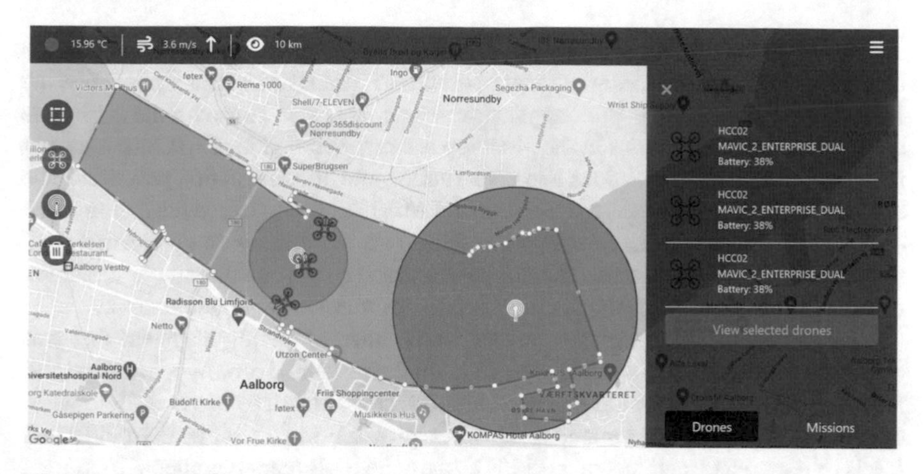

Fig. 11.3 User interface of the prototype. The area shown is the Limfjorden in Aalborg, Denmark. A fence is drawn to contain the swarm, and virtual beacons are placed to influence the search area

11.4 Five Research Challenges

The interviews with the emergency services personnel, our observations, and evaluation of an early prototype have revealed surprising tensions and concerns across a broad range of topics. In this section, we present the key topics that came up in our review of the literature and the interviews as five key research challenges for drone swarms in search and rescue operations. These challenges are visualization, situational awareness, technical issues, team culture, and public perception. We discuss initial steps to investigate these further.

We are aware that these challenges are not an exhaustive list and that there are similar efforts to take a holistic approach when developing SAR systems with drones. The US Coast Guard suggests that such systems be designed not just to satisfy technical requirements, but should be "evaluated and employed as an entire system" (UAS for SAR 2016).

11.4.1 Visualization

From our survey of existing SAR interfaces, we observed a wide variety of ways to present information. A 2D map view is often used (Agrawal et al. 2020; Rule and Forlizzi 2012; Soorati et al. 2021), and the prototypes were designed using these as models. For the design of such expert systems, there is a challenge in deciding which features should be supported and how the information should be presented. Should there be, for example, user selection of presentation style, mode selection, or efforts made toward a standardized way of representing objects and environmental characteristics? Similarly, how should the suggestions from predictive and other advanced AI features be presented? How do colors, symbols, etc., carry across countries and cultures in international operations? The DEMA drone pilots use the 'mapping missions' in the pilot application that comes standard with the enterprise drones[3].[3] The visualizations in the pilot app are not intended for SAR missions, but rather for real estate inspections and construction site documentation purposes. The application is simple and easy to use; thus DEMA has chosen to use the features that are helpful for their work, yet they have shared aspirations for an application built with SAR in mind. The visualizations they discussed go beyond the simple streaming of video and included a desire for the planning of missions and automation.

The two senior sergeants stressed the importance of having a system that was extensive yet simple to use. While they thought the presented prototype was straightforward, they feared it would become cluttered and difficult to use over time when crucial features were implemented. This is a pressing matter as there are different levels of experience and technical skills among the SAR team members. Therefore, when developing prototypes for them, it is essential to ensure simplicity and impor tant to involve less technical members of SAR missions to study the struggles and needs they face in completing their tasks efficiently.

11.4.2 Situational Awareness

In our study, we explored how advanced swarm control interfaces might impact situational awareness. We explored ways to provide predictive support to emergency personnel in which AI techniques identified possible victims and suggested a best course of action of the SAR team. It became clear that the most effective swarm interfaces would provide a balance between showing only partial details of each drone without overloading the pilot with visual indicators. Balancing the number of information elements and the form in which information is represented is an important challenge for the design of any complex interface (Oury and Ritter 2021). By adopting a user-centered approach, we have uncovered unique aspects relevant to the interfaces for supporting SAR with drones that require deeper investigation.

The interviews helped to reveal fundamental shifts in the way the operation is conducted now with single drones compared to a swarm and initial indications as to what the operator should devote their attention to. An example that illustrates this relates to the video feed from the drone cameras. The senior sergeants from our study explained that they originally had the expectation that the camera feed of all the drones would always be displayed at all times, and that the feed would have an alert indicator displayed if something of interest is found so that the operator could decide if it should be inspected more closely or ignored. Currently, a separate person who is not the pilot watches the live video feed; however, in the future imagined system with multiple video feeds, it is not certain how many simultaneous video feeds a person can realistically observe, even with the help from the alert indicator.

[3] https://youtu.be/92RgLBJcViI.

Future studies could determine if it is feasible for a person to continuously monitor multiple drone feeds and the effect this might have on their situational awareness.

One of the senior sergeants expressed his uncertainty about using a multi-drone system for SAR as it would introduce more information for a person to process. He raised the concern that the system should have some level of automation that could alleviate responsibilities from the user. Various new questions arise: Is it necessary to always display the video feed from each drone? How will the roles change among the SAR team members to work together and support the operation? The ambition of the DEMA professionals is that if a partially automated system could detect a person through the camera feed, it could be sufficient for the operator to view a feed only when a potential victim or target has been identified. A significant problem with this proposition is the fear of false negatives and the risk of missing something that a human would notice. To implement such a system, it should therefore undergo rigorous testing and evaluation to ensure that it can effectively be used alongside or as a substitute for human image processing. The culture of the SAR team seems to welcome the use of effective technology tools to aid their missions; however, accepting an autonomous system for critical tasks is a new frontier that we are exploring in the ongoing research and development.

11.4.3 Technical Issues

There are various technical challenges concerning the drone, the operator, hardware, and software. This includes, but is not limited to, sensors, communication technologies, computer vision, predictive techniques, and algorithms for controlling the swarm in a safe and efficient manner.

SAR missions can occur in harsh terrain and weather conditions. It is, therefore, a requirement for the devices to be sufficiently robust. It is also important that when a piece of equipment fails, it does not delay or interfere with the mission. If a system could be synchronized across multiple devices, it might be possible to pick up a backup device and instantly resume a mission if a device fails.

From our interview with the two senior sergeants, we found that a large part of SAR missions is spent determining which areas to search. This choice is affected by multiple factors, such as police information, as well as specific targets of value such as rivers. With multiple drones, it is more challenging to plan a route that would be appropriate for drones to take. An algorithm that suggests a course of action based on both mission and terrain-specific information could significantly help the SAR team plan and prepare a SAR mission.

When working with drones in SAR missions, the drone team instantaneously analyzes the video feed that the drones capture. However, from our initial interview with DEMA, we were informed that it is challenging for the drone operator to maintain focus, especially if they are looking at a repetitive pattern such as a field. For a drone swarm to be useful, it is necessary to develop a computer vision system that automatically detects if something has been found and alert the operator clearly and effectively.

11.4.4 Team Culture

The processes and procedures currently in place have been developed over time and tuned such that the team works tightly in unison and knows what to expect. This is vital during time-critical operations. Therefore, introducing drone swarms is not as straightforward as replacing one helicopter with a swarm of drones. The team structure will likely need to adapt with various new roles and responsibilities. The helicopter pilot and passenger are partially replaced by unmanned aircraft, but there is an increased demand in maintaining, deploying, and configuring the hardware and software and supporting it in the field. How should the transfer of responsibility from personnel that is distant to the local personnel take place and what practices will emerge? Field trials and exercises are needed to gain input from the teams and to explore alternatives.

Each SAR mission often involves professionals from multiple agencies, including police, military, and firefighters, among others. Currently, SAR missions start with a call made to the police. The police call in the DEMA resources to begin the search, providing initial details to the SAR team via telephone while the vehicles are en route to the scene. The interviewees expressed that this process works very well; however, they also imagined that a robust future system could allow the police to annotate a digital map to suggest initial search areas and to provide other key details more quickly. They described that in the initial annotations, the police could mark up areas best suited for the drones and other areas that are best suited for search dogs and personnel. This raises a lot of questions about which personnel would have access to view and explore the map and whether multiple personnel should edit and annotate simultaneously.

The military also plays a role in SAR missions, utilizing a specialized helicopter with a pilot and spotter who scans the area below. In current practice, there is direct voice communication between the helicopter and drone operator to coordinate search areas and to ensure that the drones and helicopter operate with a safe distance from each other (no closer than 200 m). In the future, with swarms of drones, there will be some level of automation. The swarm could be programmed to always maintain a safety distance to stay clear of the helicopter. To ensure the highest level of safety, it is presumed that a human drone operator remains in control and can override any movements of the swarm—and perhaps the helicopter crew should also be equipped with a view into the system so that they can understand the swarm's movements, and perhaps override the system directly and force the swarm to land should there be any imminent danger. Future studies together with the various teams are needed in order to maintain trust and understanding.

11.4.5 Public Perception

As with any novel technology, public perception and acceptance is important for the eventual adoption of this technology. Drone swarms need to provide a qualitative improvement to SAR and not just a reduction in costs. Managing the public perception and interest is important in the early stages to enable the maturing of the technology and techniques. Furthermore, the public will begin to see swarms of flying machines during emergencies, raising the question as to how this will be perceived. How will the public react to a swarm of drones flying in their area? Are there ways to communicate from the drones to the public and bystanders what is happening and what they can do to help? For example, establishing cultural symbols for bystanders to 'stand back' might be critical during operations.

Drones that actively search in public spaces are not commonly seen in residential areas in Denmark and are likely not perceived the same way as emergency vehicles such as ambulances or police cars. Studies have also shown how drones specifically used in residential areas can cause residents discomfort because of the uncertainty of their purpose and whether they can cause any harm (Bajde et al. 2017; Lin Tan et al. 2021). However, knowing that the drones are used in the context of an SAR mission is likely to be a more accepted usage of drones. It would be relevant to examine possible ways to signal to the public that drones are actively involved in an emergency, for example, by taking cues from existing emergency vehicles such as color, sound, and light signals. Taking inspiration from research on non-verbal communication with robots (Bethel and Murphy 2008), affective expression could be explored in future studies through managing the social distance between drones and people and through the use of visual and auditory cues. Drones are relatively new in the public sphere and beginning to influence cultural practices. The two DEMA senior sergeants told us that SAR inside residential areas provide some of the most challenging scenarios. This is due to multiple factors, including the presence of residents, difficult heat signatures from thermal cameras, and heightened legislation. While DEMA has the right to fly beyond standard flight rules if they deem the increased risks to be acceptable, it still does not solve the problem of the increased difficulty of the search. Drone software could play a larger role in increasing the effectiveness of drones in SAR missions in residential areas. It would be relevant for the drones to plan routes that would avoid people by, for example, exclusively flying over rooftops and avoiding flying above uninvolved pedestrians. More attention is needed to ensure that interactions with people in public contribute to feelings of safety, trust, and understanding of the drones' intentions.

11.5 Conclusion

In this chapter, we set out to share our experiences with an ongoing project in which we are designing user interfaces for controlling drone swarms for search and rescue. We conducted interviews with emergency services professionals and initial user studies with drone swarm prototypes in order to understand some of the challenges and opportunities. We learned about the current practices and procedures that are in place when utilizing a single drone to assist in SAR missions. It became clear that a more holistic view is necessary when designing in this context. Moving from manual control of one drone to automated swarms of drones will need technical advances, but also a careful consideration and involvement of the professionals. We provided highlights from the studies grouped into five key research challenges: visualization, situational awareness, technical issues, team culture, and public perception. We discussed initial steps to investigate these further. The system we are building together with DEMA is a very specialized assistive technology. We envision that it could be a tool that will help balance the mental workload of the operators while providing situational awareness. While a well-designed technical solution may be needed, our investigations thus far have shown that more focus is needed to understand the people, processes, and culture of the emergency services teams in order for swarms of drones to provide real value to search and rescue operations.

Acknowledgements We thank the Danish Emergency Management Agency professionals and Robotto for their participation. This work is supported by the Innovation Fund Denmark for the project DIREC (9142-00001B).

References

Agrawal A, Abraham SJ, Burger B, Christine C, Fraser L, Hoeksema JM, Hwang S, Travnik E, Kumar S, Scheirer W, Cleland-Huang J, Vierhauser M, Bauer R, Cox S (2020) The next generation of human-drone partnerships: co-designing an emergency response system. In: Proceedings of the 2020 CHI conference on human factors in computing systems, CHI '20, pp. 1–13. Association for Computing Machinery, New York (2020). https://doi.org/10.1145/3313831.3376825

Arnold RD, Yamaguchi H, Tanaka T (2018) Search and rescue with autonomous flying robots through behavior-based cooperative intelligence. J Int Humanitarian Action 3(1):18. https://doi.org/10.1186/s41018-018-0045-4

Bajde D, Woerman N, Bruun M, Gahrn-Andersen R, Sommer J, Nøjgaard M, Chris- tensen S, Kirschner H, Jensen R, Bucher J (2017) Public reactions to drone use in residential and public areas. Syddansk Universitetog Aalborg Universitet

Beredskabsstyrelsens droneberedskab. https://www.brs.dk/da/nyheder-og-publikationer/publikationer2/alle-publikationer/2020/beredskabsstyrelsens-droneberedskab/

Bethel CL, Murphy RR (2008) Survey of non-facial/non-verbal affective expressions for appearance-constrained robots. IEEE Trans Syst Man Cybern Part C (Appl Rev) 38(1):83–92 (2008). https://doi.org/10.1109/TSMCC.2007.905845

Bjurling O, Granlund R, Alfredson J, Arvola M, Ziemke T (2020) Drone swarms in forest firefighting: a local development case study of multi-level human-swarm interaction. In: Proceedings of the 11th Nordic Conference on Human-Computer Interaction: Shaping Experi- ences, Shaping Society. ACM, Tallinn Estonia, pp 1–7. https://doi.org/10.1145/3419249.3421239

Droner. https://www.brs.dk/da/redningsberedskab-myndighed/assistance-fra-beredskabsstyrelsen/beredskabsstyrelsens-ressourcekatalog/sarligt-materiel/droner/

Engberts B, Gillissen E (2016) Policing from above: drone use by the police. In: Custers B (ed) The future of drone use: opportunities and threats from ethical and legal perspectives. T.M.C. Asser Press, The Hague, pp 93–113. https://doi.org/10.1007/978-94-6265-132-6_5

FIRE. https://www.robotto.ai/fire

Fong T, Nourbakhsh I, Dautenhahn K (2003) A survey of socially interactive robots. Rob Auton Syst 42(3–4):143–166

Fong T, Nourbakhsh I, Dautenhahn K (2002) A survey of socially interactive robots: concepts, design. Tech. rep., and applications, Technical Report No. CMU-RI-TR-02–29, Robotics Institute

Goodrich MA, Morse BS, Gerhardt D, Cooper JL, Quigley M, Adams JA, Humphrey C (2008) Supporting wilderness search and rescue using a camera-equipped mini UAV. J Field Rob 25(1–2):89–110. https://doi.org/10.1002/rob.20226

Herdel V, Yamin LJ, Cauchard JR (2022) Above and beyond: a scoping review of domains and applications for human-drone interaction. In: CHI Conference on human factors in computing systems, CHI '22. Association for Computing Machinery, New York, pp 1–22. https://doi.org/10.1145/3491102.3501881

Hocraffer A, Nam CS (2017) A meta-analysis of human-system interfaces in unmanned aerial vehicle (UAV) swarm management. Appl Ergon 58:66–80. https://doi.org/10.1016/j.apergo.2016.05.011

Karaca Y, Cicek M, Tatli O, Sahin A, Pasli S, Beser MF, Turedi S (2018) The potential use of unmanned aircraft systems (drones) in mountain search and rescue operations. Am J Emerg Med 36(4):583–588. https://doi.org/10.1016/j.ajem.2017.09.025

Kerman S, Brown D, Goodrich MA (2012) Supporting human interaction with robust robot swarms. In: 2012 5th International Symposium on Resilient Control Systems, pp 197–202. https://doi.org/10.1109/ISRCS.2012.6309318

Kolling A, Sycara K, Nunnally S, Lewis M (2013) Human swarm interaction: an experimental study of two types of interaction with foraging swarms. J Hum-Robot Interact 2(2):103–128. https://doi.org/10.5898/JHRI.2.2.Kolling

Kolling A, Walker P, Chakraborty N, Sycara K, Lewis M (2016) Human interaction with robot swarms: a survey. IEEE Trans Hum-Mach Syst 46(1):9–26. https://doi.org/10.1109/THMS.2015.2480801

Lin Tan LK, Lim BC, Park G, Low KH, Seng Yeo VC (2021) Public acceptance of drone applications in a highly urbanized environment. Technol Soc 64:101462. https://doi.org/10.1016/j.techsoc.2020.101462

McRae JN, Gay CJ, Nielsen BM, Hunt AP (2019) Using an unmanned aircraft system (drone) to conduct a complex high altitude search and rescue operation: a case study. Wilderness Environ Med 30(3):287–290. https://doi.org/10.1016/j.wem.2019.03.004

Mishra B, Garg D, Narang P, Mishra V (2020) Drone-surveillance for search and rescue in natural disaster. Comput Commun 156:1–10. https://doi.org/10.1016/j.comcom.2020.03.012

Naghsh AM, Roast CR (2009) User interfaces for robots swarm assistance in emergency set- tings. In: Proceedings of the 23rd British HCI Group annual conference on people and computers: Celebrating people and technology, BCS-HCI '09. BCS Learning & Development Ltd., Swindon, GBR, pp 324–328

Oury JD, Ritter FE (2021) How user-centered design supports situation awareness for complex interfaces. In: Building better interfaces for remote autonomous systems. Springer International Publishing, Cham, pp 21–35. https://doi.org/10.1007/978-3-030-47775-2

Pham HX, La HM, Feil-Seifer D, Deans M (2017) A distributed control framework for a team of unmanned aerial vehicles for dynamic wildfire tracking. In: 2017 IEEE/RSJ International conference on intelligent robots and systems (IROS), pp 6648–6653. https://doi.org/10.1109/IROS.2017.8206579

Rule A, Forlizzi J (2012) Designing interfaces for multi-user, multi-robot systems. In: Proceedings of the seventh annual ACM/IEEE international conference on human-robot interaction, HRI

'12. Association for Computing Machinery, New York, pp 97–104. https://doi.org/10.1145/2157689.2157705

Scherer J, Yahyanejad S, Hayat S, Yanmaz E, Andre T, Khan A, Vukadinovic V, Bettstetter C, Hellwagner H, Rinner B (2015) An autonomous multi-UAV system for search and rescue. In: Proceedings of the first workshop on micro aerial vehicle networks, systems, and applications for civilian use. ACM, Florence Italy, pp 33–38. https://doi.org/10.1145/2750675.2750683

Silvagni M, Tonoli A, Zenerino E, Chiaberge M (2017) Multipurpose UAV for search and rescue operations in mountain avalanche events. Geomat Nat Haz Risk 8(1):18–33. https://doi.org/10.1080/19475705.2016.1238852

Soorati MD, Clark J, Ghofrani J, Tarapore D, Ramchurn SD (2021) Designing a user- centered interaction interface for human–swarm teaming. Drones 5(4):131. https://doi.org/10.3390/drones5040131

UAS for SAR (2016) https://www.dco.uscg.mil/Our-Organization/Assistant-Commandant-for-Response-Policy-CG-5R/Office-of-Incident-Management-Preparedness-CG-5RI/US-Coast-Guard-Office-of-Search-and-Rescue-CG-SAR/CG-SAR-2/UAS-for-SAR/

Part III
Creative Platforms and Their Communities

Chapter 12
Culture and Technology: Curating New Media in Collaborative Ways

Deborah Turnbull Tillman

Abstract This chapter highlights a collaborative approach to curating robotic art that can occur due to the disruptive, responsive and interdisciplinary nature of its mediums and practitioners. Each of the case studies presented speaks to the exploratory capability of the artists, designers and engineers of this contemporary art form. Due to the experiential nature of the medium, the necessity for specialist (curator) and subject (artist/artwork) in the curator-artist relationship can be disrupted in favor of working together to facilitate experiences, stage experiments, build data, and extend often tension-filled experimental public practice into part of the cultural experience. Simply put, there is more than one expert at the table when these exhibitions are designed for human engagement.

12.1 Introduction

Exhibitions featuring robotics tend to be commercial, popular, large scale, open to the public and organized around seasonal festivals or technology-based trade events. Including 'A'rt and 'D'esign in the consideration of Cultural Robotics shifts this focus, causing this practice to exist instead at the nexus of experimental practice, collaboration, and the creative use of metal, wires, and software as transdisciplinary expressive materials.

The following section explores the communities and platforms that have cropped up to support artists working collaboratively with robotic materials. These explore academic communities, studio practice, living laboratories, curator-artist relationships, and iterative and reflective modes of practice. The case studies comprise interviews with artists, curators, and technologists; encompassing the reflective practice of artists working at the cutting edge of robotic design, such as soft robotics, sound design informed by musical genres, the tension of human–robot interaction performativity, and curators exploring what it means to display these experimental

D. T. Tillman (✉)
Creative Robotics Lab, University of New South Wales, Sydney, Australia
e-mail: deborah.turnbull@unsw.edu.au

© The Author(s), under exclusive license to Springer Nature Switzerland AG 2023
B. J. Dunstan et al. (eds.), *Cultural Robotics: Social Robots and Their Emergent Cultural Ecologies*, Springer Series on Cultural Computing,
https://doi.org/10.1007/978-3-031-28138-9_12

designs as social constructs worthy of general and specialist opinion. I expand on their relationship to my own curatorial practice case by case.

The shift from specialist curator to co-producer of experience is a recently articulated one. Where many artists and curators have worked in this capacity for quite a while, it is only with the furor of data tracking and analysis that we can try to articulate how and why audiences engage with interactive works in the way they do. Where surveys used to exist within the oeuvre of marketing interns, they are now blueprints for feedback cycles in models for the engineering and coding of product testing which some of the materials necessitate.

Academically, this work has been wholistically addressed in the creative, written, and editorial works of Ernest Edmonds and Linda Candy. Together they pioneered the inclusion of audiences in the feedback cycle of iterative artworks, documenting it within the Springer Cultural Computing series, the very series this book resides in. Other books addressing this integral theme to this emerging medium of responsive art include *Interactive Experience in the Digital Age: Evaluating New Art Practice* (Candy and Ferguson 2014), *Curating the Digital: Space for Art and Interaction* (England et al. , 2016), and *Museums and Digital Culture: New Perspectives and Research* (Giannini and Bowen 2019).

Prior to these texts, Candy and Edmonds captured the lessons in the early prototype public laboratory, addressing audience evaluation in Beta_space at the Powerhouse Museum in *Interacting: Art, Research and the Creative Practitioner* (Candy and Edmonds 2011). Candy then followed these works with her own book *The Creative Reflective Practitioner* (Candy 2020) and then again editing with Edmonds in consultation with Craig Vear on *The Routledge International Handbook of Practice-based Research* (Vear 2022). Their influence on my own curatorial practice post-Beta_space and into this professional and academic inquiry is poignant and clear in the careers of the case studies highlighting the contributing practitioners in this section as well. Though these text deal with broader applications of interactivity and audience evaluation, many of the contributors to this section of the book have contributed to the Cultural Computation stream previous, now leaving their imprint within the realm of Cultural Robotics. In true Candy and Edmonds style, a new stream of Cultural Computing inquiry emerges, in the public sphere, in iterative cycles, and involving the audience as a key material. It is to them that I dedicate this section of the book.

12.2 Case Study Highlights

Leaders in the fields of experimental design look at how applied robotics as a cultural consideration allows a space for art and design to provide context to Cultural Robotics. This context can offer experimental platforms for engineering and computer science in iterative ways. It draws associations to classroom incubators, musical genres, feminism and notions of public education in terms of working across complex and ever-changing materials. These collaborations and the spaces they foster are helping to define Cultural Robotics in creative ways.

The first chapter by colleagues Anca-Simone Horvath, Elizabeth Jochum, Markus Löchtefeld, Karina Vissonova, and Timothy Merritt utilize the university classroom setting at Aalborg University in Denmark to explore the softer materials of robotics as a tactile entry point to the disciplines where art and technology intersect. In foundational classes, students follow their own designs for 'soft' robots to explore movement and materials. This provides an amazing baseline for more complex and transdisciplinary post-graduate work including studio and workshop settings. As well as contributing to guidelines for teaching and learning at a higher degree level, they also touch on the more philosophical questions of knowing and doing, and the more practical applications of engaging sustainable design principles. Interestingly, the key aim of their course is the integration of robotic technology in a socially responsible capacity.

The editors of this book perform similar tasks in UNSW Art, Design and Architecture's Creative Robotics Lab [CRL] in Sydney, Australia. What began as contributions to a stream in the discipline-wide research methodologies course for the UNSW Art and Design School came from specialist streams in the CRL now led by Brown, Dunstan, and Turnbull Tillman.

Scott Brown leads the Assistive Technology stream of research at the CRL, which aims to empower people through the creative implementation of a range of technologies and design experiences by directly collaborating with neurodiverse and disabled peoples. Workshops, such as the Autism MeetUp, hosted by Brown, link members in the lab to international design principles, using materials like LEGO™ to get to the heart of creating language around practice through design thinking methods. Inspired by a Research by Design methodology, these workshops contribute to lab culture, but are also key to areas of business such as the mission statement, rules of engagement and membership, and everyday use of the spaces for members and guests in the pursuit of research in social robotics. He lectures into the interaction course specialization in the Bachelor and Master of Design.

Belinda Dunstan is an expert in robot morphology. As a Ph.D. student, she designed a social robotics course for undergraduate students on ubiquitous tools for small scale, playful human–robot interactions utilizing platforms such as the Raspberry Pi and Arduino, titled *Social Robotics: Movement Design for Human–Robot Interaction.* The course was offered as an open elective, largely taken by double-degree media arts and computer science students, or any student wanting to understand materials, form and movement. The premise of this course was later integrated into the larger faculty offering, where Dunstan now lectures within the school of Built Environment and convenes *Computational Design: Human–Machine Interaction.* She leads the Human Futures stream of research in the CRL, which focuses on appearance design, movement planning, and the future social implications of robot design and morphology.

I am an objects and experience curator and lead the Cultural Technology stream of research in the lab. My focus is on developing criteria for the creation and evaluation of new technologies through the eyes of new media interactive art and audience engagement. I came to academia from industry, working between researchers and institutions as a key cultural liaison for the sophisticated study of human–computer interaction in the public sphere. In performing public curatorial interventions, I invite members of the CRL out of the lab and into the community to both exhibit and engage with prototype interactive artworks as part of cultural festivals in Australia and internationally. I bring a practice-led approach to reflecting on these experiences, bringing key lessons and language to my teaching across the Masters of Curating and Cultural Leadership and the media arts Honours course. I am also a key consultant on the Annual Exhibition of graduating student works at the close of each year.

The culture spoken of in the first case study by Horvath, et al., is well situated within the development of the CRL, directed by Professor Mari Velonaki, co-author on the final chapter in this section, and its influence on the broader faculty and university. Reflection on the work of the editors and the work of those at Aalborg university creates a strong argument for the emergence of academic inquiry into the materials, history, application, and future of social robotics on a global scale.

The second chapter by Wade Marynowsky, Julian Knowles, Oliver Bown, and Sam Ferguson explores the advent of socializing robotics through recognizable cultural events. In this instance, the authors explore the history of musical genres across opera, synth, and disco in relation to Marynowsky's recent works to do with musical performance. With a focus on robotic agency, these carefully choreographed works in the words of the authors, "highlight dramaturgy, choreography, robotic music gestures, and robotic musicianship are explored...in the contexts of live performance festivals and durational exhibitions" (Marynowsky, et al, abstract, p. **). The collaborative nature of the research broaches contemporary art and music, musical history, and computer science, which really emphasizes the breadth of robotic materials explored and where this kind of artwork might be platformed.

My own curatorial research has explored the work of both Marynowsky (with Knowles) and Bown (with Marynowsky and Ferguson). Marynowsky featured in my exhibition Re/Pair for the Big Anxiety Festival in 2017 (Turnbull Tillman 2017a). As the final case study of my Ph.D., the call for participation requested a work in prototype, so we could interview professional curators to build language around how they engaged with the works. To this, Marynowsky responded with the artwork *Synthesizer Robot (Synth-Bot)*, a mechanical robotic arm programmed to play a music synthesizer. This work had previously been exhibited as part of Wade's survey exhibition *Algorithmic Paraedolia* at the Incinerator Art Space (2017). I was invited to write the curatorial response to the works, which was published in the room brochure and titled *Ceci ne pas...This is not* (Turnbull Tillman 2017b). Marynowsky went on to work with Bown's algorithmic computer program Happy Brackets to create *The Ghosts of Roller Disco* (2020), a work curated into my exhibition *Never Odd or Even* at the Tin Sheds Gallery as part of international conference series TEI2020 (Tillman et al. 2020). Iterative process, as well as a deep dive into how coding as a material can either enhance or detract from the audience experience is a key piece of knowledge production that has come from working with Wade, as well as the strength that iterative process and audience feedback cycles can bring to a robotic artwork.

In the third chapter, authors Lian Loke and Dagmar Reinhardt speculate what the future role of robots might be in the performative creation of the self, mediated by robotics. As a performance artist/computer scientist and architect, respectively, together Loke and Reinhardt explore notions of intimacy and touch, of biology and machinery, of vulnerability and strength. Through the automation of an historically feminist act of applying red lipstick, this chapter features both an analysis of their artwork *code_red* (2021) and a call to action for women to participate in the design, creation, and research of social robotics. During the writing of this chapter, Reinhardt and Loke also conceived *SHErobots* with curator and author Deborah Turnbull Tillman, to which their claim of this happening in future worlds comes ever closer to arriving. Questions around what the automation of the feminine might look like, or what women might do with automation are traversed in this exhibition, in which *code_red* is featured.

Works like *code_red* are part of larger inquiry into the role of automation in the lives of women, in particular the leadership role of women in academia and industry, across practice, materials, engagement, process, and placement of robotics in society. The medium in which to take up this inquiry, being exhibition in a university gallery, allows exploration, experimentation, discussion, and publication on the topic through a feminist lens. The irony of this role falling to women is not lost on the curatorium of Reinhardt, Loke, and myself. Upon reflection on the case study of *code_red*, one can find the larger themes of tool, toy and companion as set out in *SHErobots* exhibition and written about in the exhibition catalog (Reinhardt et al. 2022). A few excerpts are captured below:

***Introduction**: We are on the brink of a new world of living with robots. Robots are moving out of factories and research labs into everyday life. Whereas digital disruption drove innovation and social change in the past two decades, the next industrial and cultural revolutions will be founded in artificial intelligence, machine learning and robotics. Now is the time to critically question who participates in the shaping of robots, and how robots will, in turn, shape humanity.* [10, p. 9].

***Tool**: Female roboticists represented in SHErobots invent and discover with industrial robots. Their research uses robots for material explorations; new material ecologies; building onsite and in outer space; communal upskilling; making their mark in industry, construction and community Female practitioners extend and deepen standard robotic makers' perspectives, and in turn, addresses alternatives. They question habitation on earth, or even in outer space. What are the resources that are available to us? What material morphologies and structural performances, what building techniques and details come into reach through the close coupling of computational design intent and workflows to advanced manufacturing and fabrication by robots? Works displayed in SHErobots stimulate a contemporary architecture and design discourse on the environment, resources, material capacities, community and habitation issues.* [10, p. 14, 15].

***Toy**: Playing with robots conceptually involves a re-imagining, and often subversion, of traditional forms of robots, what they are capable of, and how they relate to humans.*

The term toy is used here to refer to works that 'toy with' conventional stereotypes of robotics prevalent in feminist and artistic approaches, rather than the robot as a toy or plaything. It also speaks to notions of artifice, deception and subterfuge, where the surface presentation hides other motives or invisible forces. Robots are also employed in game play or creative processes of designing, making, and fabricating where the robot is either subsidiary to supporting human creativity or developed as an equal creative partner. Thinking beyond human-centric robotics to the secret life of machines opens a parallel world in which robots are free to behave as they wish, no longer subservient to human desires. In this future scenario, the robot is truly un-slaved to play. [10, p. 52].

***Companion**: As a social species, what makes us feel safe, supported, loved, or attended to? When one initially thinks of companion robots, social/sexual companions, health workers, and frontline customer service come to mind. When these are intimate in relation to bodies and our enjoyment and maintenance of them, what does the automation of cultural engagements bring to our experiences? The idea of a companion can be intonated, ambiguous, or poetic. A number of installations in SHErobots lay the groundwork for consideration of the reciprocity that is a key factor in human companionship. These artworks explore whether this is also the case for human–robot companionship.* [10, p. 84].

This exhibition is explored more fully in **the final chapter** which looks at the display of robotic materials via the exhibition platforms of a museum and an art gallery. Key case studies looking at robots as objects of curiosity, objects designed as artworks, and the hybrid works of artists like Stelarc and Elena Knox, author Wade Marynowsky and author Mari Velonaki testify to the trans-nature of robots as culture. Along with annual awards, collection objects and international festivals from the Powerhouse Museum, and the exhibition *SHErobots* at Tin Sheds Gallery, conceived with authors Dagmar Reinhardt and Lian Loke, and two key interviews by Turnbull Tillman inform the chapter. The first is with Director of Curatorial at the Powerhouse Museum, Matthew Connell in 2015, and the second with artist and author, Mari Velonaki in 2020. The display of such objects reveals a "context of collaborative making, audience engagement and notions of authenticity that make them social, and by extension, cultural." (Turnbull Tillman and Velonaki, abstract).

Below are excerpts from the interviews that were the base of the final chapter. They illustrate my own practice-based research to curating and my relationships with key industry figures in the Museums and Galleries sector of industry. This research activity was previously explored in Candy's publications, specifically in Candy and Ferguson (2014) and more broadly in Candy and Edmonds (2011), and with Velonaki in England et al. (2016).

12.2.1 Interview with Matthew Connell, Principal Curator

By Deborah Turnbull Tillman.

Powerhouse Museum, MAAS, 11 September 2015.

Robots and Culture.

DTT: [W]hen I asked you to write this chapter with me and we started thinking about previous exhibitions involving robotics, I didn't feel like we had really curated robotics; more what I would consider art or objects that start with art in mind, the materials used, and the implication [that] those materials might [encourage] some sort of other interactive engagement. But for me, it always starts and ends with art. I know that … your mind would go across disciplines immediately, right away, because of the nature of your work and the nature of my approach to your work. [Y]ou mentioned a few things to me, and I just wondered if you could talk to them. The question is: what exhibitions come to mind when I asked you about [robots and culture]?

MC: As I recall, I reminded you about Stelarc's 'Articulated Head' and I would add to that the other installation he did with the rhumbas [and Erin Gee] with the iPad screens with [their] faces on [the screens]. There was a little bit of social robotic interaction so the robots would talk to each other or talk to the [audience] in [either Stelarc's voice or Erin's opera singing].

I also reminded you that you had curated the Experimenta exhibition that had gone into ISEA2013 here at the Museum. That included Wade Marynowski's … Acconci Robot, which was one of the highlights. [D]espite [being] a robot, the work was essentially a box; a packing crate on its own floor. [This is a] contrast to his quite elaborately decorated crinolines that were so overtly spectacular.

It did beg the question and was intriguing in and of itself. Every now and then there would be a scream of delight amazement and sometime fear when [the audience] approached the robot and tried to look in little holes and work out what it was. [Then, when they] turned around to walk away, that's when the packing crate would then follow the participant [on hidden wheels]. It was like something out of Dr. Who, but beautifully done and a lovely piece of art; particularly from the viewpoint of this Museum. Its appeal was in the experience, it was contemplative, it was intriguing, and it didn't require art appreciation or an art history degree to understand it, approach it or engage with it as a piece of work.

DTT: It was also a recognizable object, as we've got lots of packing crates around the Museum.

MC: [Yes], for us it was very accessible, delightful and appealing across the board. I think it provoked a sense of wonder. Those were the two I mentioned (*Turnbull Tillman* , 2015).

12.2.2 Shifting Conversations in Robotics with Culture

Deborah Turnbull Tillman and Mari Velonaki.

Recorded 23 and 27 April, 2020.

DTT: How do you think the conversation around multi-disciplinary practice has changed between 2003 [the start of your career] to now? [What might be missing?].

***MV**: Because we're talking about social robotics now, you need to have teams made up of different disciplines. [Y]ou have people from the arts and design; you have psychologists; you have you have AI-experts; you have roboticists and mechatronic experts. When we start designing systems for the near future that have agency, that both interact with people to assist them, to protect them, to rehabilitate them; and to inject in there this element of creativity and start thinking of experiential design in the everyday. How, then, do we enhance this find?*

I'm not talking about pseudo-artistic experiences here; I'm talking about working together from different fields. When I'm designing something for a museum, someone interacts with the work for maximum 20 min, half an hour, right? Or they may revisit the work over time. This could happen with a work like Fish-Bird because its immersive, people spend time reading the writing [the poetic phrases the chairs printed out to communicate with each other]. People spend that much time because of the narrative and the storytelling and the performative aspect of the interaction. I'm giving this extreme positive scenario for [a designed] cultural or creative engagement.

It's very different to design something that people need to help them, or something that they need to share their space or the workplace with for the next 10 years. So again, you have creative systems, and that system is one that can create an experience and has a sense of renewal in there and the learning needs to be a part of it. We learn a lot from how people interact, because humans get bored very easily. Within a museum space, as a curator and an artist with works in different countries, we've learned a lot about principles. Now there are some basic principles about engagement; but surprise, comfort, and an emotional kind of elevation tend to feature.

However, it's much more than that. We start from an emotive reaction and then you realize that every work is different. That's why I find it exciting. Also, that aspect of creativity, [often realised in museums and galleries] I could prioritise as more important than [business outcomes]. But now with the [Creative Robotics] Lab and what we do at the [National Facility for HRI Research], if we talk about the new field of social robotics, it is by definition multi-disciplinary. ***I'm trying to be an advocate that art and design are not just there for decorative purposes****. We are not there to make things look better. We're not there to tick the politically correct box. We're there because we know how to create an experience. This experience is part of what is missing in social robotics.* (Turnbull Tillman, Deborah 2020).

12.3 Conclusion

Whether in a university or in the public eye, this section highlights the collaborative communities it takes to produce complex machines such as robots, and how much care it takes to contextualize them as social, and by extension cultural. As sites that humans gather around to remember, reflect, and re-imagine their histories, present and futures, museum and gallery floors as performative spaces elevate that culture and distribute it en masse. Here artists, designers, and curators can come together to contextualize these narratives. Told in the iterative stages of the materials they are comprised of, robotic art and the humans that make, experiment with and progress it, are captured in the narratives herein.

References

Candy L(2020) The creative reflective practitioner. Routledge Publishers, Oxford and New York

Candy L, Edmonds E (eds) (2011) Interacting: art, research and the creative practitioner. Libri Publications. Oxfordshire, UK

Candy L, Ferguson S (eds) (2014) Interactive experience in the digital age: Evaluating new art practice. Springer Series on Cultural Computing; Springer International Publishing, Switzerland

England D, Schiphorst T, NBryann-Kinns (eds) (2016) Curating the digital: Space for art and interaction. Springer Series on Cultural Computing; Springer International Publishing, Switzerland

Giannini T, Bowen J (2019) Museums and digital culture: new perspectives and research. Springer Series on Cultural Computing; Springer International Publishing, Switzerland

Reinhardt D, Loke L, Turnbull Tillman D (2022) SHErobots: tool: toy: companion. exhibition 20 October—10 December 2022, Published by Tin Sheds Gallery, University of Sydney, Australia, ISBN: 978-0-6455400-5-5

Turnbull Tillman D (2015) Robots and culture. Interview with Matthew Connell, Principal Curator, Powerhouse Museum, MAAS, 11 Sept 2015. Full transcript at: https://debturnbulltillman.wixsite.com/newmediacuration/post/robots-and-culture-an-interview-with-matthew-connell. Accessed 27 Jan 2023

Turnbull Tillman D (2017a) Re/Pair, The big anxiety festival, 8–10 November 2017a. Black box theatre, University of NSW, Art & Design campus. Sydney, Australia. https://www.thebiganxiety.org/events/repair/. Accessed 27 Jan 2023

Turnbull Tillman D (2017b) Ceci n'est pas...This is not. an essay on Wade Marynowsky's *Algorithmic Paraedoilia*, Algorithmic In: Paraedolia H-LC (ed) Willoughby council. Sydney Australia. Download available at: https://www.academia.edu/34794651/Ceci_Nest_Pas_This_is_NOT_an_essay_on_Wade_Marynowskys_Algorithmic_Pareidolia

Turnbull Tillman D (2020) Shifting conversations in robotics with culture. Interview with Mari Velonaki, online, 23 and 27 Apr 2020. Full transcript at: https://debturnbulltillman.wixsite.com/newmediacuration/post/shifting-conversations-in-robotics-with-culture. Accessed 27 Jan 2023

Turnbull Tillman D, Schiporst T, Cochrane K (2020) Never odd or even—Co-Curator of the TEI2020 arts track exhibition, exhibited at: Tin Sheds Gallery. University of Sydney, Sydney, 09 Feb 2020—12 Feb 2020, https://tei.acm.org/2020/participate/arts-track/. Accessed 27 Jan 2023

Vear C (2022) With consulting editors Linda Candy and Ernest Edmonds (2022). In: The Routledge International Handbook of Practice-based Research. Routledge Publisher, Oxford and New York

Chapter 13
Soft Robotics Workshops: Supporting Experiential Learning About Design, Movement, and Sustainability

Anca-Simona Horvath, Elizabeth Jochum, Markus Löchtefeld, Karina Vissonova, and Timothy Merritt

Abstract *Soft Robotics* is a class of robotics in which flexible materials make up some or all parts of the structure. Soft materials afford more flexibility than the rigid materials used in traditional robots, enabling new shapes, sizes, and interactions across a range of applications. From a pedagogical perspective, soft robotics is a good entry point for teaching robotics due to its simplicity and low production costs. Students can get started with programming a robot of their own design in a very short time, which provides a good case study for increasing accessibility in education. For the past five years, we have incorporated studio-based courses on soft robotics in an undergraduate education of Art and Technology, and most recently in a transdisciplinary workshop focused on robots and sustainability. Results from several iterations of the course help to identify key challenges and pedagogical opportunities for developing and teaching transdisciplinary courses in higher education. We contribute guidelines for conducting a successful soft robotics course that supports transdisciplinary teaching and learning activities that foster critical engagement with sustainability. We discuss the integration of different types of knowing and doing, and propose art-based and design-based research methods as useful tools for developing

A.-S. Horvath (✉) · E. Jochum
Department of Communication and Psychology, Aalborg University, Rendsburggade 14, Aalborg, Denmark
e-mail: ancah@ikp.aau.dk

E. Jochum
e-mail: jochum@ikp.aau.dk

M. Löchtefeld
Department of Architecture, Design and Media Technology, Aalborg University, Rendsburggade 14, Aalborg, Denmark
e-mail: mloc@create.aau.dk

K. Vissonova
Institute for Advanced Design Studies, non-profit, Fürj Utca 2, 1124 Budapest, Hungary
e-mail: colab@umwelten.art

T. Merritt
Department of Computer Science, Aalborg University, Selma Lagerlöfs Vej 300, Aalborg, Denmark
e-mail: merritt@cs.aau.dk

© The Author(s), under exclusive license to Springer Nature Switzerland AG 2023

B. J. Dunstan et al. (eds.), *Cultural Robotics: Social Robots and Their Emergent Cultural Ecologies*, Springer Series on Cultural Computing,
https://doi.org/10.1007/978-3-031-28138-9_13

sustainability in transdisciplinary settings. We outline activities that support learning outcomes including digital fabrication, expressive movement, embodied interaction, and design for sustainability. We conclude by mapping future development and open questions linking soft robotics, sustainability, and creative expression with the aim of integrating cultural and sustainable principles throughout the design process.

13.1 Introduction

Technology in general, and robotics in particular, is often assumed to play key roles in fighting climate change, facilitating the green transition, and enabling more sustainable futures (Anderson and Wanscher 2020; Bausys et al. 2019; Bugmann et al. 2011). However, it is not widely understood how, exactly, these technologies can contribute to sustainability, or what new sustainability challenges they might introduce. Climate change and environmental degradation require serious, committed, and cross-sectoral actions to balance technological and cultural outcomes with environmental and economic priorities. This balance, however, poses value tensions between social, economic, and environmental sustainability (Hopwood et al. 2005). A just inter- and intra-generational distribution of resources (Poel 2017) will require cultural changes that directly and immediately affect how we produce and consume goods, services and technologies. In (Dunstan et al. 2016), Dunstan et al. describe a cultural robot as a *robotic entity that participates in, and contributes to, the development of material and/or non-material culture.* Culture is recognized as a complex and integral consideration in the design, application, and advancement of social robotics (Dunstan et al. 2016). Understanding robotics as both producers and products of cultures, we consider how the design of robots in educational contexts can promote cultures of sustainability at the level of curricula in higher education. Robots materialize the tensions and complexity of sustainability. For example, new robotic technologies allow for precision farming (Galati et al. 2021) or precise fruit harvesting (Zhou et al. 2021), but these devices are often not sustainable from a material perspective because they are made from rare materials, sourced and assembled in different parts of the world. Once their end of life is reached, robot technologies advance environmental degradation through the production of e-waste. However, the use of robots can also help to minimize food waste, and thus relieves some pressure from intensive agriculture. From the standpoint of the economic sustainability of manufacturers, while it might be ecologically preferable to develop a product that is long-lasting or that follows a circular economy approach, this often requires completely new business models and manufacturing techniques to make these approaches economically sustainable for the companies (Bocken et al. 2016). At the same time, farming robots threaten to supplant low-wage laborers, and thereby could be considered unsustainable on a social equity level. Thus, we recognize how approaching the topic of sustainability in relation to designing technological artifacts, and STEM curricula more broadly, requires dealing with complexity and a systems thinking approach that leverages insights from different disciplines and perspectives.

Prioritizing sustainability when making technological artifacts is becoming increasingly important. There is a new emphasis on the use of non-renewable resources, emissions, and pollution from the types of manufacturing techniques and types of materials used (Hartmann et al. 2021), as well as concerns for the impacts of the scale of transportation of goods and accumulation of waste. Sustainability can be understood as a set of criteria applied to artifact designs and may be verified by certification bodies, such as "Cradle to Cradle" (McDonough 2002) or through methods such as life cycle assessment (Klöpffer 1997). In robotics education, however, environmental priorities still lag behind technological advancements. As Pujol and Tomás and Filho et al. show (Alves Filho et al. 2018; Pujol and Tomás 2020), development is largely technology-driven rather than sustainability-driven. Introducing sustainability principles early and often in educational contexts therefore might be a catalyst for generating new ideas and solutions.

Pujol and Tomás (2020) describe their efforts to introduce sustainability in robotics education. They describe a program where students are tasked with designing an environmentally friendly robot by choosing one of the 17 Sustainable Development Goals (SDGs) and designing a robotic solution that focuses on integrating social and economic aspects. Their approach shows how integrating sustainability aspects in technology education can act as a motivation for gender-diverse participants, with the potential to broaden engagement and participation and thus promote accessibility and a more even skill distribution. However, in their work, the participants only created concepts for robots rather than functioning prototypes.

Building on this work, we developed a transdisciplinary workshop curriculum based on experiential learning on the topic of design and development of soft robots in the context of sustainability. Soft robots are autonomous agents where significant material deformation is an integral part of the robot's function (Schultz et al. 2016). To achieve their "softness", silicone is one of the preferred materials (Rus and Tolley 2015), which is hard to recycle or reuse. While new alternative materials are currently under development, silicone remains a common material choice (Hartmann et al. 2021). This makes soft robotics a good platform for engaging with environmental sustainability focusing on engineering-related issues or material choices, as well as economic and social sustainability. We adapted a constructivist learning workshop that we ran annually for three years that focused on the intersection of soft robotics and artistic expression. The revised workshop was developed to include a focus on sustainability; in this chapter, we discuss its key outcomes.

13.2 Background

13.2.1 Soft Robotics

The defining characteristic of soft robots is that they are built from highly compliant materials with elastic properties that are close to soft biological materials (Rus and

Tolley 2015). They are often biomimetic, or directly inspired by living organisms and have a particular set of features that enable them to be employed in use cases where classical robots built with rigid materials fail. Being "soft" is a characteristic that has increasing importance in robotics, especially if the robot is meant to interact with humans or with fragile objects or environments (Dragan et al. 2013; Laschi et al. 2016; Milthers et al. 2019). One of the most prominent applications of soft robots is to be used as a robotic gripper, as their softness affords grasping irregularly shaped and fragile objects.[1] The field has seen a dramatic increase in popularity and research in the last several years, possibly linked to the wide range of potential applications from health technology in surgery (Cianchetti et al. 2014), biomedical fields (Eshaghi et al. 2021), rehabilitation (Maeder-York et al. 2014), architecture and design (Decker 2015), to exoskeletons (Awad et al. 2017), among others (Laschi et al. 2016). In the current state of soft robotics, movement is most often the primary communication channel, which accounts for the limited capabilities of soft robots to communicate their intention and future actions to human collaborators. So far, relatively little work in human–robot interaction research has focused on soft robots, with some notable examples including work from Milthers et al. (2019) and Jørgensen (2018). This complex multi-modal interplay of different interaction partners with asymmetric communication capabilities is a general issue in human–robot interaction and the design of robots in general. Therefore, while soft robots have a large potential to contribute to sustainable solutions, they also have several limitations. To overcome these limitations, new and transdisciplinary solutions are required, which necessitate the engagement of different communities of practice, which can have different understandings of what sustainability is (Vite et al. 2021).

The soft robotics community realized the need and potential for transdisciplinary collaborations early on. Yu et al. established that soft robotics is a well-suited topic for robotics education (Yu et al. 2014), and Holland et al. developed the soft robotics Toolkit in 2014 (Holland et al. 2014), which has been continuously extended with additional resources (Holland et al. 2017) that have been made available online.[2] The overall community has grown in recent years with researchers contributing material and using content in teaching. Especially successful were competitions at different conferences that invited knowledge sharing and building new collaborations (Holland et al. 2018).

From a prototyping perspective, soft robots can be built from a variety of different materials; the most common approach today is to use silicone (Ilievski et al. 2011; Rus and Tolley 2015; Zhang et al. 2017). But other approaches, including even edible actuators made from gelatin, have been proposed (Shintake et al. 2017). Hartman et al. reviewed sustainable alternatives, ranging from microbots operating in vivo, to fully biodegradable actuators (Hartmann et al. 2021). However, they also highlighted large gaps in current technologies that stretch design, materiality, and repairability properties that have sustainability at the core.

[1] https://www.festo.com/group/en/cms/12745.htm.

[2] https://softroboticstoolkit.com/about

13.2.2 Sustainability and Soft Robotics

The conviction that robots will have an important role to play in the Green Transition (Anderson and Wanscher 2020; Hartmann et al. 2021) has yet to be proven. Part of the difficulty of evaluating the potential of digitalization in general, and soft robotics in particular, is related to the paradoxes surrounding sustainability. While robotic technologies might contribute to waste reduction, minimizing the use of resources and CO_2 emissions, re-using materials, and extending the lifecycle of products, the technologies themselves are not sustainable as such. Disposable technologies make up an increasingly large proportion of our waste. As Hartman et al. (2021) point out, "the increased integration of such soft robots in our everyday life raises, in close analogy to consumer electronics, environmental concerns at the end of their life cycle". Furthermore, the introduction of soft robots into the agricultural sector, where soft actuators mean the ability to replace human labor in delicate tasks such as harvesting and fruit handling, has the potential to contribute to cascading problems that adversely affect communities who rely on this labor. The sudden introduction of new technology into a sector can have profound consequences on human and societal factors. In order to develop a holistic understanding of the effects of robot technologies, it is important to consider how they contribute to shared prosperity, not just for industry but also for the regions and people whose work is transformed.

13.3 Method

In this section, we describe the progression of the educational curricula leading up to a transdisciplinary workshop on *Sustainable soft robotics*, as well as the workshop itself, including the materials used, schedule, participants, and evaluation methods employed.

13.3.1 Soft Robotics and Art Workshops

Building upon previous work on soft robotics in education, we integrated a soft robotics workshop into a robotics and art course. We initially developed the course according to the problem-based learning (PBL) model within the context of an Art and Technology Bachelor of Arts Education. PBL (Bumblis 2005) provides a strong framework for transdisciplinary courses that combine critical thinking and problem-solving skills with hands-on experiments and practice (Jespersen 2018). The tenets of PBL relate to concepts from experiential learning (Jochum and Putnam 2015) and constructionist learning approaches (Löchtefeld et al. 2021). The earliest iteration of this course is described in Jochum and Putnam (2015).

One goal of the course is to provide students with fresh perspectives on how to engage with emerging technologies and to translate creative theories and approaches into practical results. The course deliberately integrates scientific and artistic methods, placing equal emphasis on both aesthetic and technical concerns that allow students to develop artifacts that can be used as tools for conducting research in human–robot interaction. Our decision to work with soft materials and focus on the design and fabrication of soft robot actuators, rather than traditional robotics, provides a solid foundation for a constructivist approach to active learning that emphasizes embodied knowledge, experiential learning, and the interconnectivity of seemingly dissimilar disciplines (Ackermann 2001) (Löchtefeld et al. 2021).

13.3.2 Materials

The course materials and objectives were developed using the teaching materials and activities originally developed for the soft robotics Toolkit from Holland et al. (2014) and (Holland et al. 2017) (Fig. 13.1). The course was co-conceived with a fellow researcher, Jonas Jørgensen, with experience in transdisciplinary research in art and robotics (Bering Christiansen and Jørgensen 2020). The dimension of sustainability was not initially a focus; we originally focused on the unique perspectives on form, design, movement, and context that artistic approaches bring to the emerging field of soft robotics. The course is organized around specific themes and exercises to do with artistic and computational approaches to designing expressive motion and behaviors.

Fig. 13.1 Materials used for the soft robotics workshop. Left and center: electronics and oven used for curing silicone. Right: silicone actuators and pumps

13.3.3 *Integrating* Sustainability *with Soft Robotics and Art Workshops*

Based on the robotics and art course, and in continuation of it, we conducted two workshops with the theme *Sustainable soft robotics*. We quickly became aware of how inherently unsustainable the common educational practices of designing and building soft robotics are. Therefore, we decided to bring in experts on sustainability to help us design sustainable soft robots but also to experiment with more diversity and see what results would emerge if participants with different backgrounds came together to share knowledge, conceptualize, build, and learn. This first workshop was a trial run organized by seven researchers and had five participants from industry and higher education. This trial was run in preparation of the second workshop. Figure 13.2 shows how these workshops have evolved over the years.

In this subsection, we give a detailed account of the outcomes, lessons learned, and curricula for the second iteration of the *Sustainable soft robotics* workshop and we touch upon how the trial workshop helped us to refine this curricula as well as our understanding of transdisciplinary work.

The team that conducted the workshop was comprised of researchers, with two authors who worked on the robotics and art course and five other colleagues. In total, seven researchers whose areas of investigation were: (1) constructive design, (2) computational design and architecture, (3) design philosophy with a focus on sustainability, (4) theater and performance with a focus on movement, robotics and art, (5) computing technologies and human–computer interaction, (6) robotics

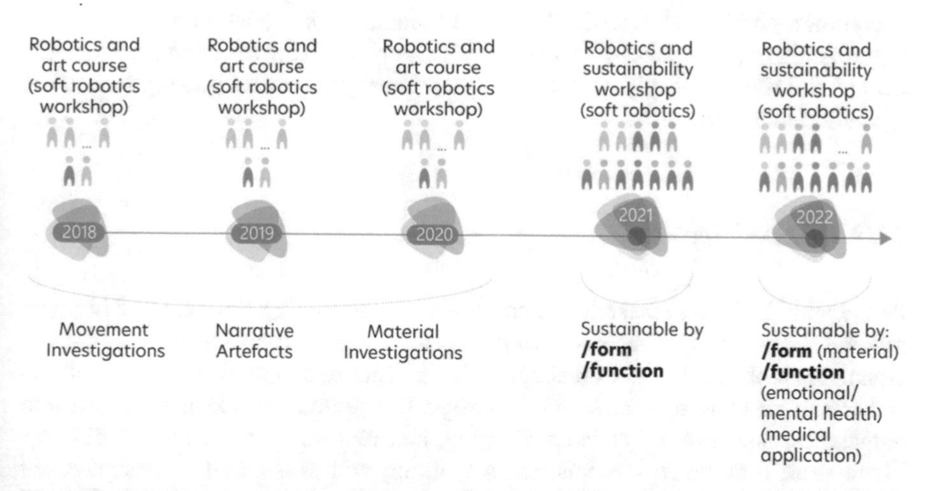

Fig. 13.2 Overall progress of the curricula and syllabus. We start in 2018 by integrating soft robotics with art, and after a few iterations of the course, we introduce sustainability in the syllabus. The first three iterations are run by two of the authors, with undergraduate students of an Art and Technology program. The fourth and fifth iterations are run by seven researchers coming from different fields with diverse student populations. Bottom: the projects developed by the students as part of the course and workshops fit under different main themes

and human-robot interaction, and (7) nature-based solutions for systems and urban design. Together, we designed the curricula and conducted a three-day workshop that focused on sustainability, soft robotics, and art. It included lectures and hands-on work as well as guided and open ideation sessions. Similar to the course, participants in the workshop worked in groups, on projects, and in a problem-based learning setup. We ended each day with 30-minute critical reflection sessions by sharing thoughts and feedback on how the day went. At the end of the three days, the groups presented their works to each other.

As a design challenge, participants were tasked with designing soft robots that engage with sustainability. In an introductory lecture on sustainability for the design of artifacts, two conceptualizations of sustainability were introduced, namely *sustainability by form* and *sustainability by function* as discussed in Vissonova (2018). In the first approach, sustainability is addressed through the qualities of an artifact's *form* or its structure. This relates to the environmental impact across the product development life cycle (Gurova et al. 2020), including the impact of sourcing materials, biodegradability, recyclability, and reusability of the materials that make up the components and the body of the artifact. Examples of such artifacts are goods made for disassembly, made from biodegrading materials, or materials that comply with infrastructures for recycling. In the case of soft robots, biodegradable actuators, molds or robot bodies would fall under this category.

A second approach is where an artifact is designed to perform a *function* that one way or another alleviates a problem dealing with social, environmental, or economical sustainability. The typical examples of this second category of artifacts are alternative renewable energy technologies, such as photovoltaic panels and wind turbines. A common problem with artifacts that are sustainable by function is that they are typically less sustainable by their form (Vissonova 2018). An example from the field of soft robotics would be the above-mentioned grippers for picking fruit in the agricultural industry.

13.3.4 Participants

Participants included researchers and students representing the fields of biology, robotics, art and design, and sustainability. A total of 30 participants including seven organizers took part in the workshop, including four undergraduate students of arts and humanities and one from biotechnology, two graduate students from arts and humanities, one from robotics engineering, and two from computational biology. There were four Ph.D. students, three working with arts-based or design-based research and one working within robotics engineering, and finally three professors: one professor of art, one of computational architecture, and one of biology. Four groups were formed to be as diverse as possible, each with at least one participant with a background in art, design, robotics technology, and biology, respectively. Each group was assigned with facilitators to guide discussions and project work. The facilitators and activities were selected from different fields.

13.3.5 *Schedule and Curriculum of the* Sustainable Soft Robotics *Workshop*

An overview of the schedule of lectures and hands-on activities conducted during the three days is shown in Fig. 13.3, while Appendix 1 contains the suggested bibliography introduced during the workshop.

There was time for group work each day as well as guided hands-on activities immersing participants in an iterative design exercise, movement exercises and a card-based ideation session on sustainability. The first ideation session was adapted from the "design cycle in an hour" (Girouard 2020). The activity focused on creating concepts, in groups, using pen and paper as well as puppets that students could use to express and discuss their ideas with each other. The higher fidelity prototyping of the concepts was carried out with the soft robot actuators that were controlled by an Arduino microcontroller. While the original design cycle exercise had a clear goal as if sponsored by a manufacturer, we introduced the participants to the challenge of problem framing in design and encouraged them to design soft robots and explore movement focused on either a specific problem or to use the exercise to find opportunities for new expression with the soft robots. Important parts of the activity involved the groups demonstrating and receiving feedback from others, which they could use to consider in further refinements or new directions. On the first day, one instructor conducted a guided session on somaesthetics (Höök et al. 2017) where we tried to think about movement through our own bodies. This involved a series of activities intended to sensitize the participants to non-verbal communication and expressive qualities of bodies, materials, space, and movement. The second guided

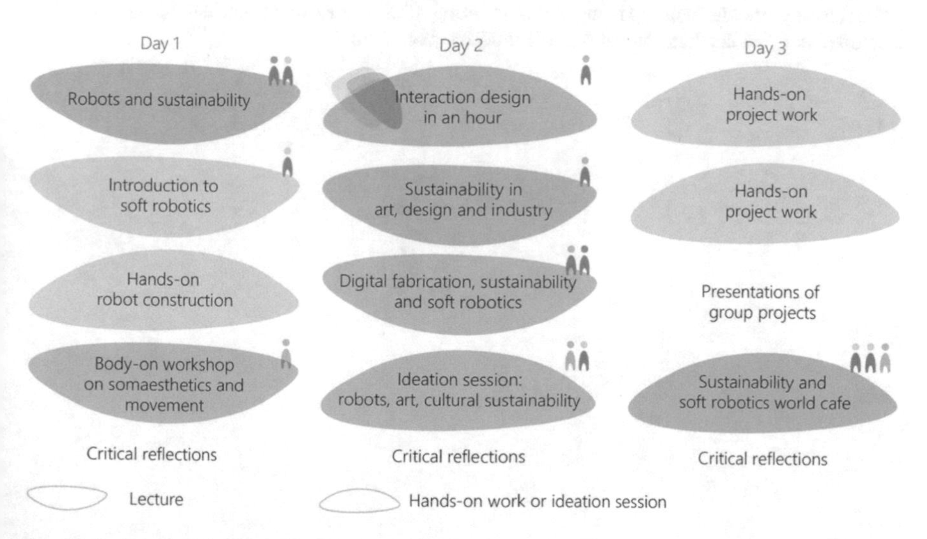

Fig. 13.3 Schedule and activities conducted during the three days of the workshop on sustainability and soft robotics

ideation session was conducted with the use of the sustainable design cards developed by Hasling and Ræbild and presented in Hasling and Ræbild (2017), Ræbild and Hasling (2019). The cards were developed with a focus on how to make sustainable development graspable in design educations and help to conceptualize sustainability by design. The groups used the cards as helpers in defining the functions and long-term and short-term effects of the design and use of their robots (Figs. 13.4, 13.5 and 13.6).

Fig. 13.4 Activities during the first day of the workshop. Top-left and top-center: body-on somaesthetics workshop, aimed at exploring movement and embodiment. Right, bottom-left, and bottom center: casting silicone in pre-existing molds, creating robot bodies, and experiencing with actuators developed in previous iterations of the soft robotics workshop

Fig. 13.5 Activities during the second day of the workshop. Top-left: ideating with the design for sustainability cards. Top center: experimenting with silicone. Top right: programming the movement of the soft robot bodies. Mid-left: exploring movement with puppets and hand actuation. Mid-center: ideation session on concept development. Bottom-left: 3D modeling the mold for a soft robot body and bottom center: 3D printing mold for a soft robot. Bottom-right: prototyping movement of the soft robot using Arduino, electronics, and puppets

Fig. 13.6 Activities during the third day. Top-left: final work on casting soft robot bodies. Center: presentation of projects. Bottom-left and right: results from the World Cafe

13.3.6 Evaluation

To evaluate the outcomes of our work we review: (1) the results of the four projects the groups developed by the end of the workshop, (2) the notes from the critical reflection sessions that took place at the end of each day, (3) the answers to the survey participants completed at the end of the workshop, and (4) the notes and feedback we received from the final reflection session, the World Cafe. In the survey, we asked open-ended questions with regard to the main lessons learned, ways to improve the curriculum, missing topics or specializations, and ways in which participants' understandings of sustainability had changed. In the last critical reflection session, at the end of the workshop, we discussed the following points: main frictions or contradictions between robotics and sustainability, connection points (themes, topics or concepts) between the different disciplines that came together, missing expertise and perspectives in defining sustainability for soft robotics, ways to avoid greenwashing[3], and finally, unified visions of possible (sustainable) futures.

13.4 Results

The results of our work include a new syllabus that integrates soft robotics with art and sustainability and a new set of activities and exercises. Through selected literature and activities our intention is to bring participants from diverse practices (from art to science to technology to design) into direct contact with the challenging intersections between soft robotics and sustainability and learn to stay with difficult/wicked problems. In this section, first, we present the four projects developed during the workshop, and second, we conduct a qualitative analysis of the notes from the critical reflections, World Cafe, and responses from the post-workshop survey, where all 30 participants in the workshop took part.

13.4.1 Sustainable Soft Robots—The Projects Developed During the Workshop

The four groups developed projects that showed different approaches to the topic of sustainable soft robots. Broadly, one project engaged with the idea of sustainability by form (the materials from which the robot was made were considered sustainable), while the other three engaged with the idea of sustainability by function (the functions the robots performed were considered to engage with sustainability).

[3] Greenwashing refers to treating the topic of sustainability superficially, making empty commitments to environmental practices for other gains.

Scrapbots: Sustainable by Form

The project that investigated sustainability by form experimented with the material of the body of the robot (Fig. 13.7). The group members integrated used silicone from other robots in the body of a newly produced robot. The group also discussed the idea of using gelatin or other biodegradable materials instead of silicone. Discussions around the durability and reliability of the material from which the robot's body would be made emerged: while gelatin is the more sustainable option (because it is biodegradable), it is also harder to control and more fragile. Gelatin would disintegrate in a few days unless kept in a cold, humidity-controlled environment. Silicone on the other hand is much more durable, easy to manipulate, and behaves in a reliable and predictable way. As one participant put it: "there is a reason we make robots from hard and durable materials". They reflected that the projected useful life of a robot is correlated to the life of the materials it is made of.

Fig. 13.7 Scrapbots—exploring sustainability by form

The Softopus: Sustainable by Function

The authors of Softopus described the project as one dealing with social sustainability. The group created a gripper that would exert a calming, soothing effect on users through a soft grip. This group spent a lot of time 3D modeling and then printing a new mold for their robot. They explored how the behavior and movement of the robot would be influenced by its external shape and not just the control algorithms that are used to actuate it. The robot could have applications in the medical industry, potentially helping with well-being, and was inspired by the third SDG (good health and well-being) (Fig. 13.8).

Tamasofty: Sustainable by Function

Tamasofty was developed as a mediating object to be applied to the care industry. The robot would mediate patients' emotions and translate them into doctor information.

Fig. 13.8 Softopus

The group tried to imagine ways in which technology could help better diagnostics and create less stressful diagnosis experiences. Conceptually, the project worked with themes such as empathy, sensory exploration, and tactile experience. Similar to the Softopus, Tamasofty aligned with the third SDG (Fig. 13.9).

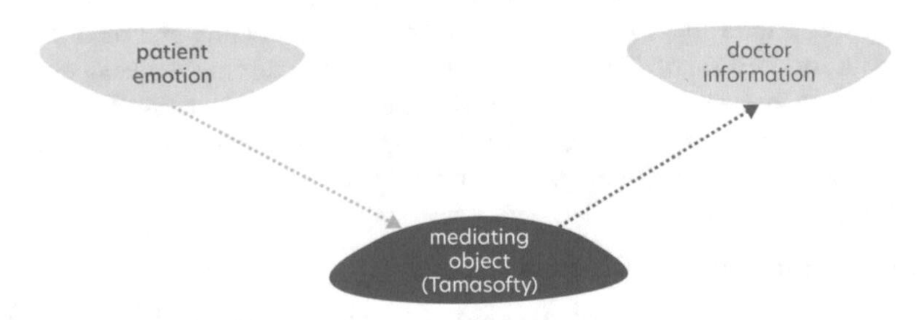

Fig. 13.9 Tamasofty is a soft robot for the care industry, sustainable by function

The Phybo Wearable Soft Robot: Sustainable by Function

The fourth project was a wearable soft robot (series/collection) functioning as a human companion. As air is blown into Phybo, it should give its user a sense of pressure, and act as a subtle reminder of "leaving your head alone and coming back to your body". Through a soft touch, Phybo would enhance the body awareness of its user thus enforcing presence through a soft touch. The form of the robot was inspired by the Fibonnaci spiral, and the group members discussed how the robot could exist at different scales. The group described their project as dealing with emotional sustainability, and similar to Tamasofty and Softopus, Phybo aligns well with the third SDG (Fig. 13.10).

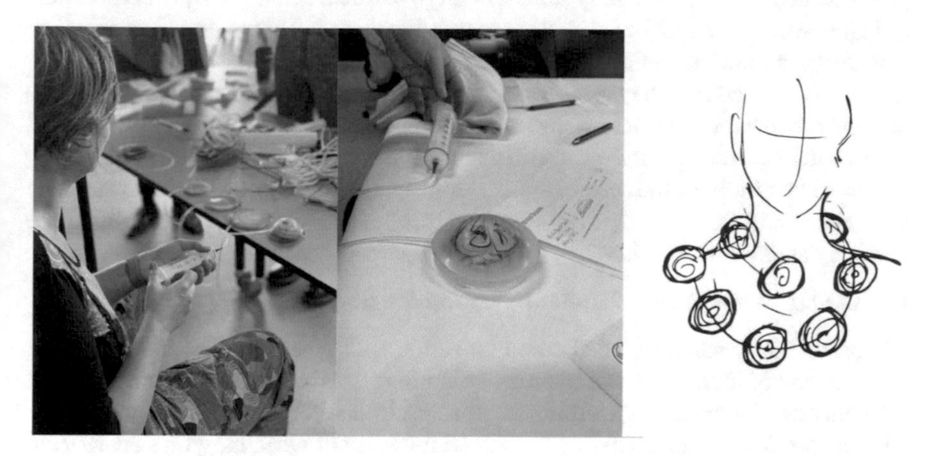

Fig. 13.10 Phybo is a wearable soft robot that deals with emotional sustainability. It promotes body awareness through a soft touch

13.4.2 Critical Reflections, Survey and World Cafe

Feedback from participants was largely positive. When asked about what were the main lessons or takeaways from the workshop, the responses related to the topics in the curricula itself (i.e., what soft robots are, how to design them, digital fabrication, somaesthetics, human–robot interaction, "that coding is not a monster"). However, in the analysis of the survey responses, coupled with our notes from the critical reflections and those from the World Cafe, we found three recurring themes: (1) a lack of consensus on what sustainability is, (2) gaining knowledge about working with others and the importance of more collaboration between fields, and (3) more time for exploration in the design phase of technological artifacts in general.

What is Sustainability?

Participants mentioned they felt "that nobody knows how to be sustainable" or that "there is no consensus about a sustainability definition", that "sustainability is a broad topic that needs to be addressed on many different levels because there are no universal solutions". Another participant stated that one of their main takeaways from the workshop was that sustainability is "complex" and about "more than environmental sustainability". Some stated that they missed "more in-depth summaries and concrete examples of sustainability", and felt that the subject was too abstract and philosophical or that "we need better language to describe what we mean by sustainability". Another participant stated that they "realized that the discussion [about sustainability] frequently moves to large political discussions, which is interesting, however, it might be interesting to consider how to contain the discussions to have focus on more specific aspects of sustainability as well as the political". Importantly, some noted that they felt that "the entire discussion about sustainability was narrowly focused on Western living and practices" and that "we think and operate with a purely academic mindset and do not put ourselves in the real world". However, it seemed that for most the sustainability design cards were a good way to ground the discussions and make the topic more graspable: "the cards invited us to think more precisely about sustainability".

Knowledge about Transdisciplinary Group Work

Almost all participants included communication with others coming from different fields as one of their main lessons learned during the workshop: "[I learned about] the concrete challenges of working in transdisciplinary groups" or "how to work with people from other fields" or "I learned more about working with people from very different backgrounds and how to adapt my communication and listening skills to fit the needs of the group" and importantly, "how cool it is to meet people from different backgrounds".

Time for Exploration

Third, many mentioned that among the lessons learned was "the need for more interaction across fields" in the design of technologies in general and soft robotics in particular, that "complex topics can be addressed via group, hands-on experimentation" but that in doing all this "the importance of time for establishing communication, and then exploration and discovery" should not be underestimated.

13.5 Discussion

Hartmann et al. (Hartmann et al. 2021) argue that becoming sustainable is the next challenge in soft robotics and that bringing diverse fields together, from material science to chemistry, engineering, biology, computer science, and robotics "will be the future challenge for autonomous robots, whether their development focuses on performance, sustainability, or both". Transdisciplinarity has been widely recommended for addressing sustainability issues (Chew et al. 2020; Hadorn et al. 2006; Keitsch and Vermeulen 2020; Vermeulen and Witjes 2020), as it is now widely accepted that no individual discipline is capable of offering a one-size-fits-all solution (Vermeulen and Keitsch 2020). To put it simply, as the Peoples Climate Movement proclaims: "To change everything, we need everyone" (https://peoplesclimate.org/), involving and giving access to decision-making to diverse stakeholders and communities of practice. However, there is much work to be done in understanding on one hand, how to scaffold collaborations between different knowledge silos and on the other, define sustainability and the steps required to support it.

Transdisciplinary education breaks down traditional disciplinary silos by integrating diverse fields and should help gain holistic knowledge, which includes input from multiple domains and stakeholders (Nicolescu 2002). Transdisciplinarity is a multi-perspective endeavor that transgresses boundaries of disciplines and bridges theory and practice. However, conducting transdisciplinary research is challenging, as many studies have shown (Pohl et al. 2021, 2020; Sellberg et al. 2021). Transdisciplinary knowledge is difficult to achieve and implement in current educational settings because it challenges the status quo of academic institutions (Jantsch 1972) and frequently falls in between departments and faculties and of current knowledge dissemination practices (Beavis and Gibbs 2020). While many people recognize the value of transdisciplinary research and teaching, there remain very real obstacles and institutional barriers that prevent biology students from taking courses in applied philosophy, or art students from following computer science courses. There are many value spheres involved in sustainability issues (Murphy 2012), and there are also many value spheres involved when undertaking transdisciplinary work. A number of studies attempt to outline best practices for conducting transdisciplinary research (Abrams 2006; Hoffmann et al. 2017; Pohl et al. 2020). The process of bringing multiple, diverse stakeholders together to creating new knowledge based on these interactions is dubbed *integration* (Bammer 2005; Pohl et al. 2020).

In order to conduct sustainability research in transdisciplinary settings, Hadorn et al. (Hadorn et al. 2006) highlight the need to examine three main areas, namely (a) the problem field in which change is needed, (b) identification of more sustainable practices, and (c) how existing practices might be transformed. Here, we identify the problem field as robotics in higher education, and used an established problem-based course to uncover how we could identify more sustainable practices within teaching and research, experiential learning about robotics and art. We synthesize the workshop outcomes and discuss how we brought together diverse views, values, and disciplinary expertise to begin breaking down some of the structures of siloed thinking and approaches. We hope our experience can prove useful for educators and practitioners in the fields of cultural robotics and human—robot interaction who want to entangle sustainability and transdisciplinarity in their curricula or practices.

Knowledge sharing takes on many forms, including details about the design or use of artifacts and also more generalized forms of intermediary knowledge such as guidelines and best practices (Löwgren 2013). Pedagogical best practices often emerge from extensive meta-analyses across a wide set of research examples (McGee and Reis 2012). We take initial steps by providing guidelines and suggestions based on our results and experiences. We identify three main considerations for conducting transdisciplinary courses: (1) integrating different ways of knowing, learning, and doing, (2) using artistic and designerly methods to foster participation and process-based inquiry, and (3) staying with the trouble and embracing transdisciplinary tensions.

13.5.1 Integrate Different Ways of Knowing, Learning and Doing

In conducting the workshop on soft robotics, art, and sustainability as a transdisciplinary setup, the development of a shared vocabulary is one of the chief priorities. The language used should be aligned so that all participants and stakeholders are on equal footing and are capable of communicating with one another. It is important to refrain from using discipline-specific jargon and identify instances where people from different fields might use different words to express similar ideas, or alternately when they use similar words to express different ideas. For example, the concept of agency has a particular semantic and latent meanings in biology that differ from how this concept is commonly understood in social sciences or robotics. Beyond language, similar issues can be investigated from different disciplinary perspectives, e.g.,, the concepts of efficiency and optimization have relevance for artistic practice as well as mechanical production, although knowledge about optimization is achieved through different means. On the other hand, knowledge that has been created in one field can help to solve problems in another field, e.g., the movement techniques developed in puppetry and theater can inform control systems for robots (Jochum and Murphey 2014). Sometimes using the modes of working from one field can help to unlock new ideas in another field (Höök et al. 2017; Horvath et al. 2020).

Transdisciplinary research takes time. We recognize that new material takes time to be absorbed, especially if the concepts are totally unfamiliar to participants. Peer-to-peer presentations and discussions among participants with diverse backgrounds help activate this knowledge in lively and meaningful ways. In our workshop we found it useful to anchor our discussions in specific themes (e.g., agency, complexity, or sustainability) and share how different fields and actors understand and approach these themes. From our experiences, we learned to allow ample time and different opportunities to learn about the different fields coming together as ultimately mapping what each field knows and how it develops new knowledge is especially important in order to ensure effective communication and meaningful collaboration.

During and following our workshops, we identified the need for skilled transdisciplinary facilitators who are able to recognize when there is a need to emphasize the diversity of perspectives and techniques during discussions and to stage situations that encourage discussion. These facilitators can also provide opportunities for different actors open to different ways of learning, knowing, conceptualizing, and problem-solving. Facilitators can be instrumental in fostering a willingness to get out of one's comfort zone and help encourage participants to lean in toward others. That our workshop made little distinction between teachers and learners in the group work and exercises helped create an atmosphere of collaboration and openness. This is similar to what (Vermeulen and Keitsch 2020) suggest when they proposed facilitators should involve stakeholders from various backgrounds. By encouraging the co-production of knowledge through linking case-specific examples with abstract understanding, participants become more open to change and understanding.

Pohl et al. (2021) call this integration of different "thought styles" and management of "thought collectives". They suggest that integration should go beyond the cognitive dimensions and that emotional and social integration are equally important to fostering transdisciplinary learning. Admittedly, this can be hard to achieve and difficult to maintain, especially during an accelerated learning or design process. One of our workshop participants commented: "the social dynamic could have been more balanced". People from similar fields that share common ground (e.g., human–computer interaction and human–robot interaction, or architecture and structural engineering) might naturally form clusters, as they already share a common vocabulary, worldview and ways of working. Close collaborations between participants who are far apart might even create tensions. However, learning to stay with the trouble might be an important factor for knowledge transfer and transdisciplinary work.

13.5.2 Use Artistic and Design Methods to Foster Participation and Process-Based Inquiry

A number of studies have suggested that using the arts in sustainability science can contribute to new understandings and new modes of engaging audiences in conversations about climate change and environmental sustainability (Curtis 2011; Heinrich and Kørnøv 2021; Heras et al. 2021; Trott et al. 2020). Others have discussed the role of design-based research (Chew et al. 2020) in transdisciplinary education. These studies show how co-design and participatory action research can foster collaborative and inclusive dialogs and exploration. Design and the arts can help to communicate with broad audiences by engaging "head, hands, and heart" (Sipos et al. 2008). Chew et al. argue for using design both as a *problem-solving logic* and as a *method* in transdisciplinary work (Chew et al. 2020).

We argue for involving the problem-solving methods of design-based research and art-based research in educational curricula that engage with robotics and sustainability. Sustainability has been described as a highly complex or wicked problem (Murphy 2012; Blok et al. 2016) and as a systemic issue: an improvement in one part of the system can result in a deterioration of another part (Murphy 2012).

Apart from distinct vocabularies, different fields have particular approaches to formulating and solving problems. Rittel and Webber describe two types of problems:

tame problems and wicked problems (Rittel and Webber 1973). Tame or benign problems are expressible as numerical data, they are definable and separable, they have clear missions, and it is clear if and when the problem has been solved. Wicked problems, on the other hand, cannot be definitively described and have no optimal solutions (Rittel and Webber 1973). Material science or mathematics are fields that can deftly work with tame problems. Urban planning, the social sciences, art, and design often work with wicked problems. This is important because when bringing diverse practitioners together in problem-based setups, there might be vastly different understandings of what constitutes a problem, how it is formulated, and whether it can be solved fully or if simply offering a new perspective, critique, or incremental improvement is enough.

It has been widely discussed that designers and artists have different approaches to solving problems compared to researchers from the natural sciences or those working within engineering (Cross 1982). Whereas in the sciences, controlled experimentation and rigid classification or analysis are expected, in the design disciplines, there may be a focus on visual representations, transformation, modeling, pattern formation, and synthesis or construction. Artists may approach their work through analogy, metaphor, narrative, and criticism. This is also connected to the nature of the problems these fields deal with. This was also obvious and expressed during our workshop by all the scientists working together with artists. One of the biologists stated: "I am not sure what we are doing [in the project], we started working on something, but we don't know where we are going, or whether we are on the right path. This is not how we work in biology, and it is very hard for me to let go and work in this way".

If we understand sustainability as a wicked problem, and robotics education and practices as usually dealing with tame problems, it might be important to engage with the exploratory methods from art and design when dealing with sustainability issues and robotics. Using art and design as methods would mean starting workshops on robotics, art and sustainability by first defining problems and common understandings of what sustainability is. It could also mean starting to work and experiment with robot design without a clear plan or output in mind. The output would then not have to be a new robot but could be a map of ways in which robot design can engage with some aspects of sustainability. If the output is a robot, it would not have to perform typical functions such as controlled movement but could be used to communicate issues of climate change and environmental or social sustainability.

In Rittel's definition, identifying or defining the problem is the same thing as finding the solution for wicked problems; the problem can't be defined until the solution has been found. In other words, *"the process of solving the problem is identical with the process of understanding its nature"*. (Rittel and Webber 1973). In our context, the process of exploring the topics of sustainability and soft robotics helped participants uncover a host of problems concerning soft robot design and application, from the inherent unsustainability of the materials used to make them, to the hollow claims that equate robotics with the green transition without specifying exactly how these technologies contribute meaningfully to environmental or social sustainability. As one participant noted, their one take-away message was that *there is not much research into robotics and sustainability. This is a question that needs further exploring*. Frequently, participants questioned whether and how robotics has an actual role to play toward more sustainable futures. This type of ongoing questioning challenges the notion that robots (or any technology) are silver bullets for ameliorating the effects of climate change, or the fallacy that efficiency and optimization are inevitably sustainable.

By bringing sustainability, understood as a value and as a set of limitations in the curricula and in the discussions, soft robots were designed “in a far more complex, and critical frame” than they are normally considered (Fry 2005). The hands-on robot prototyping, building, and fabrication grounded the topics and concepts in material practices that stimulated continuous reflection and discussion.

In our workshop, we allowed equal time and value spent on concept development, constructive approaches of sketching and modeling, as we did for discussions about ways to systematically design movements and study task performance through experiments. We also gave considerable focus to the artistic perspectives and encouraged a critical look at the materials and discussions about robotics in culture, and welcomed deeper discussions about how soft robotics can support artistic expression. To encourage aesthetic expression through movement, we incorporated movement exercises (grounded in dance, somatics, and puppetry) and used this to sensitize participants to the nuances of movement when considering how to program their robots to move.

In short, we suggest using methods from design and the arts in crafting curricula on robotics and sustainability in the following ways: (1) to facilitate participation and co-creation in transdisciplinary work, which is essential in sustainability studies, (2) as means of communicating issues related to sustainability and soft robotics and fostering critical engagement from diverse audiences, and (3) as methods of inquiry, in process-based explorations rather than outcome-based explorations. Outcome-based explorations are more common in the natural sciences and engineering.

13.5.3 Stay with the Trouble in Transdisciplinary Work

Similar to Vermeulen and Keitsch (2020), we also saw that questions on “who now holds the truth?” and “is some knowledge more valuable than other?” could emerge in transdisciplinary teaching, learning, and knowledge creation. As Haider et al. explain, “to be trained as a sustainability scientist then requires new ways of engaging with each other, with the world around us, and of reflection within our own scientific processes” (Haider et al. 2018), or in Donna Haraway’s words: “Alone, in our separate kinds of expertise and experience, we know both too much and too little” (Haraway 2015).

While we did not experience this in our work, it is important to note that in transdisciplinary setups discrimination can become a problem. Representatives of some fields might think that their knowledge and ways of thinking and doing are more legitimate than others. Academics might look down on industry representatives as not being "theoretical enough" or for being ahistorical, while industry representatives might call academics "too theoretical". As one workshop participant put it: "some of the activities were really eye-opening [...] what I fear is that they could be very divisive and create friction". As (Guimarães et al. 2019) argue, engaging in transdisciplinary work can challenge the identities of participants. Therefore, special care should go into creating a space of mutual respect because as Constanza et al. explain, "no discipline has intellectual precedence in an endeavor as important as achieving sustainability" (Costanza 1993).

We found it important to engage with systems thinking and complexity science when discussing and dealing with sustainability for robotics. It might be useful to consider sustainability in the context of paradox theory, as defined in corporate management studies: applying a paradox lens to sustainability acknowledges tensions among different desirable, interdependent and sometimes conflicting sustainability goals (Carmine and Marchi 2022; Hahn et al. 2018).

In Fig. 13.11, we attempt to give a visual representation of these intersections and possible tensions for different understandings of sustainability. This could serve as a canvas for problem formulations when starting to engage with sustainability in the design of technological artifacts, and might prove useful both for educators and practitioners. For example, as detailed previously, economical sustainability might not align with social sustainability (i.e., automating low-skilled jobs can be economically efficient for a company, but might cost jobs; designing products for longer use will ease the pressure on the environment, but might not align with current economic models), therefore social justice can be a pressure point between social and economical sustainability.

Nonetheless, the positive side of possible disciplinary tensions is that they foster critical engagement with one's own field. Ultimately, the workshop on sustainability, soft robotics, and art gave participants the chance to explore sources of unsustainability, as Murphy calls them (Murphy 2012) of soft robot design. Perhaps one of the more important outcomes of the workshop was that it led everyone to engage critically with their own practices and ways of doing and knowing as well as ask questions about technology-driven development. Designers reconsidered their pen and paper ways of ideation, technologists considered whether technology has a role in fighting climate change and if so, then what that role is. Finally, artists became more reflective of their creative practices and on ways to systematically assess the effects of their work while the scientists learned how to "let go" and work with methods and within frameworks for creative exploration.

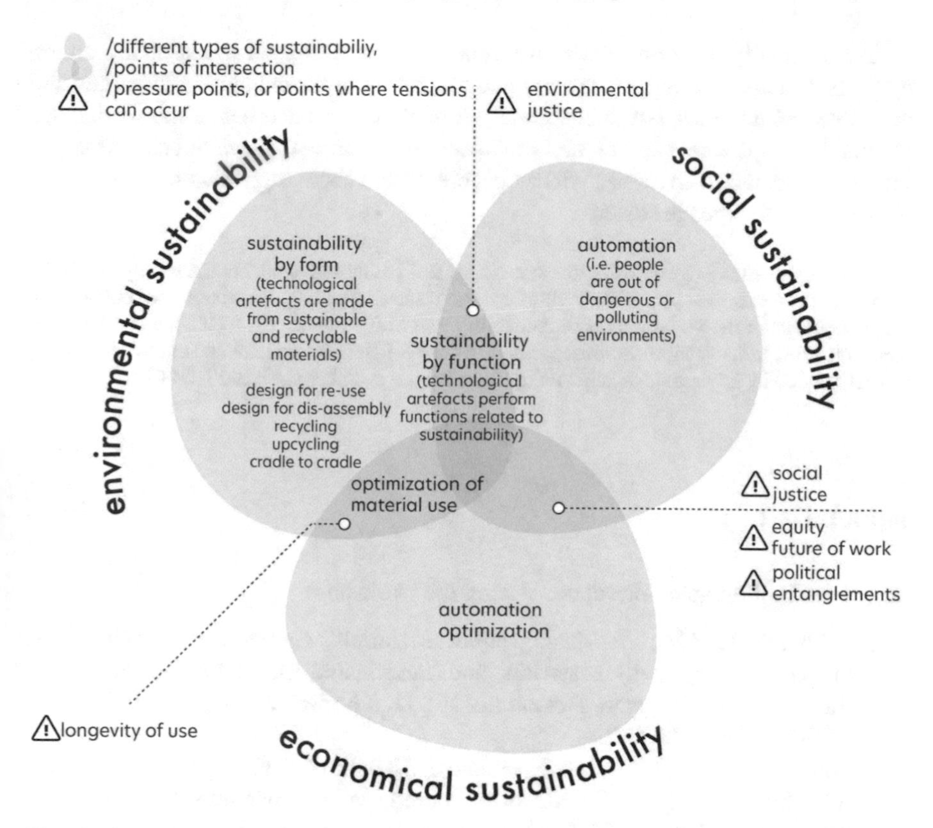

Fig. 13.11 Different types of sustainability, points of intersection, and possible pressure points

13.6 Conclusion

In this chapter, we provided highlights from soft robotics workshops we conducted over five years and described how we expanded to focus on sustainability through a transdisciplinary approach in a revised workshop. The lessons learned have been distilled into guidelines and suggestions so that educators and practitioners interested in conducting similar workshops might benefit from our experiences. In short, first we share lessons we learned about integrating different types of knowing and doing in transdisciplinary setups needed in sustainability research. Second, we show how using art-based and design-based research methods can be useful (a) in facilitating participation, (b) as a means of communication to broad audiences, fostering critical engagement about sustainability, and (c) as methods of inquiry in process-based explorations engaging with sustainability understood as a wicked, complex problem. Finally, we propose that systems thinking and paradox theory might be useful lenses when staying with the trouble that transdisciplinary tensions for sustainability bring. While the workshops we conducted were generally well-received, there are various aspects that can be improved. As teachers and educators, it will be incumbent on us to continually re-examine our curricula in a deliberate and intentional manner, not only updating our knowledge and understanding of the changes in materials and available technologies, but to continually reflect on the outcomes of the workshops we conduct in the near future, and to experiment with new ways to bring students into direct contact with messy, wicked problems and encourage them to seek out the important challenges in society.

Acknowledgements We would like to thank the Art and Technology students at Aalborg University who, over the years, took part in the Multimedia Programming course. We would also like to thank the participants in the workshop on Sustainability and Soft Robotics for their time and valuable feedback. We also thank our consortium partners in the Artificial Biology, Robotics, and Art (ABRA) project, funded by Erasmus+ Strategic Partnership Project Number KA203-2020-003.

Appendix 1

Suggested bibliography introduced during the workshop:

1. Ashby, M.F.: Chap. 1—background: Materials, energy and sustainability. In: M.F. Ashby (ed.) Materials and Sustainable Development, pp. 1–25. Butterworth-Heinemann, Boston (2016). DOI https:// doi.org/10. 1016/B978-0-08-100176-9.00001-3.
2. Elmoughni, H.M., Yilmaz, A.F., Ozlem, K., Khalilbayli, F., Cappello, L., Atalay, A.T., Ince, G., Atalay, O.: Machine-knitted seamless pneumatic actuators for soft robotics: Design, fabrication, and characterization. Actuators 10(5), 94–(2021)
3. Ghedini, F., Bergamasco, M.: Robotic creatures: Anthropomorphism and interaction in contemporary art. In: 19th International Symposium in Robot and Human Interactive Communication, pp. 731–736. IEEE (2010)
4. Girouard, A.: Design Lifecycle in an Hour (2020). URL: https://medium.com/@audreygirouard/design-lifecycle-in-an-hour-59017d8d10c1
5. Guerlac, S.: Livingness, information, and the really real. Theory event 24(1), 131–157 (2021)
6. Hakio, K., Mattelmäki, T.: Future skills of design for sustainability: An awareness-based cocreation approach. Sustainability (Basel, Switzerland) 11(19), 5247– (2019)
7. Hamidi, F., Baljko, M.: Engaging children using a digital living media system. In: Proceedings of the 2017 Conference on Designing Interactive Systems, DIS'17, p. 711–723. Association for Computing Machinery, New York, NY, USA (2017). https://doi.org/10.1145/3064663.3064708.
8. Hamidi, F., Stamato, L., Scheifele, L.: Turning the invisible visible: Transdisciplinary bioart explorations in human-dna interaction. In: Conference on Human Factors in Computing Systems—Proceedings (2021)

9. Hartmann, F., Baumgartner, M., Kaltenbrunner, M.: Becoming sustainable, the new frontier in soft robotics. Advanced materials (Weinheim) 33(19), e2004413–n/a (2021)
10. Hasling, K., Ræbild, U.: Sustainability cards: Design for longevity. In: Product Lifetimes and the Environment 2017—Conference Proceedings, PLATE'17, pp. 166–170. Delft University of Technology and IOS Press (2017). https://doi.org/10.3233/978-1-61499-820-4-166
11. Heider, F., Simmel, M.: An experimental study of apparent behavior. The American journal of psychology 57(2), 243–259 (1944)
12. Hoffman, G., Ju, W.: Designing robots with movement in mind. Journal of human–robot interaction 3(1) (2014)
13. Hopwood, B., Mellor, M., O'Brien, G.: Sustainable development: mapping different approaches. Sustainable development (Bradford, West Yorkshire, England) 13(1), 38–52 (2005)
14. Jochum, E., Millar, P., Nuñez, D.: Sequence and chance: Design and control methods for entertainment robots. Robotics and autonomous systems 87, 372–380 (2017)
15. Jørgensen, J.: Appeal and perceived naturalness of a soft robotic tentacle. In: Companion of the 2018 ACM/IEEE International Conference on Human–Robot Interaction, HRI'18, p. 139–140. Association for Computing Machinery, New York, NY, USA (2018). URL https://doi.org/10.1145/3173386.3176985
16. Jørgensen, J.: Interaction with soft robotic tentacles. In: Companion of the 2018 ACM/IEEE International Conference on Human–Robot Interaction, HRI'18, p. 38. Association for Computing Machinery, New York, NY, USA (2018). https://doi.org/10.1145/3173386.3177838.
17. Kac, E.: Foundation and development of robotic art. Art journal (New York. 1960) 56(3), 60–67 (1997)
18. Margheri, L., Rossiter, J., Laschi, C., Iida, F., Cianchetti, M.: Soft Robotics: Trends, Applications and Challenges: Proceedings of the Soft Robotics Week, April 25–30, 2016, Livorno, Italy, Biosystems Biorobotics, vol. 17. Springer International Publishing (2016)
19. Merritt, T., Hamidi, F., Alistar, M., DeMenezes, M.: Living media interfaces: a multi-perspective analysis of biological materials for interaction. Digital creativity (Exeter) 31(1), 1–21 (2020)
20. Milthers, A.D.B., Bjerre Hammer, A., Jung Johansen, J., Jensen, L.G., Jochum, E.A., Löchtefeld, M.: The helpless soft robot—stimulating human collaboration through robotic movement. In: Extended Abstracts of the 2019 CHI Conference on Human Factors in Computing Systems, CHI EA'19, p. 1–6. Association for Computing Machinery, New York, NY, USA (2019). URL https://doi.org/10.1145/3290607.3312807
21. Rafsanjani, A., Bertoldi, K., Studart, A.R.: Programming soft robots with flexible mechanical metamaterials (2019).
22. Reller A., Diesenbacher J.: Are there enough resources for our lifestyle?. In: Stebbing, Tidchner (eds.) Changing Paradigms: Designing for a Sustainable Future. Cumulus Think Tank, pp. 154–167 Aalto university press. (2015).

23. Rus, D., Tolley, M.T.: Design, fabrication and control of soft robots. Nature (London) 521(7553), 467–475 (2015).
24. Ræbild, U., Hasling, K.M.: Experiences of the sustainable design cards: Evaluation of applications, potentials and limitations. Fashion Practice 11(3), pp. 417–442 (2019).
25. Sachyani Keneth, E., Kamyshny, A., Totaro, M., Beccai, L., Magdassi, S.: 3D printing materials for soft robotics. Advanced materials (Weinheim) 33(19) (2021).
26. Shah, D., Yang, B., Kriegman, S., Levin, M., Bongard, J., Kramer-Bottiglio, R.: Shape changing robots: Bioinspiration, simulation, and physical realization. Advanced materials (Weinheim) 33(19), (2021).
27. Shintake, J., Sonar, H., Piskarev, E., Paik, J., Floreano, D.: Soft pneumatic gelatin actuator for edible robotics. In: 2017 IEEE/RSJ International Conference on Intelligent Robots and Systems (IROS), pp. 6221–6226. IEEE (2017)
28. Sánchez-Somolinos, C.: 4D Printing: An Enabling Technology for Soft Robotics, Chap. 14, pp. 347–362. John Wiley Sons, Ltd (2020). DOI 10.1002/9783527822201.ch14.
29. Veenstra, F., Jørgensen, J., Risi, S.: Evolution of fin undulation on a physical knifefish-inspired soft robot. In: GECCO 2018—Proceedings of the 2018 Genetic and Evolutionary Computation Conference, GECCO'18, pp. 157–164. ACM (2018)
30. Vissonova K.: Effects of Design and Sustainable Design of Technical Artefacts. In: Vermaas P., Vial S. (eds) Advancements in the Philosophy of Design. Design Research Foundations. Springer, Cham. pp. 433–451 (2018)
31. Zhang, J., Jackson, A., Mentzer, N., Kramer, R.: A modular, reconfigurable mold for a soft robotic gripper design activity. Frontiers in robotics and AI 4 (2017)

References

Abrams DB (2006) Applying transdisciplinary research strategies to understanding and eliminating health disparities. Health Educ Behav 33(4):515–531. https://doi.org/10.1177/1090198106287732

Ackermann E (2001) Piaget's constructivism, Papert's constructionism: what's the difference?

Alves Filho SE, Sa STdL, Burlamaqui AMF, Aroca VR, Goncalves LMG (2018) Green robotics: Concepts, challenges, and strategies. IEEE Lat Am Trans 16(4):1042–1050 (2018). https://doi.org/10.1109/TLA.2018.8362135

Anderson K, Wanscher C (2020) Robotics technology: a key enabler for the green transition. Odense Robot 1–51. https://www.odenserobotics.dk/robotics-technology-a-key-enabler-for-green-transition/

Awad LN, Bae J, O'Donnell K, De Rossi SM, Hendron K, Sloot LH, Kudzia P, Allen S, Holt KG, Ellis TD et al (2017) A soft robotic exosuit improves walking in patients after stroke. Sci Transl Med 9(400)

Bammer G (2005) Integration and implementation sciences: building a new specialization. Ecol Soc 10(2). http://www.ecologyandsociety.org/vol10/iss2/art6/

Bausys R, Cavallaro F, Semenas R (2019) Application of sustainability principles for harsh environment exploration by autonomous robot. Sustainability 11(9):2518
Beavis A, Gibbs P (2020) Transdisciplinary knowledge-an emergent concept. Springer International Publishing, Cham, pp 7–15
Bering Christiansen M, Jørgensen J (2020) Augmenting soft robotics with sound. In: Companion of the 2020 ACM/IEEE international conference on human-robot interaction, HRI '20. ACM, New York, pp 133–135. https://doi.org/10.1145/3371382.3378328
Blok V, Gremmen B, Wesselink R (2016) Dealing with the wicked problem of sustainability the role of individual virtuous competence. Bus Prof Ethics J. https://doi.org/10.5840/bpej201621737
Bocken NM, De Pauw I, Bakker C, Van Der Grinten B (2016) Product design and business model strategies for a circular economy. J Ind Prod Eng 33(5):308–320
Bugmann G, Siegel M, Burcin R (2011) A role for robotics in sustainable development? In: IEEE Africon'11. IEEE, pp 1–4
Bumblis J (2005) Apollo 11 revisited—An example of problem-based learning. In: Arabnia H, Reza H (eds) SERP '05: Proceedings of the 2005 international conference on software engineering research and practice, vols 1 and 2. CSREA, International Technology Institute; World Academic Science & Information Technology; HPC wire; GRID today, pp 590–595
Carmine S, De Marchi V (2022) Reviewing paradox theory in corporate sustainability toward a systems perspective. J Bus Ethics. https://doi.org/10.1007/s10551-022-05112-2
Chew J, Lee JJ, Lehtonen M (2020) Towards design-driven transdisciplinary education: navigating the challenges and envisioning the role of design as a facilitator (2020). https://doi.org/10.21606/drs.2020.344
Cianchetti M, Ranzani T, Gerboni G, Nanayakkara T, Althoefer K, Dasgupta P, Menciassi A (2014) Soft robotics technologies to address shortcomings in today's minimally invasive surgery: the stiff-flop approach. Soft Robot 1(2):122–131
Costanza R (1993) Ecological economics: the science and management of sustainability. Am J Agr Econ 75:1–20. https://doi.org/10.2307/1243998
Cross N (1982) Designerly ways of knowing. Des Stud 3(4):221–227. https://doi.org/10.1016/0142-694X(82)90040-0
Curtis DJ (2011) Using the arts to raise awareness and communicate environmental information in the extension context. J Agric Educ Ext 17(2):181–194. https://doi.org/10.1080/1389224X.2011.544458
Decker M (2015) Soft robotics and emergent materials in architecture. In: Material studies—real time eCAADe conference, pp 409–416
Dragan AD, Thomaz AL, Srinivasa SS (2013) Collaborative manipulation: new challenges for robotics and hri. In: 2013 8th ACM/IEEE international conference on human-robot interaction (HRI). IEEE, pp 435–436
Dunstan BJ, Silvera-Tawil D, Koh JTKV, Velonaki M (2016) Cultural robotics: robots as participants and creators of culture. In: Koh JT, Dunstan BJ, Silvera-Tawil D, Velonaki M (eds) Cultural robotics. Springer International Publishing, Cham, pp 3–13
Eshaghi M, Ghasemi M, Khorshidi K (2021) Design, manufacturing and applications of small-scale magnetic soft robots. Extreme Mech Lett 44:101268. https://doi.org/10.1016/j.eml.2021.101268
Fry T (2005) The scenario of design. Des Philos Pap 3(1):19–27. https://doi.org/10.2752/144871305X13966254124158
Galati R, Mantriota G, Reina G (2022) Mobile robotics for sustainable development: two case studies. In: Quaglia G, Gasparetto A, Petuya V, Carbone G (eds) Proceedings of I4SDG workshop 2021. Springer International Publishing, Cham, pp 372–382
Girouard A (2020) Design lifecycle in an hour (2020). https://medium.com/@audreygirouard/design-lifecycle-in-an-hour-59017d8d10c1
Guimarães MH, Pohl C, Bina O, Varanda M (2019) Who is doing inter- and transdisciplinary research, and why? An empirical study of motivations, attitudes, skills, and behaviours. Futures

112:102441. https://doi.org/10.1016/j.futures.2019.102441. https://www.sciencedirect.com/science/article/pii/S001632871830483X

Gurova O, Merritt TR, Papachristos E, Vaajakari J (2020) Sustainable solutions for wearable technologies: mapping the product development life cycle. Sustainability 12(20). https://doi.org/10.3390/su12208444. https://www.mdpi.com/2071-1050/12/20/8444

Hadorn GH, Bradley D, Pohl C, Rist S, Wiesmann U (2006) Implications of transdisciplinarity for sustainability research. Ecol Econ 60(1):119–128

Hahn T, Figge F, Pinkse J, Preuss L (2018) A paradox perspective on corporate sustainability: descriptive, instrumental, and normative aspects. J Bus Ethics 48:235–248. https://doi.org/10.1007/s10551-017-3587-2

Haider LJ, Hentati-Sundberg J, Giusti M, Goodness J, Hamann M, Masterson VA, Meacham M, Merrie A, Ospina D, Schill C, Sinare H (2018) The undisciplinary journey: early- career perspectives in sustainability science. Sustain Sci 13:191–2014. https://doi.org/10.1007/s11625-017-0445-1

Haraway D (2015) Staying with the trouble: making kin in the Chthulucene. Duke University Press

Hartmann F, Baumgartner M, Kaltenbrunner M (2021) Becoming sustainable, the new frontier in soft robotics. Adv Mater 33(19):2004413. https://doi.org/10.1002/adma.202004413

Hasling K, Ræbild U (2017) Sustainability cards: design for longevity. In: Product lifetimes and the environment 2017—Conference proceedings, PLATE '17. Delft University of Technology and IOS Press, pp 166–170. https://doi.org/10.3233/978-1-61499-820-4-166

Heinrich F, Kørnøv L (2021) Art and higher education for environmental sustainability: a matter of emergence? Int J Sustain High Educ. https://doi.org/10.1108/IJSHE-01-2021-0012

Heras M, Galafassi D, Oteros-Rozas E, Ravera F, Berraquero-Díaz L, Ruiz-Mallén I (2021) Realising potentials for arts-based sustainability science. Sustain Sci 16:1875–1889. https://doi.org/10.1007/s11625-021-01002-0

Hoffmann S, Pohl C, Hering JG (2017) Methods and procedures of transdisciplinary knowledge integration: empirical insights from four thematic synthesis processes. Ecol Soc 22(1). http://www.jstor.org/stable/26270124

Holland DP, Park EJ, Polygerinos P, Bennett GJ, Walsh CJ (2014) The soft robotics toolkit: shared resources for research and design. Soft Rob 1(3):224–230

Holland DP, Abah C, Velasco-Enriquez M, Herman M, Bennett GJ, Vela EA, Walsh CJ (2017) The soft robotics toolkit: Strategies for overcoming obstacles to the wide dissemination of soft-robotic hardware. IEEE Robot Autom Mag 24(1):57–64

Holland DP, Berndt S, Herman M, Walsh CJ (2018) Growing the soft robotics community through knowledge-sharing initiatives

Höök K, Hummels C, Isbister K, Marti P, Márquez Segura E, Jonsson M, Mueller F, Sanches PA, Schiphorst T, Ståhl A, Svanaes D, Trotto A, Petersen MG, Lim YK (2017) Soma-based design theory. In: Proceedings of the 2017 CHI conference extended abstracts on human factors in computing systems, CHI EA '17. Association for Computing Machinery, New York, pp 550–557. https://doi.org/10.1145/3027063.3027082

Hopwood B, Mellor M, O'Brien G (2005) Sustainable development: mapping different approaches. Sustain Dev 13(1):38–52

Horvath AS, Rühse V, Raptis D (2020) Soundsculpt: a design framework for 3d modelling and digitally fabricating sound patterns. In: 32nd Australian conference on human-computer interaction, OzCHI '20. Association for Computing Machinery, New York, pp 572–581. https://doi.org/10.1145/3441000.3441017

Ilievski F, Mazzeo AD, Shepherd RF, Chen X, Whitesides GM (2011) Soft robotics for chemists. Angew Chem 123(8):1930–1935

Jantsch E (1972) Inter- and transdisciplinary university: a systems approach to education and innovation. High Educ 1(1):7–37. https://doi.org/10.1007/BF01956879

Jespersen L (2018) Problem orientation in art and technology. J Probl Based Learn High Educ 6(1):1–14. https://doi.org/10.5278/ojs.jpblhe.v6i1.2341

Jochum E, Murphey T (2014) Programming play: puppets, robots, and engineering. https://doi.org/10.4324/9781315850115
Jochum EA, Putnam LJ (2015) Robot aesthetics: practice-based research in robotic art and performance. In: AAU conference on how to do things with art
Jørgensen J (2018) Appeal and perceived naturalness of a soft robotic tentacle. In: Companion of the 2018 ACM/IEEE international conference on human-robot interaction, HRI '18. ACM, New York, pp 139–140. http://doi.acm.org/10.1145/3173386.3176985
Keitsch MM, Vermeulen WJV (2020) Transdisciplinarity for sustainability: aligning diverse practices. Routledge. https://doi.org/10.4324/9780429199127
Klöpffer W (1997) Life cycle assessment. Environ Sci Pollut Res 4(4):223–228
Laschi C, Rossiter J, Iida F, Cianchetti M, Margheri L (2016) Soft robotics: trends, applications and challenges. Livorno, Italy
Löchtefeld M, Milthers ADB, Merritt T (2021) Staging constructionist learning about energy for children with electrochromic displays and low-cost materials. In: 20th international conference on mobile and ubiquitous multimedia. ACM, Leuven, p 17. https://doi.org/10.1145/3490632.34906541
Löwgren J (2013) Annotated portfolios and other forms of intermediate-level knowledge. Teractions 20(1):30–34. https://doi.org/10.1145/2405716.2405725
Maeder-York P, Clites T, Boggs E, Neff R, Polygerinos P, Holland D, Stirling L, Galloway K, Wee C, Walsh C (2014) Biologically inspired soft robot for thumb rehabilitation. J Med Devices 8(2). https://doi.org/10.1115/1.4027031. URL https://doi.org/10.1115/1.4027031.020933
William McDonough MB (2002) Cradle to cradle design: remaking the way we make things. North Point Press
McGee P, Reis A (2012) Blended course design: a synthesis of best practices. J Asynchronous Learn Netw 16(4):7–22. https://eric.ed.gov/?id=EJ982678. Publisher: Sloan Consortium
Milthers ADB, Bjerre Hammer A, Jung Johansen J, Jensen LG, Jochum EA, Löchtefeld M (2019) The helpless soft robot—stimulating human collaboration through robotic movement. In: Extended abstracts of the 2019 CHI conference on human factors in computing systems, CHI EA '19. ACM, New York, pp 1–6. https://doi.org/10.1145/3290607.3312807
Murphy R (2012) Sustainability: a wicked problem. Sociologica 1–23. https://doi.org/10.2383/38274
Nicolescu B (2002) Manifesto of transdisciplinarity. State University of New York Press
A Note from Peoples Climate Movement. https://peoplesclimate.org/
Van de Poel I (2017) Design for sustainability. Philos Technol Environ 121–142
Pohl C, Pearce B, Mader M, Senn L, Krütli P (2020) Integrating systems and design thinking in transdisciplinary case studies. GAIA-Ecol Perspect Sci Soc 29(4):258–266
Pohl C, Klein JT, Hoffmann S, Mitchell C, Fam D (2021) Conceptualising transdisciplinary integration as a multidimensional interactive process. Environ Sci Policy 118:18–26. https://doi.org/10.1016/j.envsci.2020.12.005
Pujol FA, Tomás D (2020) Introducing sustainability in a robotic engineering degree: a case study. Sustainability 12(14). https://doi.org/10.3390/su12145574
Ræbild U, Hasling KM (2019) Experiences of the sustainable design cards: evaluation of applications, potentials and limitations. Fash Pract 11(3):417–442. https://doi.org/10.1080/17569370.2019.1664026
Rittel HWJ, Webber MM (1973) Dilemmas in a general theory of planning. Policy Sci 4:155–169. https://doi.org/10.1007/BF01405730
Rus D, Tolley MT (2015) Design, fabrication and control of soft robots. Nature 521(7553):467
Schultz J, Mengüç Y, Tolley M, Vanderborght B (2016) What is the path ahead for soft robotics? Soft Rob 3(4):159–160. https://doi.org/10.1089/soro.2016.29010.jsc
Sellberg MM, Cockburn J, Holden PB, Lam DPM (2021) Towards a caring transdisciplinary research practice: navigating science, society and self. Ecosyst People 17(1):292–305. https://doi.org/10.1080/26395916.2021.1931452

Shintake J, Sonar H, Piskarev E, Paik J, Floreano D (2017) Soft pneumatic gelatin actuator for edible robotics. In: 2017 IEEE/RSJ international conference on intelligent robots and systems (IROS), IEEE, pp 6221–6226

Sipos Y, Battisti B, Grimm K (2008) Achieving transformative sustainability learning: engaging head, hands and heart. Int J Sustain High Educ 9(1):68–86. https://doi.org/10.1108/14676370810842193

Trott C, Even T, Frame S (2020) Merging the arts and sciences for collaborative sustainability action: a methodological framework. Sustain Sci 15(4):1067–1085. https://doi.org/10.1007/s11625-020-00798-7

Vermeulen WJ, Keitsch M (2020) Challenges of transdisciplinary research collaboration for sustainable development. Routledge, pp 200–208

Vermeulen WJ, Witjes S (2020) History and mapping of transdisciplinary research on sustainable development issues: dealing with complex problems in times of urgency. Routledge, pp 6–26

Vissonova K (2018) Effects of design and sustainable design of technical artefacts. Springer, Berlin, pp 433–451

Vite C, Horvath AS, Neff G, Møller N (2021) Bringing human-centredness to technologies for buildings: An agenda for linking new types of data to the challenge of sustainability. In: De Angeli A, Chittaro L, Gennari R, De Marsico M, Melonio A, Gena C, De Russis L, Spano L (eds) CHItaly '21. Association for Computing Machinery, pp 1–8. https://doi.org/10.1145/3464385.3464711

Yu X, Nurzaman SG, Culha U, Iida F (2014) Soft robotics education. Soft Rob 1(3):202–212. https://doi.org/10.1089/soro.2014.0009

Zhang J, Jackson A, Mentzer N, Kramer R (2017) A modular, reconfigurable mold for a soft robotic gripper design activity. Frontiers Rob AI 4:46

Zhou H, Wang X, Au W, Kang H, Chen C (2021) Intelligent robots for fruit harvesting: recent developments and future challenges

Chapter 14
Sonic Robotics: Musical Genres as Platforms for Understanding Robotic Performance as Cultural Events

Wade Marynowsky, Julian Knowles, Oliver Bown, and Sam Ferguson

Abstract This chapter examines how artist Wade Marynowsky's recent robotic performance art projects are framed within musical genres: Opera, in *Robot Opera* (2015), Ambient/Glitch, in *Synthesiser-Robot* (2017); and Disco, in *The Ghosts of Roller Disco* (2020). By positioning the projects within known music genres, the research expands the canon of Cultural Robotics by providing platforms that allow wider communities to understand the presentation of robotic performance as cultural events within a historical context. Notions of robotic agency, dramaturgy, choreography, robotic musical gesture, and robotic musicianship are explored across three case studies, which are presented in the contexts of live performance festivals and durational exhibitions: (1) *Robot Opera*, a dramaturgically designed, interactive opera for eight, larger than life-sized robots; (2) *Synthesiser-Robot*, a solo autonomous robot performance for a repurposed industrial robot arm, theUR3, and a hardware-software interface, the Ableton Push; and (3) *The Ghosts of Roller Disco*, a choreographed performance for eight robotic roller skates. The research highlights the importance of robotic agency by applying autonomous and interactive movement, localised sound, and surround sound design in creating immersive and engaging robotic performance art experiences.

W. Marynowsky (✉)
University of Technology Sydney, Sydney, Australia
e-mail: wade.marynowsky@uts.edu.au

J. Knowles
Macquarie University, Sydney, Australia
e-mail: julian.knowles@mq.edu.au

O. Bown
Interactive Media Lab, University of New South Wales, Sydney, Australia
e-mail: o.bown@unsw.edu.au

S. Ferguson
Creativity and Cognition Studios, School of Computer Science, Faculty of Engineering and IT, University of Technology Sydney, Sydney, Australia
e-mail: samuel.ferguson@uts.edu.au

© The Author(s), under exclusive license to Springer Nature Switzerland AG 2023

B. J. Dunstan et al. (eds.), *Cultural Robotics: Social Robots and Their Emergent Cultural Ecologies*, Springer Series on Cultural Computing,
https://doi.org/10.1007/978-3-031-28138-9_14

14.1 Introduction

In the *Cultural Robotics* chapter *Robot Opera: A Gesamtkunstwerk for the twenty-first Century* (Marynowsky et al. 2016) the notion of robotic performance agency is detailed through the history and theories surrounding representations of the robot in popular culture and art history. Thematics including the Uncanny, the Camp; the Robot as High Culture, and Reciprocity are examined in relation to Marynowsky's previous robotic artworks.

For example, *The Hosts*: *A Masquerade of Improvising Automatons* (2009) (Fig. 14.1); featured five larger than life-sized, and human-like, robot characters in ornate dresses. The work draws connections to the Venetian masquerade balls of the sixteenth century Renaissance and the ornate dresses constructed by Jaquet-Droz, such as The Musician [circa 1768 (Voskuhl 2013)]. The characters in *The Hosts* wear sumptuous, embroidered ball gowns and have individual masquerade guises: a clown in a black and white harlequin print, a princess in a pink-ribboned bodice, a military officer with stars and stripes, and a cowboy-hatted cowboy. Similarly, the sound design for each character is matched to the guises. Three female vocalists (Christina Harrison, Kusum Normoyle, and Debra Petrovich) were asked to perform abstract vocals and/or spoken word, which was recorded in the studio and transferred to the robot's internal computer. The robots' sound samples are randomly played when an object is detected by the ultrasonic range finding sensors, or when a particular programmed behaviour is activated over Wi-Fi communication sent from the control computer. The voice samples are intended to increase the emotional impact of affective gestures (Bramas et al. 2008), such as driving towards a human. It is intended that if the robots sound human, the audience experiencing them will understand the robots as being more like them (Eyssel et al. 2012).

The localised on-board sound coming from the robots' speakers had a strong influence on the main finding, which was that people had the sense that the robots were following and or approaching them individually, when in fact the robots were acting of their own accord. As a viewer of the work, Melody Willis recalled, "They all turned and gathered around me. I felt psychically powerful, like a child with extrasensory perception (ESP)". The main behaviour of the robots was being able to automatically navigate the space without colliding with any obstacles, including humans, so the experience that Willis describes was purely dependent on random circumstances. The artificial intelligence programme that allowed the robot to navigate is, in one sense, a sleight-of-hand tool that artists may draw from their bag of magic tricks to suggest that the inanimate is now animate, a definition of the Uncanny (Jentsch 1997). As the robots glide across the floor gracefully, they danced a completely automated, sensor-based choreography. The choreography was constructed of various scenes, such as slowly awakening from the darkness, coming to life one by one and navigating, spinning in unison, and so on. The duration of the performance was 10 min; the sequence then looped around in a cycle over the course of a day. These scenes of the overall arching choreographed sequence informed the dramaturgy of the later work, *Robot Opera* (2015) (Fig. 14.2).

Fig. 14.1 *The Hosts: A Masquerade of Improvising Automatons*, Mediations Biennale, (2019), The 2nd International Biennale of Contemporary Art, Poznan, Poland, 2010

Fig. 14.2 *Robot Opera* (2015), National Kaohsiung Centre for the Arts, Taiwan, 2016

The representation of robots in film has had a strong influence on the audience's mental model of how a robot should behave (Latupeirissa and Bresin 2020). In *The Hosts,* the ambient background sound created by Marynowsky was influenced by György Ligeti's *Atmosphères*, used in Kubrick's soundtrack for the film *2001: A Space Odyssey* (Kubrick and Clarke 2001). The genre of music in *The Hosts* is not directly referred to but through the atmospheric sound design, an outer space and science fiction film world is suggested.

The dramaturgical platforms explored in the earlier works mainly focused on the Uncanny (Hover et al. 2021) and the Uncanny Valley (Mori et al. 2012), which predominantly explores the visual representation of a robot's appearance, whilst in *Robot Opera*, the focus was on the importance of sound, music, and choreography to communicate a sense of artificial life in a stripped-back machine aesthetic. Whilst the earlier chapter defined the conceptual framework of *Robot Opera* as a Gesamtkunstwerk for the twenty-first century, a detailed inspection of the sound design and the use of operatic form is first presented here. The conference paper "'The Ghosts of Roller Disco', A Choreographed, Interactive Performance for Robotic Roller Skates" details the technical research and development of the work and was published in TEI 2020-Proceedings of the 14th International Conference on Tangible, Embedded, and Embodied Interaction (Marynowsky et al. 2020), whereas this chapter details the sound design for the work and the importance of musical genre.

14.2 Musical Genre and Its Role in Robotic Performance

Much previous work has focused on robotic musicianship from the perspective of mechatronics, algorithmic musical models, machine learning/listening, and the technical aspects of human–robot interaction (Hoffman and Weinberg 2010; Weinberg et al. 2020), but to date, little work has been done to critically reflect on the role of performance context and musical genre in robotic musicianship. On the basis of the three robotic performance works outlined in this chapter, we propose that this overlooked dimension of robotic performance is critical in how works are interpreted and received by audiences. Moreover, when we speak of musical genre, it is important to consider that this manifests itself in a wide range of performance practices around the music as well as in the musical materials themselves.

In the field of human musical performance, it is axiomatic that genre and the related concept of performance context have a profound bearing on the presentation and reception of musical works (Fig. 14.3). In order to understand how genre operates as a vehicle, it is important to recognise that genre is not only expressed through musical style and materials but also through the extra musical or performative conventions that operate around the musical materials. In this way, we might understand genre to be an expansive multi-modal, multidimensional concept.

At the highest level, there are specific relational structures and an established set of ritualised practices between performers and the audience that are associated with the notion of genre. In specific terms, there are established performance

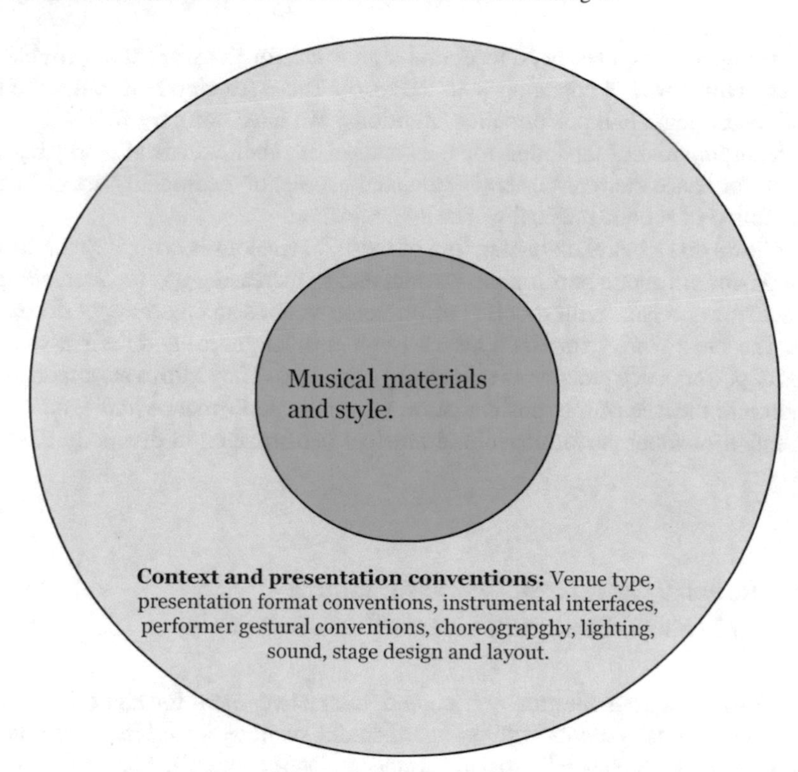

Fig. 14.3 Musical genre and performance context

settings and trends in the approaches to the mediation of performances via audio-visual technologies, common gestural conventions, and choreographic styles, not only in respect of instrumental interfaces but also beyond them. Beyond instrumental gestures, there are choreographic dimensions to musical genres that are core to performance, production, and reception, which operate as a shared language or set of expectations between performers and the audience. For example, in the classical music and operatic traditions, there are concert hall-based rituals governing performer placement and movement on stage and the formal acknowledgement of performers by the audience. Within the genres of popular music, there are similar conventions and rituals that differ from the concert hall traditions. The audience is more participatory both physically and vocally in an apparently less formal sense than in concert hall traditions, but arguably similarly ritualised. In rock, the instruments are distributed in similar or standard formations in a space. In electronica genres, the performers interact with electronic interfaces and there are less obvious connections between gestural inputs and sonic outputs. The proposition is that when composers compose, performers perform, and audience members observe, they are intuitively drawing upon this stabilised set of shared norms.

Genres carry with them a set of structural attributes which shape the work in space and time (scenes, interludes, sets, brackets, 'drops', etc.). These structuring attributes operate as a set of 'conceptual signposts' for the work that provide the audience with a way of engaging with the work. These structural attributes are tied to notions of genre and performance traditions. We have not only found that they provide an important foundation for the creation of robotic musical works but that these performance conventions and associated aspects of theatricality are critical to the reception of robotic musical works by audiences.

It follows that a better understanding of these dimensions is critical to the further development of robotic performance works and to increasing our understanding of how audiences engage with robotic performance work as an emerging performance form. The three works examined here all use musical genres and their associated rituals of performance practice as organising principles. They form a sustained investigation into the role of the musical genre in robotic performance and examine the intersection of robot performance and musical performance in distinctly different ways.

14.3 *Robot Opera*: Robotic Performance as Musico-Dramatic Gesamtkunstwerk

Robot Opera is a dramaturgically designed, interactive opera for eight, larger than life-sized (2.3 m tall) robots with on-board sound, choreographed in a large theatre space with an 8–16 channel surround sound projection system. The project investigates the notion of the *Gesamtkunstwerk*, a term coined in 1927 by the German philosopher Karl Friedrich Eusebius Trahndorff (Trahndorff 1827) to describe the concept of the 'total artwork', that is, a work that synthesises all art forms into a single unified multidisciplinary work.

In the research and development stages of *Robot Opera*, Marynowsky and Julian Knowles (music and sound designer) worked in collaboration with the performance group Branch Nebula (Lee Wilson and Mirabelle Wouters) to develop a dramatic arc for the large-scale performance piece. The arc started with all the robots at one end of the space, all lined up in a row, equally spaced all lights and sound off, except for a low hum. Reveal—the robots slowly turn around to face the audience, with lights on robots coming on; then Sequencer—a beat sequence broken across eight robots, with local LED lights pulsing in response to the sound in various shaped pattern. Slowly moving forward, one robot comes forward and detects a human. A local controlled (DMX) moving head spotlight flashes in blue, while a robotic vocoder voice sample says 'Human detected'. This signifies detection and all robots move forward and interact, avoid and wander, intermingling with the crowd. After 10 min, they return to a rectangular formation. In the noise section, robots spin around fast as if they are out of control, before stopping all of a sudden. One robot returns to avoid and wander, a red light traversing the space and speaking in a female vocoder voice. All lights off, darkness.

In *Robot Opera*, the title of the work explicitly invites the audience to engage with the work as a musico-dramatic narrative that is located within a European performance tradition that predates robotic performers, stretching back to the late sixteenth century. Beyond its title, *Robot Opera* adopts a range of operatic conventions in its execution. These include scene-based dramatic structures that structure the work through time, a musical division between the soloists and accompaniment (in this case, sound produced by robots and accompaniment produced by surround speaker arrays), alternation between soloist and chorus structures (in this case solo robotic performance contrasted with full ensemble performance), and so on.

The intention of this work is to deploy well understood theatrical and musical devices as a structuring principle for the comparatively unfamiliar context of an entirely robotic performance work. In this way, the established traditions of performance can operate as a platform through which the audience can engage with an entirely non-anthropomorphic robot performance ensemble.

Whilst *Robot Opera* directly draws on operatic conventions, it also subverts them once they are established. *Robot Opera* manifests staging differences from conventional opera in that the work is presented in a large open theatre space where, for the second half of the performance, the audience is free to enter the stage area and interact with the robots. The second half of the work, therefore, sees the breaking of the 'fourth wall' between performers and the audience. This is a radical departure from the operatic tradition but at the same time introduces an element of theatrical surprise and a use of theatrical space that is more consistent with contemporary experimental performance traditions (Bailes 2011).

On a musical level, the relationship between the musical materials in *Robot Opera* and the opera tradition are conceptual and structural, rather than direct or imitative. The *Robot Opera* score does not sound like opera, yet it adopts a great many of the compositional structures from it. Musically, the eight sound-enabled robots operate as an ensemble and as a collection of soloists. At various times, they sound together as a tight ensemble. At other times, there is a soloist, or a soloist is surrounded by supporting sound from the other robot performers. At the perimeter of the space, there is a multichannel loudspeaker array that provides an accompaniment to the performers on stage. On a conceptual level, this loudspeaker array can be understood to be the equivalent of the orchestra pit. The loudspeaker 'orchestra' provides atmosphere, accompaniment and generally supports the on-stage sound from the robot performers and the dramatic action.

Speech and text play important roles in *Robot Opera* as they do in the opera tradition, with the robot performers vocalising at various points. The vocalisations from the robots consist of single words randomly hocketed around the robot ensemble who are wandering randomly across the stage area. These are supplemented by nonverbal speech-like electronic utterances, (Robinson et al. 2021) referencing the robotic languages of popular culture film. The result is a spatial cloud of words and utterances from the robot ensemble as they move through the audience. The use of speech in this context resembles operatic recitative but stops short of traditional narrative function. Its purpose is an affective speech layer of evocative words rather than a vehicle for the delivery of narrative content. In the latter half of the work, the robots seek out humans, engage them, and verbalise material via on-board loudspeakers, introducing a level of performer/audience interaction that is not part of the opera tradition. The closest the robot text comes to meaningful speech is the utterance of the words 'Human detected', which occurs when a robot detects an audience member in front of them. This text is delivered in a 1:1 relationship with a proximate audience member as opposed to the typical 'one to many' format of opera recitative and it creates a state of heightened dramatic tension in an intimate connection with a single audience member.

Given that there were no human performers in the production, the audience looked for agency in the robot performers, viewing the robot performers as autonomous performance agents, and projecting a number of qualities onto them as a result of the mise-en-scène and performance context. Rich data was collected on this work, including audio and video documentation of audience interactions and research surveys of audience members. These data clearly shows that audience members projected qualities onto the robot performers that arose from an imagined view of robots as agents, some of which were not programmed or present in software. These imagined qualities suggest that the performance context and musical genre are critical in how an audience reads a robotic performance work. The technical reality of agency or interaction in this instance was less important than the imagined agency or interaction. Much as we understand theatrical illusion to be a critical part of all human performance, so it is with robot performers.

The work invites the audience to consider robots as performance agents within a performative and dramaturgical system based on the operative form. The title, references to genre, mise-en-scène, choreography, dramatic structure, and musical score all support this reading. The work shows how musical genre, and its performance conventions can be explicitly used and subsequently subverted within the same work. The key insight is that genre and the expectations of performance practice play important roles in the exchange between the audience and robot performers.

14.4 *Synthesiser-Robot*: Expressive Robotic Gesture and Emotion in Robotic Music Performance

Synthesiser-Robot is a semi-anthropomorphic robotic musician (Kemper 2021) that performs a solo autonomous performance for a repurposed industrial robot arm, Universal Robots' UR3, and a hardware-software interface, the Ableton Push. Mounted to the end of the UR3 is a Robotiq adaptive gripper normally used to pick and place in a factory setting. In this instance, the gripper is used to turn knobs and press the buttons of the Ableton Push, much like a musician or DJ would in a live performance context. A custom-built tripod stand was created to raise the UR3 to table height. A custom-built lectern was also created as a stand for the Push and placed in front of the UR3, evoking the notion that the UR3 is about to perform a piece of music (Fig. 14.4).

Fig. 14.4 *Synthesiser-Robot* (2017). UR3 robotic arm, custom stand, Ableton Push, Mac mini, speakers, audio, installation, and duration variable

The Ableton Push pads were configured, in Ableton Live, to have sound samples assigned across the 64 pads. The pads were velocity sensitive, allowing touch sensitivity via the applied force of the UR3 gripper. Individual ambient/glitch sound samples were designed for each of the 64 pads and a musical composition was created whereby the UR3 would play the musical events from the Push controller.

Whilst installed in a gallery environment, *Synthesiser-Robot* references electronic music performance in its musical materials, performance interface, and performance gestures. Like *Robot Opera*, it uses these reference points as a set of conceptual signposts for the audience to invite them to engage with a robot as the primary performer. Unlike *Robot Opera*, however, the musical score has a closer sonic relationship to the genre it references, being ambient electronica and modern DJ performance. Within these genres, performers typically manifest a range of non-functional, expressive gestures, such as dancing or head nodding, that are not directly connected to the production of sound but are purely expressive in the visual domain. These gestures can equally be thought of as a form of performative expression and an invitation from the performer to the audience, encouraging them to respond to the music.

Musically and performatively, this work quite closely models itself on the conventions of electronica composition and performance. In keeping with the norms of electronic dance music performance, at the beginning of the performance, *Synthesiser-Robot* presses the play button on the Push to trigger a predetermined sequence of ambient background tracks. After the background music starts, the UR3 begins its

7 min 30 s performance, adding live electronic elements over the top of the background by playing the Push controller. This is a very common performance practice in electronic music genres. The performance is driven by a manually created pre-programmed sequence of single and repetitive loops in which the UR3's arm moves up and down pressing on the Push's sound pads in varying gestures. In between bursts of playing the Push controller, the UR3 reacts to the music by moving freely, performing dance-like movements. Similar to the way, a DJ might nod their head or fist pump the air to encourage the audience to dance, and the UR3 moves gracefully to the music, communicating to the audience that they may enjoy the music in the same way. In this way, *Synthesiser-Robot* demonstrates the expressive potential of dance-like movement and expressive gestures to expand the meaning of 'robotic musical gesture' beyond the context of direct instrumental performance (Hoffman and Weinberg 2010).

Synthesiser-Robot is a paradoxical title as both a robot and a synthesiser are human constructs that cease to exist without human interaction. In this case, both elements exist by interacting with each other in an automated and synthesised, self-contained programmed performance loop. As the robotic performance continues throughout the course of a day (during the month of the exhibition), humans are invited to the exhibition during opening hours and may encounter the robotic performance of *Synthesiser-Robot* at any time within its loop cycle. This highlights that the context and construction of the premise of the work relies heavily on the cultural understandings of the communities of the art gallery going attendees.

In respect of its musical materials, *Synthesiser-Robot* draws on the tradition of ambient electronic music and glitch electronica. Ambient music is a genre of music developed in the 1960s and'70s that emphasises tone and atmosphere over traditional musical structure or rhythm. Its history may be traced back to Erik Satie's notion of furniture music (Potter 2016), minimal music (Warburton 1988), and musique concrète (Deutsch 2009) and was popularised by composer Brian Eno's album *Ambient 1: Music for Airports* (1978). Ambient music has been associated with spiritual and new age movements due its use in relaxation and reducing stress. The genre has evolved to include a wide range of styles from electronic music using synthesizers and acoustic instrumentals using flutes and drums, singing bowls, to world music and spiritual chanting from indigenous cultures. In contrast, Glitch music emerged as a purely electronic genre of music in the 1990s. Pioneers include German labels such as Mille Plateaux and the work of Ryoji Ikeda in Japan. Glitch has been described by Kim Cascone as having an 'aesthetic of failure' (Cascone 2000) and may be distinguished by the deliberate use of glitch-based sonic artefacts such as clicks and cuts and deep digital signal processing. Oval's studio album *Wohnton*, produced in 1993, helped to define the genre by adding some of the aforementioned ambient aesthetics. The tradition of glitch music embraced to a certain extent the prior tradition of ambient music in that a large body of work responds to both aesthetics. It is in this convergence that the music for *Synthesiser-Robot* can be located.

The genre of ambient/glitch music is used here as a platform for expressing relaxing, contemplative and spiritual music, which is an unexpected form of music to be played by an industrial robot arm. Traditionally, one would expect an industrial robot arm to play repetitive industrial music at high speed, but by programming the arm with slow moving gestures we invite the audience to consider the robot's movements as a response to the slowly evolving ambient music, which the UR3 is creating in real time. Glitching digital audio artefacts such as stuttering, decimation, and sonic bleeps are used to represent the "Ghost in the Machine" (Clarke 2011) in the Arthur C. Clarke sense by referring to the virtual consciousness inside the robot and its processes in operation. They symbolise and emphasise the digital nature of robotic performance as it exists in the popular imagination.

The contextual platform of the art gallery helps to frame these gestural and musical cues to guide viewer's response in considering the robot as a sentient live performer. The non-humanoid appearance of an industrial robot arm takes on anthropomorphic qualities as a result of its behaviours within this well-known mis-en-scène. Additionally, *Synthesiser-Robot* proposes a future in which humans might accept robots as solo musicians, pay money to see robotic bands, and purchase music created by algorithms. It was envisioned that with further research and development, the robot could be programmed with machine learning algorithms to have a greater knowledge of Abelton's Push grid button configuration. Machine learning algorithms could also be used to allow the robotics system to compose music, which questions the notion of originality in creative practice. When faced with artificial intelligence (AI) algorithms that can generate artwork that is sold at auction (Kinsella 2018), we can ask ourselves, what is the role of the artist in the age of computation reproduction embedded with machine intelligence? The role of the artist is to engage with the tools and times they live in. For example, the first AI generated portrait made by the French collective Obvious, which sold at Christie's in 2018 for US$432,000, is an important milestone in the history of AI art.

As detailed in the book *Beyond the Creative Species* by Bown (2021), artists have been using generative algorithms to create art and music to some extent since the 1980s (such as AARON by Harold Cohen [1985–86]) (Cohen 2017). More recently, a relevant musical example includes Artificial Intelligence Virtual Artist (AVIA), an electronic composer that was trained, using deep and reinforcement learning algorithms, to read and compose music, originally classical in genre. Since 2019, the company offers a commercial product, Music Engine, which is capable of generating compositions in various styles, such as rock, pop, jazz, fantasy, shanty, tango, and twentieth-century cinematic (Barreau 2019). The main difference between AI generated music and a highly skilled professional is that the human artist makes creative decisions to produce music that is currently at a much higher level in terms of originality and creativity. Importantly the artwork produced is assigned value by existing in a complex techno-social-cultural environment.

14.5 Sound Design for *The Ghosts of Roller Disco*: Deconstructed Fragments

Disco, a genre of dance music that emerged in the 1970s in the United States, developed in opposition to the rock and punk music of the time. Its sound is typified by four-on-the-floor beats, syncopated basslines, string sections, horns, electric piano, synthesisers, and electric rhythm guitars. Some of its most popular musicians include Donna Summer, the Bee Gees, Chic, etc. Roller Discos were a popular teenage alternative to adult discos and for a short time in the 1980s in Sydney, Australia, they were a highly popular weekend pastime.

In *The Ghosts of Roller Disco*, the theme is the genre of disco within the context of a roller disco, but this time treated in a more nostalgic way. In contrast to the previous works and their futuristic machine aesthetic, a set of eight robots were created and embodied in the form of autonomous retro'80s roller skates. Each robot used sensors and motors to drive the skates around a performance space with simple autonomous behaviours, such as coordinating path planning and object avoidance.

In addition, each robot was equipped with a built-in speaker and synthesis toolkit enabling it to play music in real time, coordinated over Wi-Fi communication.

Musically, the goal was not to be faithful to the original musical style but to use the music and context as referential material to reconfigure and play with, situated in a dreamlike setting where the roller skates themselves have come to life as ghosts, detached from human bodies. The result is an ambient and experimental musical performance that is constructed from musical materials derived from disco themes but driven by this original performance context and its affordances. The array of fast-moving small speakers moving around on the skates provides the potential for a distributed choral ensemble of sound, coordinating with and reacting to each other. The real-time rendering of the sound was derived from the movement of the skates. Originally, the goal was to use the robots' movement control data to control parameters in the music, which could then be coordinated across the array of speakers to be musically in time. However, it proved more technically feasible, and as effective, simply to use data taken directly from an accelerometer built into each robot to control the sound—an appropriate proxy for the robot's movement control data.

We invited two professional musicians, Adrian Lim-Klumpes (keyboards) and Eve Klein (voice), to prepare audio samples inspired by specific disco tracks to create a 'sound world' of musical phrases with which to work; two solo voices that could be set on top of a fixed backing track played over a regular PA. The musical material was prepared so that it could be freely remixed in real time. Originally the goal was to have those phases played in a precisely timed way, but we found that allowing them to play in a temporally loose manner was effective in creating a more perceptually complex soundscape, which set up a rich interplay between the robots' sound and the strong pulse of the backing track. This lent itself to the sense of nostalgia and dreamlike feeling associated with the ghost-skate setting. This in turn also set up an aesthetic juxtaposition with the more comical presence of the darting, critter-like skates, suggesting more complex interplay between the characters. The

Fig. 14.5 *The Ghosts of Roller Disco*, (2020)

main sound used was Eve Klein's vocal samples that were stretched in time using the granular synthesis software installed and emended into the hardware inside the boots. An accelerometer sensor was used to map the movement of the skate to the granular pitch of the sample, creating a vocal sound stream across an elongated time frame. The installation of the work in the show *Never odd or eveN*, (TEI2020 Arts Track Exhibition, at Tin Sheds Gallery, University of Sydney), featured a large black light installed in the ceiling, making everything that was white—the skates, people in white clothes and teeth—illuminate, adding another experiential layer to the abstracted disco environment. Here, the memory of attending roller discos as a teenager is suggested in an installation of ghost-like robots, autonomously navigating and waiting in a parallel world for the cycle of repeating fashions and fads to reignite (Fig. 14.5).

14.6 Future Work

Each of the works has strong potential to be further developed and combined in multiple ways. Multiple *Synthesiser-Robots* could be incorporated into the structural frames of the robots in *Robot Opera* to further investigate the performative nature of individual robotic agents. This would enable the robots to have one to two arms, so that in addition to driving around, detecting, and following humans the robots could point in multiple directions as well as performing 'live' music on their Push pads.

It is intended that *The Ghosts of Roller Disco* project be developed so that two robo-skates are bound together by legs and a lower torso. This would enable the potential to develop human-like figure-skating manoeuvres such as circle work, figure eights, jumps, flips and mid-air spins. When these developments come to fruition, it is intended that the robo-skates enter the Robot Olympics. A social robot version could be a waiter or waitress that, with a tray on top of the lower torso, could deliver food and drinks to art gallery opening attendees.

With additional research, *Synthesiser-Robot* could be developed into a robotic band of four musicians along with a visual projection accompaniment, to create an audio-visual and robotic spectacle that could tour the world stage. To date, there have been multiple robotic musical bands that have toured internationally, such as Compressorhead (Davies and Crosby 2015) and Z-Machines, however, these bands are strongly situated and framed within the genre of rock.

The question of the future remains: will these types of robot performances be interesting enough to entertain a human audience and for how long will it last before the next robotic band comes along? Which genre of music will robots perform and in which contexts? Importantly, how can genres be expanded by robotic performers to create new art forms?

14.7 Conclusion

This chapter examines the role of genre and performance context in the reception of robotic performance works and, by extension, how such works can be understood as cultural events within a historical context. Whether the robot is remote controlled, tele-operated, autonomous, or programmed with artificial intelligence algorithms, the performance context, references to genre, and theatrical conventions play a significant role in the way in which works are read. Once a robot is placed within the context of a musical or narrative genre, it may be accepted to be a believable robotic agent due the suspension of belief and theatrically of the presented illusions.

Audiences must understand the language of the theatre, opera, music, and or cinema in order to follow the narrative construction of the form. The use of thematics and theatrical illusion may also be used to sustain a sense of empathy with the characters on stage, in film or a piece of live music. In cinema, the mise-en-scène, along with the cinematography and editing of a film, influences the believability of a film in the eyes of its viewers. Similarly, in robotic performance, thematic constructs such as autonomous and interactive movement, sound and light influence the audience's believability of a robotic agent with a robotic performance artwork.

As demonstrated in *The Hosts* and *Robot Opera*, to automatically navigate a given space without colliding with any obstacles is an essential component of expressing base level artificial intelligence. Additionally, when a robot interacts with humans by following them and stating 'Human detected' when they are in close proximity acknowledges that the robot is aware that a human is nearby, enhancing the believability of the robotic agent. Simple lighting effects such as pulsing the light in relation to the volume of the local robot's audio or spoken word acting as a visual representation of speech has proven effective in communicating to the audience that the robot is speaking.

When using multiple robots, an audience may project social formations and hierarchies onto the robot group, particularly when they autonomously form recognised configurations out of what may seem to be a chaotic formation. For example, in *Robot Opera,* we used multiple formal arrangements to represent uniformity after intermingling within large audience groups. In the introduction and reveal sections, all the robots were lined up in a row, a metaphor for an army line of soldiers. Again, in the middle section, all the robots regrouped (after intermingling) to line up in groups of four and across from each other to form a rectangle in the middle of the space. Within the sequencer section, we referenced the left-to-right animated light patterns of electronic drum machines, by creating a beat sequence broken up across eight individual robots, with local LED lights pulsing in various shaped patterns. Above is a brief list of theatrical techniques that were employed to enhance Human–Robot Interactions with artistic contexts. It is clear that the future for robotic performance situated within these artistic contexts has strong potential for further engagement with audiences across a wide range of experiences and narratives.

14.8 Credits and Exhibition Listings of the Works

The Hosts: A Masquerade of Improvising Automatons (2009). Artist and sound: Wade Marynowsky, Electrical engineer: Aras Vaichas, Programmer: Jeremy Apthorp, Lighting design: Mirabelle Wouters, Costumes: Sally Jackson. Exhibited at the Performance Space, Sydney, 2009; *Beyond Mediations*, Mediations Biennale, The 2nd International Biennale of Contemporary Art, Poznan, Poland, 2010; and in the retrospective, *Nostalgia for Obsolete Futures*, Ian Potter Centre, National Gallery of Victoria, 2014.

Robot Opera (2015). Artist: Wade Marynowsky, Music and Sound Design: Julian Knowles, Lighting Design: Mirabelle Wouters, Dramaturgy: Lee Wilson, Electrical Design: Ben Nash, Programmer: Imran Khan. The world premiere of *Robot Opera* was co-presented by Performance Space and Carriageworks in Sydney. It was the key work in the *Liveworks* festival of live and experimental arts. In 2016, the project's international debut was at the National Performing Arts Centre, National Kaohsiung, Taiwan, 2016. *Robot Operetta* was exhibited in *Dream Machines*, Hazelhurst Regional Gallery and Arts Centre, 2017.

Synthesiser-Robot (2017). Artist: Wade Marynowsky, Music and Sound Design: Julian Knowles. Exhibited in *Algorithmic Pareidolia*, Incinerator Art Space, Willoughby, September–October 2017; *Re/Pair*, part of the Big Anxiety festival, Black Box Theatre, UNSW Art and Design, 2017; and in *Data Life, Mechanical Life, Synthesise Life*, The Beijing Media Arts Biennale, Central Academy of Fine Arts Museum, Beijing, China, 2018.

The Ghosts of Roller Disco (2020). Artist: Wade Marynowsky, Hardware: Angelo.

Fraietta, Brendan Lamb, Nicholas Welsh, Software: Michael Gratton, Sam Ferguson, Alex McClung, Angelo Fraietta, Sound Design: Oliver Bown, with additional music and sound design by Eve Klein and Adrian Lim-Klumpes. Exhibited at *Never odd or eveN*, TEI2020 Arts Track Exhibition, Tin Sheds Gallery, University of Sydney, NSW, Australia, February 9–12, 2020.

References

Bailes SJ (2011) Performance theatre and the poetics of failure. Routledge. https://doi.org/10.4324/9780203846179

Barreau P (2019) Artificial intelligence virtual artist. https://www.aiva.ai/

Bown O (2021) Beyond the creative species. MIT Press. https://doi.org/10.7551/mitpress/10913.001.0001. http://direct.mit.edu/books/book/5026/Beyond-theCreative-SpeciesMaking-Machines-That

Bramas B, Kim YM, Kwon DS (2008) Design of a sound system to increase emotional expression impact in human-robot interaction. In: International conference on control, automation and systems. https://doi.org/10.1109/ICCAS.2008.4694222

Cascone K (2000) The aesthetics of failure: "post-digital" tendencies in contemporary computer music. Comput Music J 24(4):12–18. http://www.jstor.org/stable/3681551

Clarke AC (2011) The collected stories of Arthur C. Clarke. Hachette, UK

Cohen P (2017) Harold Cohen and AARON. AI Mag 37(4):63–66. https://doi.org/10.1609/aimag.v37i4.2695

Davies A, Crosby A (2015) Compressorhead: the robot band and its transmedia storyworld. In: Cultural robotics: first international workshop, CR 2015, Held as Part of IEEE ROMAN 2015, Kobe, Japan, 31 Aug 2015. Revised Selected Papers, pp 175–189. https://doi.org/10.1007/978-3-319-42945-8_14

Deutsch S (2009) A concise history of western music for film-makers. Soundtrack 2(1):23–38. https://doi.org/10.1386/st.2.1.23_1

Dunstan BJ, Silvera-Tawil D, Koh JTKV, Velonaki M (2015) Cultural robotics: robots as participants and creators of culture. In: Koh JTKV, Dunstan BJ, Silvera-Tawil D, Velonaki M (eds) CR 2015: Cultural robotics—Lecture notes in computer science, vol. 9549 (2015). https://doi.org/10.1007/978-3-319-42945-8_1

Eyssel F, Kuchenbrandt D, Bobinger S, De Ruiter L, Hegel F (2012) 'If you sound like me, you must be more human': On the interplay of robot and user features on human-robot acceptance and anthropomorphism. In: HRI'12—Proceedings of the 7th annual ACM/IEEE international conference on human-robot interaction. https://doi.org/10.1145/2157689.2157717

Hoffman G, Weinberg G (2010) Gesture-based human-robot jazz improvisation. In: 2010 IEEE international conference on robotics and automation. IEEE, pp 582–587. https://doi.org/10.1109/ROBOT.2010.5509182. http://ieeexplore.ieee.org/document/5509182/

Hoffman G, Weinberg G (2010) Shimon: an interactive improvisational robotic marimba player. In: Proceedings of the ACM international conference on human factors in computing systems, pp 3097–3102. https://doi.org/10.1145/1753846.1753925

Hover QR, Velner E, Beelen T, Boon M, Truong KP (2021) Uncanny, sexy, and threatening robots: the online community's attitude to and perceptions of robots varying in humanlikeness and gender. In: ACM/IEEE international conference on human-robot interaction, pp 119–128. https://doi.org/10.1145/3434073.3444661

Jentsch E (1997) On the psychology of the uncanny (1906). Angelaki 2(1):7–16. https://doi.org/10.1080/09697259708571910

Kemper S (2021) Locating creativity in differing approaches to musical robotics. Front Robot AI. https://doi.org/10.3389/frobt.2021.647028

Kinsella E (2018) The first AI-generated portrait ever sold at auction shatters expectations, fetching $432,500—43 times its estimate (2018). https://news.artnet.com/market/first-everartificial-intelligence-portrait-painting-sells-at-christies-1379902
Kubrick S, Clarke AC (2001) A space odyssey (1968)
Latupeirissa AB, Bresin R (2020) Understanding non-verbal sound of humanoid robots in films. In: Workshop on mental models of robots at HRI 2020. Cambridge
Marynowsky W, Knowles J, Frost A (2016) Robot opera: a Gesamtkunstwerk for the 21st century. In: Jeffrey TKVK, Dunstan BJ, Silvera-Tawil D, Velonaki M (eds) Cultural robotics. Springer, Berlin, pp 143–158. https://doi.org/10.1007/978-3-319-42945-8_12
Marynowsky W, Ferguson S, Fraietta A, Bown O (2020) 'The ghosts of roller disco', a choreographed, interactive performance for robotic roller skates. In: TEI 2020—Proceedings of the 14th international conference on tangible, Embedded, and embodied interaction (2020). https://doi.org/10.1145/3374920.3375284
Mori M, MacDorman K, Kageki N (2012) The uncanny valley [From the field]. IEEE Robot Autom Mag 19(2):98–100. https://doi.org/10.1109/MRA.2012.2192811
Potter C (2016) Erik satie: a Parisian composer and his world. Boydell & Brewer
Robinson FA, Velonaki M, Bown O (2021) Smooth operator: tuning robot perception through artificial movement sound. In: ACM/IEEE international conference on human-robot interaction. https://doi.org/10.1145/3434073.3444658
Trahndorff KFE (1827) Aesthetik oder Lehre von der Weltanschauung und Kunst. Maurer, Berlin
Voskuhl A (2013) Androids in the Enlightenment. University of Chicago Press. https://doi.org/10.7208/chicago/9780226034331.001.0001
Warburton D (1988) A working terminology for minimal music. Intégral 2:135–159. http://www.jstor.org/stable/40213909
Weinberg G, Bretan M, Hoffman G, Driscoll S (2020) Robotic musicianship: robotic musicianship embodied artificial creativity and mechatronic musical expression. Springer Nature. https://www.springer.com/gp/book/9783030389291

Chapter 15
Rouge and Robot: The Disruptive Feminine

Lian Loke and Dagmar Reinhardt

Abstract In future worlds, robots will participate and collaborate in human activities, and deeply interact with humans through personal, intimate and immediate actions around and even on the human body. Taking feminine rituals of grooming the body and highlighting body attributes as a focus, we investigate the intersection of biological and machine moving bodies, exploring vulnerable moments of touch. Introducing a robot into an entrenched cultural act that references the universal female is a disruptive, critical tactic for rethinking dominant approaches to human–robot collaboration. Through a Deleuzian analysis of our video artwork *code_red* (2021), we propose an alternative theoretical interpretation of the interaction between an industrial robotic arm and a female figure. We then turn to a broader consideration of how this particular analysis ties back into current thinking in collaborative, industrial and social robotics, through a discussion of concepts of breaching the intimate zone, resonance, and micro-gestures in human–robot relations.

15.1 Introduction

Imagine a robotic arm approaching the vulnerable skin of a woman's lips. Trembling in anticipation of receiving strokes of rich, red, creamy fluid, she wears her lipstick as warpaint.

The red lipstick—a war cry of prostitutes, coquettes, and harlots (the original crushed blood of lice replaced by concoctions of modern chemistry)—announces their readily available sexuality to the universe as much as to the potential buyer. Interpreted simultaneously as a poster board and red flag that signals the woman, the connection, the embrace and the consumption, the ecstasy, and with it the (temporary) loss of self through exultation. The red of the mouth signals a promise of

L. Loke (✉) · D. Reinhardt
The University of Sydney, Sydney, Australia
e-mail: lian.loke@sydney.edu.au

D. Reinhardt
e-mail: dagmar.reinhardt@sydney.edu.au

© The Author(s), under exclusive license to Springer Nature Switzerland AG 2023
B. J. Dunstan et al. (eds.), *Cultural Robotics: Social Robots and Their Emergent Cultural Ecologies*, Springer Series on Cultural Computing,
https://doi.org/10.1007/978-3-031-28138-9_15

desire and fulfilment. But the red also announces the universal female that can be constructed and expressed through an application of paint, the proposition that a role/characteristic is independent from our given body attributes—and with this, a social contract/agreement that is irrespective of gender but an attribute that can be borrowed or worn like a coat or a dress. She dresses in red.

Yet this red is also a warpaint, a warning. The female persona takes a stand and voices an opinion, unrequited. The wearing of lipstick is a sociocultural, feminine act inflected with the political. The red mouth that is framed, calling for rights to vote, rights to equality, and rights to own time/money/personal space. Crying out loud for the right to self-determination. Women who roar. As a contradiction, the act of applying lipstick is highly personal, delicate, and discreet. A subtle careful gesture, private, when she opens the lipstick, rotates the tip, traces her lips in front of a mirror, and thus slides into an alternate version of herself.

Introducing a robot into this highly personal ritual invites a critique of how biological and machinic bodies could be reconfigured in the intimate zone of close encounter. Deploying an industrial robotic arm is a defamiliarising tactic intended to disrupt and open up thinking on how we weave robots into our daily lives and private, intimate acts. It is also part of our arsenal of creative practise of the disruptive feminine, a questioning of the construction of the universal female, with its culturally entrenched feminine modes.

The vision of industrial robots is of machines programmed to function, focused on control and towards repetition, enabling serial or customised production, high-end precision fabrication and manufacturing of modules and parts, and consequently supporting production and relieving human workers of arduous labour. Current approaches to humans working and collaborating with industrial robotic arms are primarily concerned with how to safely interact at close quarters. While increasingly equipped with sensor-based feedback and thus becoming responsive to surroundings or moving entities, these are not yet sophisticated machines in the sense of a full capability for interacting with or acting for humans, or on human bodies. A human body in the scenario of industrial workshops is often conceived of in mechanical and ergonomic terms, with the robot and the human considered as separate, discrete entities that work on a shared task, often in parallel or subdivided tasks rather than in a seamless, fluid interaction. How the relation between machinic and biological bodies can be understood as direct and reciprocal, or even as a coupling or merging of bodies is little explored in a context of industrial robotics, and only to some extent in creative or social robotics.

Consequently, our research explores human–robot relationships through dimensions of body, sensation, and interaction. We proffer a different understanding and interpretation of human–robot collaboration when applied to the provocation of an industrial robot drawing lipstick on human lips. Documented in our video artwork *code_red* (2021), we investigate a female figure and robot interacting in a vulnerable moment of touch (Fig. 15.1). Here, the robot is explored as the frame by which the figure is held, opening up the tension within the conditions of the human figure, and introducing the issue of the boundary and breach of the intimate zone between human and robot through application of the intimate gesture of caress.

Fig. 15.1 *code_red* video still

In this chapter, we provide a background introduction to the technical setup and videography for our creative artwork, *code_red* (2021). The myriad of relationships and perspectives are explored through the camera frame to highlight and amplify nuances of actions and reactions between human and machine. In the section A Deleuzian Framework, we then introduce a series of theoretical concepts derived from Deleuze's (1981/2003) *Francis Bacon: The Logic of Sensation* that we apply as an analysis of *code_red*. We explore concepts of the Figure and Fact, Sensation as Intensification and Spasm, Coupling as Merging and Resonance, and Painting Invisible Forces, to illustrate and explore the parallels and divergences between Deleuze's interpretation of the paintings of Francis Bacon and our female/machine framework. In the discussion, we expand upon the analysis to juxtapose and link the theoretical concepts to contemporary issues and ideas in human–robot interaction, psychology, philosophy, and choreography that are informing the field at the intersection of creative, industrial and social robotics. Finally, in the conclusion, we highlight how future research in robotics can be adopted to extend human–robot interactions beyond choreographic practises or functional programming as we commonly understand it, towards investigating a merging and sensation (through a lens as offered by Deleuze), and to enquire deeper into our intimate relationships and exchanges with the machines and artificial intelligences that form part of our lifeworld.

15.2 Code_red

Code_red (1 m 50 sec, 2021) is a video artwork[1] that addresses ethical issues of intimacy, agency, vulnerability, and trust in human-robot relations. It portrays a solitary female figure awaiting the painting of her lips in red by a robot. In what seems like a simple, perfunctory action sequence by the robot, through a series of images that are deliberately intense and ambiguous, we aim to unsettle the viewer and call into question the integration of a robot in established cultural rituals of feminine grooming and presentation.

Its technical dimension explores the way in which an industrial, six-axis articulated robotic arm (KUKA KR10) and customised end effector can be programmed to enter the safety zone around a situated participant to then apply lipstick to the human lips. A simple robotic protocol serves to generate a basic path and incrementally builds up factors including participant posture, tool position, path of approach, and path of deposition (see Fig. 15.2). This is based on an initial facial scan, and transfers of data through a series of computational softwares bridging point cloud data to parametric modelling in Grasshopper GH program for lip contours, and robot programming in KUKA|prc for several types of robot motion. The robot motion packages include distinct gestures for applying the lipstick, including a 'dabbing' (with partial deposition at point locations), drawing lines from the middle section of the lips, a continuous contour tracing as a complete loop from start to end point as motions relative to the human as an object. A female participant is positioned in the designated KUKA KR10 robotic workspace and zone of reach. Calibrations for participant and robot location were undertaken as 'freeform' setup (devoid of restrictive securing and fixing of the participant's head or shoulders for safety reasons). The robot paints the bottom lip twice in a motion from left to right, and then draws on the top lip from the centre to one side and then the other. The angle of the lipstick is varied along the path to achieve a better coverage in a first approximation of a real-life scenario of lipstick application.

The subsequent videographic dimension explores several setups between representations of a human face, and the interaction between the industrial robotic arm and a human subject. The video was developed for a programmed robotic motion of an industrial robotic arm, with an initial application of a lipstick on a facial substitute. These studies are run over a 3D printed face mounted within the robotic workspace in order to confirm the conditions of the setup, including incremental release of the tooltip, positioning of a participant, robotic movement line over mouth, robot general trajectory before and after accessing the lips for drawing and to trial perceived expressive performance of the robot arm. The robotic movement is highlighted as slowly approaching the face, with a rotational and complex choreography that at times echoes animal motion. The cinematography opens scenarios of rupture and breach, where the lipstick is deposited and tracked, smudged as a smear on the face proxy. Through videographic investigation, the relationships between robot and human, the

[1] *Code_red* video viewable at https://vimeo.com/636034376.

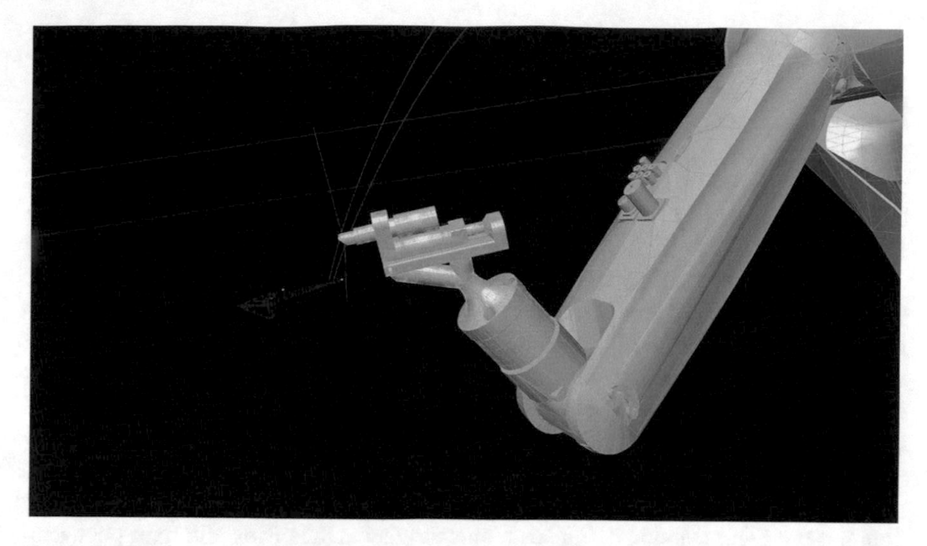

Fig. 15.2 Simulation of robot programmed to draw on human lips

concept of programming, the use of material on the surface of a living entity, and the construction of boundaries (and breach thereof) are called into question.

Moreover, the analytic review of the movie sequence revealed a human response to the robotic touch that had not been anticipated, in the form of small actions, gestures, and affective responses (sensations) that the female experienced. The video traces these discrete responses of the participant to the approaching robot, where she adjusts her face and body position, retracts or moves into the robot, and additionally actively forms her lips, in correspondence to the effective robot position and movement, stage by stage. To understand this better, we reviewed *code_red* through the critical lens of French philosopher Gilles Deleuze's reading of British artist Francis Bacon's paintings.

15.3 A Deleuzian Framework for Body, Gesture, and Sensation in Human Robot Interactions

We adopted Deleuze's *Francis Bacon: A Logic of Sensation* (1981/2003) to derive a framework for analysis of *code_red* and its particular impacts and pathways for new potentials in human-robot relationships. This framework provides a conceptual anchor for reconstructing a female/machine framework. The following section is divided into four parts: *Figure and Fact: Bodies Kept in Motion*; *Sensation: Intensification and Spasm*; *Coupling: A Merging and Resonance*; and *Painting Invisible Forces: A Topography of Lips*. Each of these sections is closely aligned with key moments in Deleuze's analysis, and is correlated to moments in *code_red*.

15.3.1 Figure and Fact: Bodies Kept in Motion

A round area often delimits the place where the person—that is to say, the Figure—is seated, lying down, doubled over, or in some other position. [...] [These techniques] do not consign the Figure to immobility but, on the contrary, render sensible a kind of progression, an exploration of the Figure within the place or upon itself (Deleuze 1981/2003, p. 1–2).

From the start, the Figure has been a body, and the body has a place within the enclosure of the round area. But the body is not simply waiting for something from the structure, it is waiting for something inside itself, it exerts an effort upon itself in order to become a Figure. Now it is inside the body that something is happening; the body is the source of movement. This is no longer the problem of the place, but rather of the event (Deleuze 1981/2003, p. 15).

Deleuze outlines a placement of the Figure, a positioning of the body in Bacon's paintings that isolates the Figure. There is a spatialising material structure that contains the Figure, and in doing so creates the fact, a localising and fixing determination of the body in space. The figure is expressed primarily through a movement by which it is confined, thus confining the figure itself. This movement is actioned by the figure and takes place within the dimensions set by the fact (an action programme) or the framework (as a spatial device, the area, ring, or frame). Through this, dependencies are constructed between Figure and Fact as tensions that are held until a moment of violent release, where the figure temporarily experiences/is overcome by an internal/internalised sensation. The triptych painting *Three Studies of Lucian Freud* (1969)[2] by Bacon aptly demonstrates these ideas.

Code_red translates the dynamics described in Bacon's paintings to a performative and explorative sequence of human–machine interactions. As can be argued, the round area of Bacon's paintings holds the figure in the same way as a workspace for interaction setup would, delineating robot reach and intersection with workspaces, planes it is programmed to address, or a human's personal zone (see Fig. 15.3). In the setup with a female and a six-axis industrial robotic arm, the robot thus becomes the fact by which the female figure is held, both as a literal mechanic device and through the motion trajectory the robot executes, thus opening up the tension within the conditions of the figure towards the choreography between the female and the machine. The robot is the physical fact upon which the female is oriented, in expectation of the process that is about to be executed, and so the robot provides a frame. However, this frame is, more accurately, determined by the robot motion trajectory and movement path.

[2] Francis Bacon, *Three Studies of Lucian Freud* (1969), https://www.francisbacon.com/artworks/paintings/three-studies-lucian-freud.

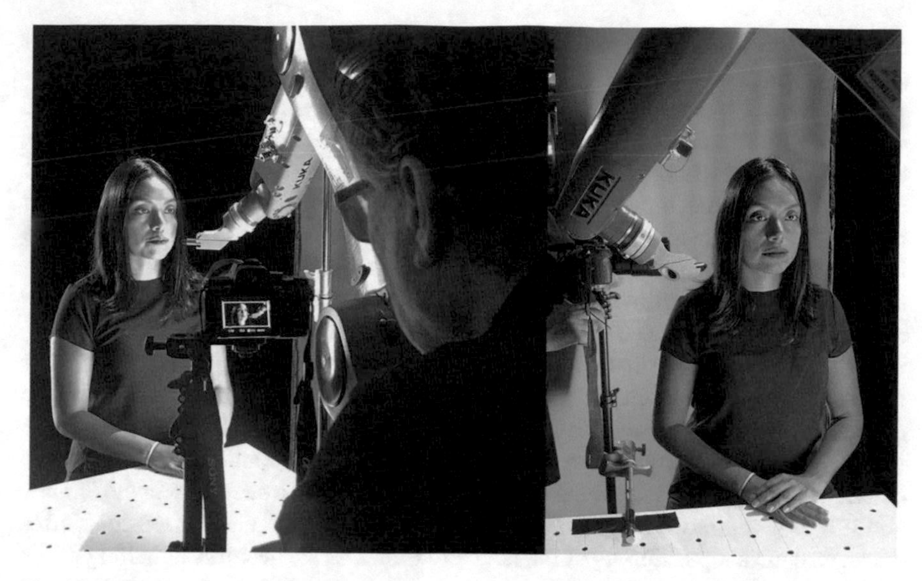

Fig. 15.3 Shoot for *code_red*: the human is precisely positioned in relation to the robot workspace to ensure accurate calibration. Or, in Deleuzian terms, the human Figure is captured in the Frame of the robot

15.3.2 Sensation: Intensification and Spasm

> The Figure is the sensible form related to a sensation; it acts immediately upon the nervous system, which is of the flesh, whereas the abstract form is addressed to the head, and acts through the intermediary of the brain, which is closer to the bone. [...] Sensation is the opposite of the facile and the ready-made, the cliché, but also of the "sensational," the spontaneous, etc. Sensation has one face turned towards the subject (the nervous system, vital movement, "instinct," "temperament" [...] and one face turned towards the object (the "fact," the place, the event) (Deleuze 1981/2003, p. 34)
>
> Sensation is what is painted. What is painted on the canvas is the body, not insofar as it is represented as an object, but insofar as it is experienced as sustaining *this* sensation (Deleuze 1981/2003, p. 35)

Bacon depicts figures being captured in the frame. Yet far from rendering a passivity of the Figure, his techniques of painting sensation convey an intensification of feeling that acts directly on the nervous system of the viewer. As Deleuze explains, it is through "a movement 'in-place,' a spasm" (Deleuze 1981/2003, p. 41) produced in part by the painting of bodily deformations that "*the action of invisible forces on the body*" (Deleuze 1981/2003, p. 41) is revealed. In *Study for Bullfight*

No. 1 (1969),[3] the bodily deformations of man and bull contain the vital movement of their violent encounter.

In *code_red* (with reference to Fig. 15.4), the apparent stillness or passivity is required by the female in order for her body to be in the precise location for the robot to draw on her lips, and continuing across this topography through the programmed motion. Furthermore, amidst this being-held-tight, there is a moment of quiver and twitch. This intrinsically agitated micro movement (or 'spasm') calls into question the female as the object ready to be drawn upon. The quiver interrupts the determined formation of the lips, a small movement that precedes the moment of physical contact—and which is open to interpretation. Is it an expression of fear or hesitation, with the full knowledge that with a fast and forceful move the robot might damage the tissue and thus breach the programmed contract? Or is the woman in a state of anticipation, with the internalised sensation running through the captured body unable/unwilling to retract from the approaching robot? It could be argued that the externally oriented body movement of the robot introduces a sensation in the female, and it is at that point that her body is escaping through the relieved tension, introduced through a repositioning of the field and frame, or the gestural signalling of the other moving body, the robot touch.

It is here that Deleuze offers another hypothesis, whereby "the levels of sensation would be like arrests or snapshots of motion, which would recompose the movement synthetically in all its continuity, speed, and violence." (Deleuze 1981/2003, p. 30). In that sense, the relationship between the two bodies of robot and female enters a decomposition of motion and movement into temporal frames that both hold the body postures, but also an intensive, intensified feedback on the internal ongoings of the person subjected to the robot motion trajectory.

15.3.3 Coupling: A Merging and Resonance

It is a characteristic of sensation to pass through different levels owing to the action of forces. But two sensations, each having their own level or zone, can also confront each other and make their respective levels communicate. Here we are no longer in the domain of simple vibration, but that of resonance. There are thus two Figures coupled together (Deleuze 1981/2003, p. 65).

What produces the struggle or confrontation is the coupling of diverse sensations in two bodies, and not the reverse, so that the struggle is also the variable Figure of two bodies sleeping intertwined, or which desire mixes together, or which painting makes resonate (Deleuze 1981/2003, p. 69)

In Bacon's striving to paint portraits that were neither figuration nor abstraction, he eschews narrative to get at pure sensation. When two figures are painted, there is a coupling of figures that are couplings of sensation(s). It is the confrontation of two

[3] Francis Bacon, *Study for Bullfight No. 1* (1969), https://www.francisbacon.com/artworks/paintings/study-bullfight-no-1.

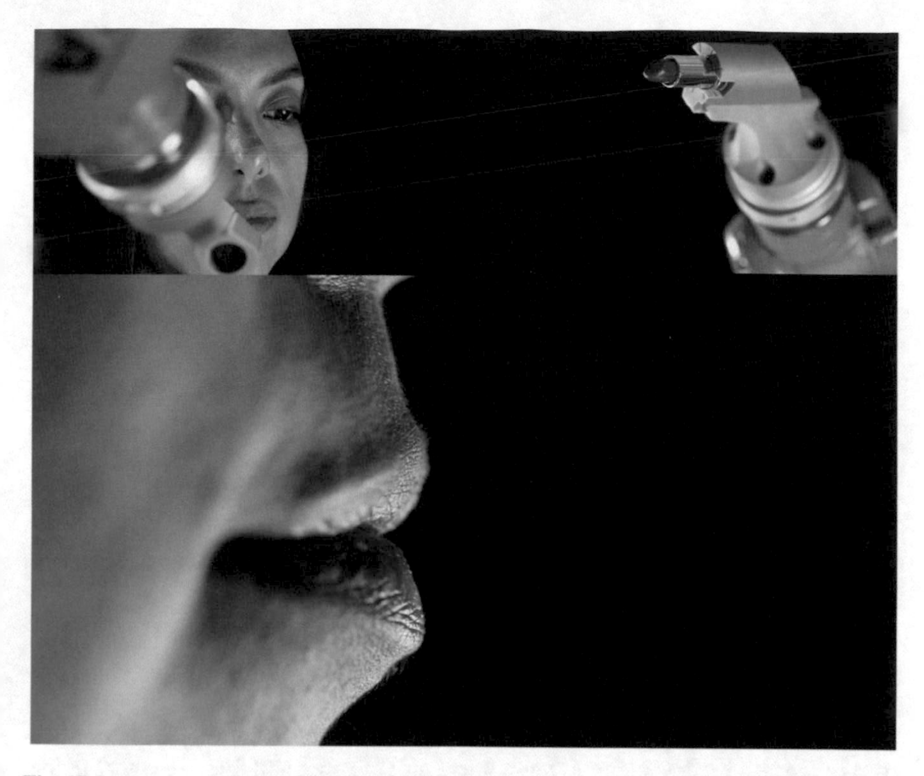

Fig. 15.4 *code_red* video stills depicting the spasm of sensation that belies the anticipation of approach by the female figure as the robot advances

sensations that produces resonance. According to Deleuze's logic, within the actions and interactions of two figures that are both juxtaposed and interrelated with each other in the programmed and choreographed movement, and through space and time, then vibration becomes resonance. In the triptych painting *Three Studies of Figures on Beds* (1972)[4] by Bacon, it is difficult to discern where one body starts and the other begins. There is an interplay of sensations and energies between bodies that resonates in a rhythmic complexity.

With *code_red*, a coupling of bodies takes place between the robot and the human. The robot is no longer solely the frame, but is part of the figure, or more accurately, a second figure. The two figures operate at different levels of sensation: the robot through its expressive motion, the human through an illusion of stillness (see Fig. 15.5). The robot is the dynamic, animated entity at this scale. It is programmed to move with behaviours that imply surveying, circling, and approaching the human figure. A predator and prey. Flexing and gyrating, it takes its time before launching in

[4] Francis Bacon, *Three Studies of Figures on Beds* (1972), https://www.francisbacon.com/artworks/paintings/three-studies-figures-beds.

Fig. 15.5 *code_red* video stills depicting juxtaposed actions/reactions of female figure and robot operating at different levels of sensation

on its target. The female figure—the anchor, the pivot point. Already here a resonance and rhythm are established.

Moreover, the juxtaposed actions/reactions, the joint movement and countermovement of the two bodies render a (temporary) unit of robot and female, where in that movement and in the response both become one, in the same way that two dancers become one entity, temporarily, in the dimensions of time, space, and process. Both bodies cannot be considered absolutes; they are not static in space or even stable entities. Female body and robot body exist as gradient conditions that expand and contract across space and time, and across multiple versions and combinations in reference to the other. This is where the machine setup exceeds the potential of the painting: the dynamic capture of the robotic framework allows for multiple dimensions across which the female figure is both captured and constituted in a frame and released through an infinite number of actions. A coexistence of two independent but synchronised bodies, the biological and the machinic, for interaction with each other, where in a single connected movement these bodies become one organisational unit.

A tension builds as the moment of contact is drawn out. The close-up shots of the female figure's lips intensify the vital forces that reside internally and occasionally flicker into visible form. She leans forward into the robot in an almost imperceptible gesture, into the machine movement to get closer to the applicator. In a very literal sense, the contact between robot and female figure is one of touch. In merging, contact can be conceived of as one body-entity, no longer a distinct separation between the two bodies, a giving over to the other, a resonance. An act of intimacy that allows something, someone, the other to cross that boundary and enter into your private zone. What is this act of approach and contact—is it occupation? Is it the danger of penetration of the hard and cold steel into the soft and quivering flesh? Is there an element of volatility to this violence, or is it simply the imagination of what else could happen in an instant that is unguarded? Its soft application, a tender caress … or the danger that would imply that what is touched is ultimately destroyed in this moment.

What is less visible but revealed in the programming of the robot and interview with the female participant, is the active movements of the muscles of the lips to produce a necessary resistive force in receiving the application of the lipstick. Despite the miniscule scale of movement required—a micro-gesture—this active agency by the human is significant. We return to this idea of the micro-gesture in the Discussion.

15.3.4 Painting Invisible Forces: A Topography of Lips

It is as if invisible forces are striking the head from many different angles. The wiped and swept parts of the face here take on a new meaning, because they mark the zone where the force is in the process of striking. This is why the problems Bacon faces are indeed those of deformation and not transformation. These are two very different categories. The transformation of form can be abstract or dynamic. But deformation is always bodily, and it is static, it happens at one place; it subordinates movement to force, but it also subordinates the abstract to the Figure. When a force is exerted on a scrubbed part, it does not give birth to an abstract form, nor does it combine sensible forms dynamically: on the contrary, it turns this zone into a zone of indiscernibility that is common to several forms, irreducible to any of them, and the lines of force that it creates escape every form through their very clarity, through their deforming precision (Deleuze 1981/2003, p. 58–59)

In rendering invisible forces visible, Bacon uses deformation of the Figure, often the head, to illustrate "the forces of pressure, dilation, contraction, flattening, and elongation that are exerted on the immobile head" (Deleuze 1981/2003, p. 58). The origin of these forces may be attributed to "the most natural postures of a body that has been reorganised by the simple force being exerted upon it: the desire to sleep, to vomit, to turn over, to remain seated as long as possible" (Deleuze 1981/2003, p. 59). The deformed face in *Three Studies of Henrietta Moraes* (1969)[5] by Bacon is less about an apparent violence, and more a record of the dynamic forces at work in that life.

Unlike the majority of human–robot collaborations where the human and robot work together on an object or surface, in our case of the lipstick, the human body becomes the surface upon which the robot draws. The lips are a topography, a canvas on which force is exerted. But being human, the lip surface is a muscle and thus a mobile topography, at once a canvas and an active agent. The landscape of the mouth is a sensitive topography, and at the same time the area with the highest muscle control. It is this area that becomes an arena of exchange, where the mouth meets

[5] Francis Bacon, *Three Studies of Henrietta Moraes* (1969), https://www.francisbacon.com/artworks/paintings/three-studies-henrietta-moraes-1.

the machine, and where this area of muscle control actively receives the approaching agent that would place the undetermined liquid material onto her lips.

While Bacon paints the invisible forces occurring on and represented by damaged bodies that dissolve through motion, the body drawn upon in *code_red* is equally endangered (see Figs. 15.6 and 15.7). The robot drawing on this pair of lips is considered a breach of regular standards for human–machine interactions. It impedes on the standard zone that is commonly reserved for ensuring the safety of the easily damageable human body relative to force and speed, but more importantly the non-negotiable trajectory of the robotic arm. The nature of the soft and vulnerable flesh stands here in stark contrast to the dominant precision of the machine. The lips are an interface at which external and internal forces meet; the stroke of lipstick by the robot, the internal quiver of the body. Despite the apparent passivity of the human locked into the frame by the motions of the robot whilst awaiting the application of lipstick, it is through the micro-gestures of the lips that an active agency is evident. And here the lips also carry another dimension of danger or endangerment—situated at the mouth, an orifice (where the robot is close to entering the female body itself).

In fact, the female can be understood as objectified but as a multiple—as but one of many in a factory of robotic repetition, marked towards a universality, and lack of difference. Here the lips become a territory for domination. Through technologies of digital scanning and classification of facial features, machine learning algorithms

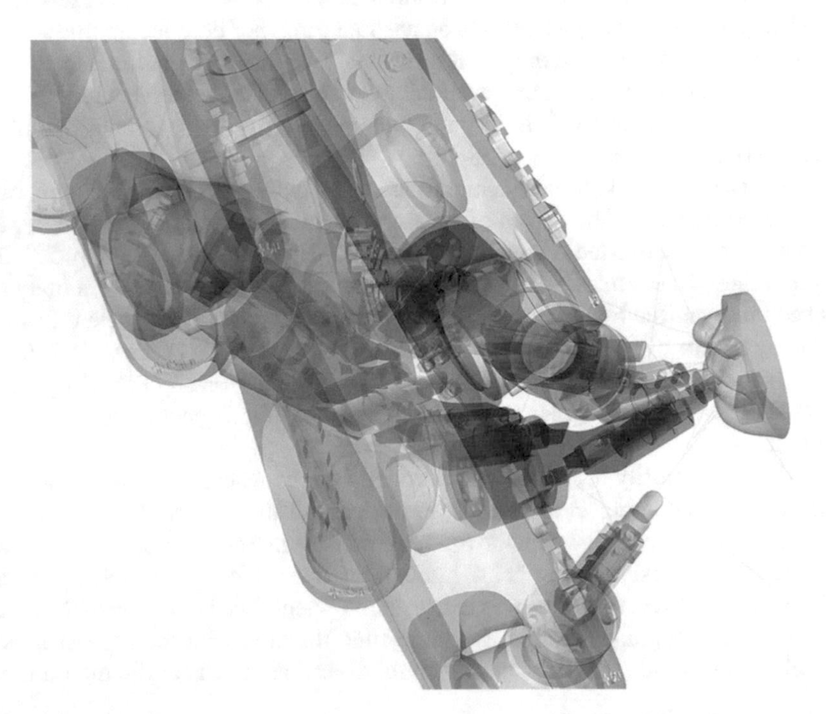

Fig. 15.6 Simulation of the process of robotic motions in drawing upon the lips reveals what looks like a swarming attack of forces

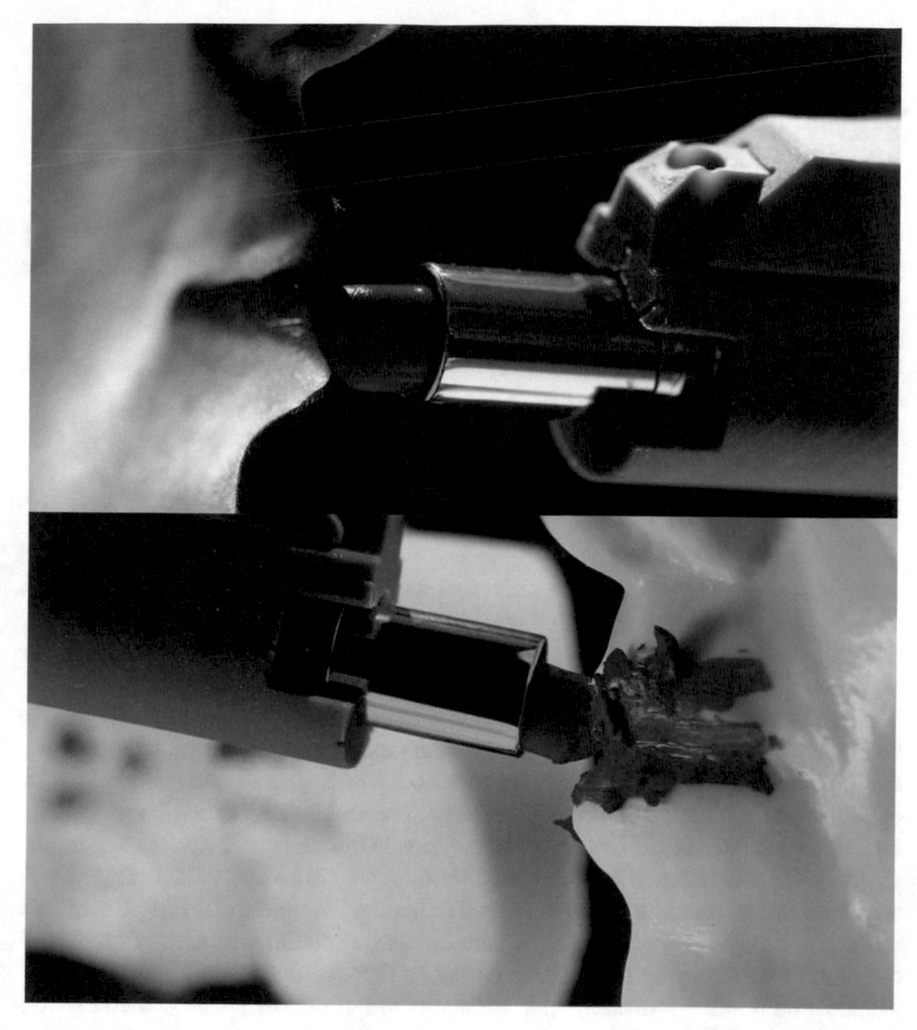

Fig. 15.7 *code_red* video stills depicting the painting on human lips and factory faces by the robot; the contrast between a delicate caress and a violent force

driving computer vision sort and classify lips of action and desire into a standard typology, stripping them of their unique plasticity and agency.

And finally, whereas from a robot programmer's technical perspective the geometry of the human lips is a complex three-dimensional topography, the application of colour through the lipstick is a variable in itself, a diminishing source where material is layered. This topography of the lips is an interface where external and internal forces meet—the drawing force of the robot and the compressed muscle activated by the female. Smudging the colour over the contours, moving over this soft hill, moving into areas beyond the boundary, leaving the domain of representation, going beyond where the lipstick should be applied, towards other areas within the face, literally turning the colour into the paint for war.

15.4 Discussion

Now that we have worked through an analysis of *code_red* using the theoretical framework from Deleuze, we turn to a broader consideration of how this particular analysis ties back into current thinking in collaborative, industrial, and social robotics. We frame this discussion through concepts of breaching the intimate zone, resonance, and micro-gestures.

15.4.1 Breaching the Intimate Zone: Safety and Danger

When robots were introduced into industrial workplaces, the goal was robotic assistance of manufacturing tasks with the human as operator (Colgate et al. 1996). More recently, the shift to collaborative robots (cobots) places the robot in close proximity to the human, triggering a rethinking of what is considered safe (Martinetti et al. 2021). Whereas in manufacturing contexts the robot works on a physical object, in the domestic scenarios of social robotics for personal grooming that we are exploring, the human body becomes the object of the robot's task. The human is no longer the operator but a recipient of the robot's actions. The soft, fleshy human body poses an interesting challenge to human-robot interaction in the context of repurposing industrial robotic arms so that they are more explicitly geared to address a human's personal space. Here, the coupling of two bodies (the human/biological and robot/machinic) through action is far from simple, as the industrial robotic arm is usually considered to remain in a safe distance so as not to damage the human body. Whereas in industrial robotics considerations for safety are of utmost priority, social robots are commonly built to interact with people through inbuilt sensor systems and clad in protective materials (which keep safe humans as much as the machinery of the robot). However, here we posit safety as an approach, not a physical distance, cage, or cladding. Unlike benign social robots that "have to comply to rules of safety, friendliness and legibility in order to facilitate interaction with the humans" (Granjon 2017, p. 4), artist Paul Granjon makes non-benign robotic art machines that bite and kick, and thus pose a danger to humans. The kinetic behaviour of these robots establishes thresholds of safety and danger.

How is this robot motion decoded and interpreted by the human participant? The expressive character of the robot motion on its approach as it nears in proximity to the face of the human is critical in establishing a sense of safety in the person. The qualitative dynamics of the robot motion (its path, speed, and acceleration) contribute to the affective impact on the human (Saerbeck and Bartneck 2010; Venture and Kulić 2019). Too fast and direct may be read as violent and aggressive—a confrontation, an attack, depending on the context of the gesture (Chartier et al. 2017). Alternatively, slowing down on approach may be read as non-threatening, as caring and considerate.

Besides the affective experience of robot motion, the concept of affective touch in social robots (Kerruish 2017) requires consideration for cases of personal grooming by robots where the robotic tool comes in contact with the soft, malleable surface of the human body. Questions arise as to how to choreograph the machine actions through space, as well as the gestural/tactile interaction with the human body (in our case, the lips). The transition from large gestures and motion paths of the robot in approaching the person to small, delicate gestures and pressure on human skin raises issues of invasion versus integration of personal space with the robotic arm and its workspace. The creative practise of project:Galath3a explores similar issues of trust and consent in human–robot interaction and personal grooming, with the face as a canvas for the robotic brush strokes.[6] The artists put their own bodies on the line to explore and test how it feels to receive robotic touch as a potential breach of the intimate zone, that might transform into a synergistic coupling of human and machine.

15.4.2 *Resonance as Kinaesthetic Empathy*

Returning to the Deleuzian concept of resonance, where the interplay of sensations between bodies leads to a merging or blurring of boundaries between human and machine, we note a correspondence with contemporary theorising on human-machine relations that decentre the human and argue for non-human agency. In the language of agential realism, Gemeinboeck uses the term "intra-bodily resonances" (Gemeinboeck 2021) to refer to an embodied empathy (Despret 2013), an attuning to bodily energies, vibrations, rhythms, or a relational dynamics between humans and machines. At this level of embodied or kinaesthetic empathy (Foster 2010), it is less about modelling robots after humans—one of the primary criticisms of dominant approaches to social robotics (Gemeinboeck 2017)—than acknowledging the shared primordial movement language across species (Sheets-Johnstone 1999). Qualitative movement dynamics are a dance between entities at varying scales and intensities. Processes of attunement take place in sensing, feeling, adapting and sustaining vital relationships between humans, animals and machines. In Gemeinboeck's relational-performative aesthetics approach to the design of human-robot interaction, "positioning oneself in the middle of the encounter and the relationships it produces deliberately undermines focusing on the individualism of interacting agents and, instead, promotes attending to the crisscross of perceptual flows, movement dynamics, and emergent effects that give rise to meaning" (Gemeinboeck 2021, p. 2).

Drawing on the concept of kinaesthetic empathy from dance (Foster 2010) and the associated theory of mirror neurons (Gallese et al. 1996), along with recent work

[6] https://awards.mediaarchitecture.org/mab/project/338.

on the relationship between kinetic objects (non-human entities) and kinaesthetic empathy (Miyoshi 2019), from the human perspective the robot motions convey expressive qualities, forces, tensions, and rhythms that are felt internally by the observer. The internal sensation of observed movement enables the human to feel and recognise the movement patterns of the robot in a qualitative and affective sense, for example, as aggressive or nurturing. In future living with robots in close proximity and intimate scenarios, where bodily intertwining and boundary crossing occurs, programming of robotic motion behaviours must go beyond functional programming towards a choreography of flows, sensations, intensities and resonances between bodies, human and machine.

15.4.3 Micro-Gestures as Active Agency and Collaboration

The quivers and twitches of the female figure in *code_red* were interpreted through the Deleuzian concept of sensation as the outward expression of the internal intensification and vibration of the body; a release of tension or suspension as a spasm, a micro-expression. The capture of the female figure in the frame of the robot generates a set of forces producing this affect. Affect as understood by Deleuze is not equivalent with common conceptions of affect that reduce it to categories of emotion found in psychology and human-robot interaction studies.

In psychology, a micro-expression is defined as an involuntary facial movement that is connected to an emotional state and of very short, almost imperceptible duration (Haggard and Isaacs 1966). The detection of micro-expressions can be clues to hidden emotions and deceptive behaviour (Ekman and Friesen 1969), although recent studies contest claims that micro-expressions can be reliably used in lie detection systems that analyse facial expressions (Burgoon 2018). More recently, in the research domain of robotics, micro-gestures of the entire body are proposed as an extension of micro-expressions of the face for detection of emotional state from body movement (Chen et al. 2019).

The fleeting twitches of the lips in our study could be viewed as micro-expressions, belying an emotional response to the approaching robot. However, our use of the term *micro-gesture* is not one of affect, but one of active agency. The close-up details of the moving image reveal the movement within that landscape, the micro-gestures that flex and pull the muscles in this topography of flesh—an active movement on behalf of the female participant reaching towards the lipstick to make contact. And in meeting (there can be no shying away), in providing an active resistance to the oncoming force, the micro-gestures can be interpreted as collaborative. Active yet almost imperceptible work is done at this small scale of haptic interaction by the human agent to collaborate with the robot and enable the successful application of the lipstick. It should be noted that the haptic relationship of the robot-lipstick-lips is a muted and asymmetric version of the reciprocal interaction of the human's hand-lipstick-lips, in which the human hand lipstick and the lips engage in a haptic choreography as equal partners. This dance of the lips with the lipstick is unique for each woman, as materially documented in the sculpted shape of each woman's lipstick, and captured in the photographic series *Lipstick* (1992) by Stacy Greene.[7]

15.5 Conclusion

While the original term for and function of robots stems from *R.U.R.—Rossums's Universal Robots* by the Czech playwright Čapek (1920), where robots in human form are manufactured in support of human labour, we expand in *code_red* the robot-human relationship to the universal female—the woman supported by the machine, in interaction with the machine, in response to the mechanic touch, in preparation for future encounters. As we have argued, this interaction is by no means subservient but a highly intense, at stages relieved, motion-driven, and importantly internalised and sensuous choreography of movements, where the female participant holds agency rather than being subjected by the machine.

Taking feminine rituals of bodily grooming and presentation as a focus, we explore the intermingling of biological and machinic moving bodies, allowing vulnerable moments of touch. Our methodology of robotic programming and filmmaking makes visible conceptual and ethical challenges in introducing an industrial robotic arm to apply lipstick to human lips. From the robot perspective, the human body is rendered as object and surface, yet the human maintains agency in a micro-gesture of the lips, and more importantly, in the internal sensation that she experiences that is instilled by the robotic touch.

Through creative artwork and cinematographic tracing of human–robot relationships in *code_red*, our focus here resides not primarily with classic tropes of cultural storytelling; nor of technical solutions to the robot applying lipstick to a woman's lips. Instead, we pursue dimensions of sensation and intensity, affect, and resonance within physical, and consequentially emotional, interactions with machines that can contribute to the current discourse. Like Deleuze's analysis of Bacon's work, our female-robot narrative is subordinated to a series of images (visual and aural) that, in our case, attempts to produce a disturbance, an ambivalence, regarding how to interpret the relations between robot and human, and so question the way in which we construct agency, tasks, and functions for both.

In a forward trajectory, we thus speculate on the way in which in future worlds an ecology of social or assistive robots, or creative robots, will participate in, determine, contradict, or enrich human activities of personal, intimate actions that situate the body and its emotions, its internal world, through small acts of touch. *Code_red* opens a new understanding of the female activist and female body as object of desire or control, as agent provocateur, with implications for the role of robots in direct interactions with the innumerous, amorous, and armoured bodies that humans possess over their lifetime, from childcare to assisted elderly living. By integrating the robot into highly individual and personalised rituals, new forms of collaboration can

[7] https://stacygreene.com/home/projects/lipstick/

emerge that couple human agency with care databases expressed through the data-linked robot body, traversing the affective boundary between humans and machines. Then, as this chapter argues, the disruptive feminine can become a springboard for thinking through how robots can counter the taboo of touch entrenched in cultural narratives and practises.

Acknowledgements We are grateful for the contributions and collaborations to produce the *code_red* film and project by Paul Warren (cinematography, videography), Lindsay Webb (sound design), Susana Alarcon (model and DMAF technician), Lynn Masuda (robot programmer), and the support of DMAF, Sydney School of Architecture, Design and Planning, University of Sydney.

References

Burgoon JK (2018) Microexpressions are not the best way to catch a liar. Front Psychol 9:1672. https://doi.org/10.3389/fpsyg.2018.01672

Čapek K (1920) R. U. R.: Rossum's universal robots. Aventinum

Chartier G, Berthoz A, Brian E, Jaisson M (2017) Violence and uncertainty: interactional sketches for a cognitive analysis of violent actions. Soc Sci Inf 56(2):198–219. https://doi.org/10.1177/0539018417694772

Chen H, Liu X, Li X, Shi H, Zhao G (2019) Analyze spontaneous gestures for emotional stress state recognition: a micro-gesture dataset and analysis with deep learning. In: 2019 14th IEEE international conference on automatic face gesture recognition (FG 2019), pp 1–8. https://doi.org/10.1109/FG.2019.8756513

Colgate J, Wannasuphoprasit W, Peshkin M (1996) Cobots: robots for collaboration with human operators. In: Proceedings of the 1996 ASME international mechanical engineering congress and exposition; conference. pp 433–439, 17–22 Nov

Deleuze G (1981/2003) Francis Bacon: the logic of sensation. Continuum

Despret V (2013) Responding bodies and partial affinities in human–animal worlds. Theory Cult Soc 30(7–8):51–76. https://doi.org/10.1177/0263276413496852

Ekman P, Friesen WV (1969) Nonverbal leakage and clues to deception. Psychiatry 32:85–105

Foster S (2010) Choreographing empathy: Kinesthesia in performance. Routledge, London. https://doi.org/10.4324/9780203840702

Gallese V, Fadiga L, Fogassi L, Rizzolatti G (1996) Action recognition in the premotor cortex. Brain 119:593–609

Gemeinboeck P (2017) Creative robotics: Rethinking human machine configurations introduction (editorial). Fibreculture J 28:1–7. https://doi.org/10.15307/fcj.28.203.2017

Gemeinboeck P (2021) The aesthetics of encounter: a relational-performative design approach to human-robot interaction. Front Robot AI 7:217. https://doi.org/10.3389/frobt.2020.577900

Granjon P (2017) This machine could bite: On the role of non-benign art robots. Fibreculture J 28. https://doi.org/10.15307/fcj.28.208.2017

Haggard E, Isaacs K (1966) Micro-momentary facial expressions as indicators of ego mechanisms in psychotherapy. Appleton-Century-Crofts, pp 154–165

Kerruish E (2017) Affective touch in social robots. Transformations J 116–135

Martinetti A, Chemweno PK, Nizamis K, Fosch-Villaronga E (2021) Redefining safety in light of human-robot interaction: a critical review of current standards and regulations. Front Chem Eng 3:32. https://doi.org/10.3389/fceng.2021.666237

Miyoshi K (2019) What allows us to kinesthetically empathize with motions of non-anthropomorphic objects? 4:52. https://doi.org/10.5278/ojs.jos.v4i2.2447

Saerbeck M, Bartneck C (2010) Perception of affect elicited by robot motion. In: Proceedings of the 5th ACM/IEEE international conference on human-robot interaction, HRI '10. IEEE Press, pp 53–60
Sheets-Johnstone M (1999) The primacy of movement. John Benjamins Publishing Company
Venture G, Kulić D (2019) Robot expressive motions: a survey of generation and evaluation methods. J Hum -Robot Interact 8(4). https://doi.org/10.1145/3344286

Chapter 16
On Display: Robots as Culture

Deborah Turnbull Tillman and Mari Velonaki

Abstract Robots are necessarily transdisciplinary things. Like everything that occupies that space in-between our taxonomies, or that third space, they can evoke strong feelings of curiosity or fear. The elements of variance and verisimilitude they can embody create a distance, another space, wherein curators can draw attention to the cultural aspects of robotics by researching and displaying the 'stuff' of robotics in cross-disciplinary contexts, such as exhibitions. This chapter will focus on the exhibitions of artist Mari Velonaki and Deborah Turnbull Tillman (in collaboration with fellow curators) whereby elements of robotics have come into proximity with exhibitions on art, design, computers and engineering. Their display in the context of collaborative making, audience engagement and notions of authenticity makes them social, and by extension, cultural.

16.1 Introduction

Museums and galleries are spaces that have helped establish clear lines across disciplines in the mind of the public. At a time when cultural platforms are entering an interdisciplinary phase, the exhibition of robotics is paving the way to crossing these disciplines, particularly in relation to new research. Introducing layers of information technology to the exhibition floor has made it so that these categories are able to become more malleable, more permeable. The ability to de-silo strict taxonomies has become possible.

This chapter examines curatorial and creative relationships that author Turnbull Tillman has established around the display of robotic materials. Conversations with curators Matthew Connell, Dagmar Reinhardt and Lian Loke, with artist and author 2 Mari Velonaki, bring into focus the current understanding of how robotics operate,

D. T. Tillman (✉) · M. Velonaki
Creative Robotics Lab, University of New South Wales, Sydney, Australia
e-mail: deborah.turnbull@unsw.edu.au

M. Velonaki
e-mail: mari.velonaki@unsw.edu.au

© The Author(s), under exclusive license to Springer Nature Switzerland AG 2023

B. J. Dunstan et al. (eds.), *Cultural Robotics: Social Robots and Their Emergent Cultural Ecologies*, Springer Series on Cultural Computing,
https://doi.org/10.1007/978-3-031-28138-9_16

who designs them, how they function and what the future may hold as robots become less spectacle, and more tool, toy or companion. The experimental nature of the art gallery holds a different function than the historical receptacle of a museum, but in the case of relational research with Turnbull Tillman as an intermediary, researchers can work across sites, across ideas and across materials to gain a clearer understanding of robots as culture.

In trying to define robotics, cultural professionals tend to think in terms of engineering and ingenuity. Robots are strong, they are accurate, they are fast, they have incredible repeatability and they don't get tired, bored, disobedient or sick. They tend to be considered, first and foremost, as ideal extensions beyond our human limitations. But they are also *other* things. They were initially created as automata, as objects of wonder and speculation by genius clockmakers. They ask us to contemplate what is beyond our limitations as humans by posing philosophical questions, such as what does it really mean to be alive? What does it mean to have agency? What does it mean to be human and possess human traits, both positive and negative? Robots later emerged as mechanistic abstractions of ourselves, where our humanness stops short but technology picks up the slack, as with telescopes, corrective eyewear, calculators and computers. In this chapter, authors Turnbull Tillman and Velonaki will discuss robotic ideas and objects for cultural audiences, and also how the transition through intent and into exhibition creates a kind of categorisation that becomes either relatable or rejectable but is always intriguing.

16.2 Inter-, Cross- and Transdisciplinary 'Things'

> Contexts for research are often forged at the edges of disciplines
>
> (Muller et al. 2015)

In examining the context in which robots come to exhibition floors through different avenues, one can trace them as interdisciplinary 'things' in that they "represent more than one branch of knowledge" (Magnusson 2013; Latour 2004). In existing across defined categories as examples of different disciplines, there is a relational aspect to the categorisation of cross-disciplinary things. They can represent but also "*relate to* more than one branch of knowledge." Transdisciplinarity exists specifically in a research context, where different disciplines are actively working together across categories to create new knowledge, methodologies and ideas because of the related aspects of their different fields. Robotics research exemplifies this approach.

As an example of inter-, cross- and transdisciplinary *things*, robots affect most people's lives in a way that gives them a social aspect. This folds into a larger culture of mechanical and technological pursuits, but ultimately returns to the philosophical, about what it means to be human; about what it means to be able to engage and think, even to decide and act. A good example of how robotics become about more than one thing is the understanding of what robotics are. Where Turnbull Tillman initially would not have classified herself as intentionally curating robots, rather contemporary art made from robotic materials, author Velonaki works in collaboration with experts in software engineering, mechatronics and cultural institutions.

This precedent, where interdisciplinary teams of artists, technologists and curators work together to generate new research specifically across art-science, was solidly set at the Powerhouse Museum with the Beta_space project (Turnbull and Connell 2011) and continues to be explored through the work of Lizzie Muller at the National Institute for Experimental Arts, University of New South Wales (UNSW) (Muller et al. 2015), Matthew Connell at the Museum of Arts and Applied Sciences (Turnbull and Connell 2014) and author Turnbull Tillman through her research initiative New Media Curation and the Creative Robotics Lab, UNSW (Turnbull Tillman et al. 2015).

The third space is a place put forward as a psychosocial innovation to collaborative research in a public context where creative cognition (and recollection) is valued as part of the evaluative process of art-science research. Its roots challenge a 'two-culture approach' flagged as far back as the 1950s by C.P. Snow, which were collated and tested more recently by the Psychosocial Research Group (PRU) at the University of Central Lancaster (UCLAN) in the UK, and most recently by Muller at the UNSW Galleries at the University of New South Wales in Sydney, Australia (Muller et al. 2015). The third space is explored most recently by an international research team comprised of Lizzie Muller, Jill Bennett, Lynn Froggett and Vanessa Bartlett. In short, the third space is a research platform where a visual matrix methodology that prioritises art-sensitive research; that involves scientific inquiry, namely HCI or HRI and can exist in a public space with artists, scientists and the general public working together (Ibid). Previously, only the specialists were consulted, but more recently, through the work of Muller, Ernest Edmonds and Linda Candy (Alarcon-Diaz et al. 2014) and authors Turnbull and Connell (2011, 2014), audiences at interdisciplinary cultural institutions have become the medium through which experience can be gauged. Below are some curatorial examples in which authors Velonaki and Turnbull Tillman work with curators Connell, Reinhardt and Loke.

16.3 Autonomy and Characterisation: Robotics and Culture

Robots as hybrid objects are desirable because they represent a time, space and culture in which the character of a person is imbued on an object (Magnusson 2013) that presumably has agency and the capacity to mimic human behaviour in a technological way. It is a mnemonic device, in a way—a self-reflective object. The types of human behaviour of particular interest would be the ability to mimic thinking or responding to one's own environment. Where there are also machines that do this, the point at which a robot crosses over from being a machine is when it appears to be making decisions in response to the environment it is sensing. The ability for a machine to act *autonomously* characterises it as a robot. It doesn't always have to be humanoid in appearance, but that does pose a question that nags author Turnbull Tillman. Does authenticity factor into a positive engagement with an autonomous system (for her, the audience's engagement with interactive art) (Turnbull Tillman et al. 2015)?

Roboticist and artist Hiroshi Ishiguro has collaborated with the Creative Robotics Lab Director, artist Mari Velonaki (Author 2 and Turnbull Tillman's Ph.D. supervisor). Specific to the work of Ishiguro is that the characterisation he is imbuing his authentic replicas with are those of himself, his daughter, and the cultural (very gendered) stereotypes of Japanese males and females, as found in Gemenoid HI-4, HI-2 and F. The larger context in which Ishiguro designs and realises his hyper-realistic robots is the Uncanny Valley, the curve with which audiences react to animated objects that are clearly not alive, and the repulsion that most people still feel when a non-human humanoid robot too closely mimics human behaviour. Ishiguro's creations are examples of what he articulates as being so close to the real thing that the feeling of discomfort or revulsion characterised by the Uncanny Valley is due to a failure to accurately and authentically mimic human behaviour in robotics (MacDorman and Ishiguro 2006). Indeed, his humanoid robots are close enough to warrant a second look when Ishiguro or his daughters are in photos with them (Fig. 16.1).

Author Velonaki has worked closely with Hiroshi Ishiguro. She hosted his staff and his Gemenoid robots at the Creative Robotics Lab in 2003, and again in 2014. Here Velonaki considered elements of reality and authenticity in her 2009 artwork, *The Woman and the Snowman.* In this installation, Velonaki compares two fictitious

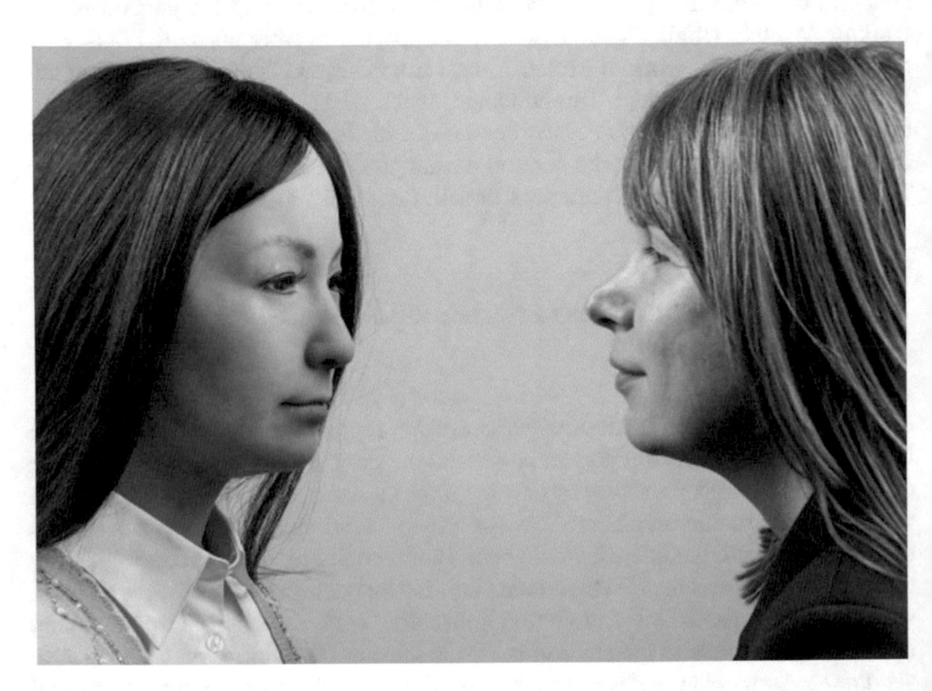

Fig. 16.1 Mari Velonaki with Gemenoid F, 2009

Fig. 16.2 Mari Velonaki, *The Woman and the Snowman*, 2009

characters, a snowman and a woman. Through sound and an abstracted kinetic object, Velonaki explores how technology has encroached on and changed the way people relate to objects and to each other. In showing an obviously fictitious character of a snowman alongside a woman who ends up being a robot, the idea of reality is overturned, left open to contemplation, and exposed (Fig. 16.2).

When installing the *SHErobots* exhibition at Tin Sheds Gallery (Sydney, Australia) in October 2022, Velonaki commented that instead of showing *Fish-Bird* (2002–3), perhaps she should have shown *The Woman and the Snowman.* When asked why, she was contemplating Elena Knox's *Pathetic Fallacy* (2022), which had been staged and filmed at Velonaki's Creative Robotics Lab. Knox's work was compiled as part of her Ph.D. project and includes one of Hiroshi Ishiguro's Gemenoids in the film. It portrays an elderly woman and a young female robot. The elderly woman grooms the younger robot, both admiring and bemoaning her beauty and how she will never age. Author Turnbull Tillman wonders if this exchange could be reminiscent of any intergenerational exchange in her introductory essay to *SHErobots* (Reinhardt et al. 2022, p. 86). Velonaki thought perhaps it would have created a nice discussion between the two pieces, both contemplating what is real and how far the range of 'the Other' extends (Fig. 16.3).

In many ways, these two works explode the traditional roles of women as both decorative and care-giving companions. They toy with the notion of how women are displayed and considered in society, and perhaps that by making strange this relationship, alternate identities might be considered, alternate autonomies reached. There is a similar consideration of the work that Velonaki included in *SHErobots.* In *Fish-Bird* (2002–3), first created at the University of Sydney's Australian Centre for Field Robotics with mechatronics expert David Rye, two wheelchairs behave as companions to each other, and to the audience members that engage with them. As you walk into a designated space, they could be stationary, be caught in a choreographed

Fig. 16.3 Elena Knox, *Pathetic Fallacy*, 2022

dance or printing out messages to each other. Once they sense you, they immediately approach, seeking contact. They don't behave as even electronic wheelchairs might, but instead seek to create a connection with each other and their human visitors, following them and each other around. In Velonaki's own words, "Fish and Bird…fall in love but cannot be together due to technical difficulties. In their shared isolation [they] communicate intimately with each other and their visitors via movement and text" (Reinhardt et al. 2022, p. 91). Although these words have been spoken and written countless times since the artwork's inception, the limited autonomy with which these objects try to connect, looking as they do, in very human ways, has an even stronger impact post-COVID-19 pandemic, when most people remained confined to small spaces over two distinct lockdowns with only one or two other humans for companionship and reliant on text communications for outside contact (Fig. 16.4).

In discussion on the topic, the authors articulate what is missing from conversation around interdisciplinary arts and authenticity:

> **DTT: I notice that a lot of your works have a trajectory.** They don't just show once, but there's an iterative quality to them that makes learning within research possible. And even if you don't know that you're doing it, it's kind of an automatic reflection, and then a shift in perception and a shift in making and exhibiting it again in a different way as part of another conversation. This is, I suspect, how your robots become social as well, is that they're involved in multiple levels of social commentary at any given time.
>
> **MV:** It is important and it's a good point, but because, for example, like with *Fish-Bird*, every time we exhibited, we made it site specific for the location, for the museum or the gallery where it was going to be installed. For example, we connected the robots to online maps, so they possess information about their surroundings, their vicinity. We included vocabulary from the local language. The last time we installed *Fish-Bird* at the Bilbao in Spain, there were many opportunities to include indigenous language samples, so the robots learned a

Fig. 16.4 Mari Veloanki, *Fish-Bird* (2004)

new environment, a new language. This integration to their environment through language gave the sense of current or real time and would keep the [kinetic agents] connected by printing something from the local newspapers every morning.

I feel it's important to give more back, to learn more, to use the platform for other people, to learn, to create, to improve. Our robots have parallel lives outside of the galleries and museums they are exhibited in. We use the [robots] as both demonstrators and research platforms in the labs when they're not in exhibitions. Now we're working on the sound component with *Diamandini*, but there are all these other experiments that can happen in parallel which are very different to the exhibition [scenario]. It's important to show something different to what has come before. So, after all these years *Diamandini* now has a new component, but I would like to incorporate a different sound component that she, that the woman from the *Red Armchair Series*, that *Fish-Bird*, that the *Woman [and] the Snowman* didn't have before; that improves reciprocal interaction in a new way. (Turnbull Tillman and Velonaki 2020)

Each time Velonaki iteratively progresses a robotic artwork, the more it has a chance to learn from humans about human behaviour and the more humans learn about themselves.

16.4 Design and Functionality: ISAAC Versus BAXTER

When looking at the progression of displaying robots, the Cyberworlds exhibition at the Powerhouse Museum curated by Matthew Connell does a wonderful job. One such robot has occupied space in the Cyberworlds galleries for some time. ISAAC the robot was collected by Connell in 1999 for the launch of the Cyberworlds exhibition. It [he] was on display, save for maintenance, for 16 years. His two primary modes were dancing for and playing a game with audience members. He was successful as a robotic agent largely because he is programmed to mock human behaviour. The more rude, mocking or disrespectful his is, the more popular he was with the museum audience. In these engagements, ISAAC personifies perceived negative human traits in a way that made the audience empathise with their own humanity, their own frailty and weakness. Where the end result of engaging with ISAAC was fun and entertaining, even challenging at times, there were elements of his display that protected the audience from their engagement with him. ISAAC was only ever powered on, or live, if his glass case was shut and locked. People were not permitted inside the case or near ISAAC when he was 'alive.' The strength and obedience with which he responded to his programming were so responsive that he wasn't yet aware of things like 'being careful,' 'minding others' or that care for human life might be more important than performing the tasks he had been programmed to do. ISAAC functioned as an obedient responsive system, in that he responds to his programming with industrial strength and obedience. He did so with grace, timing, accuracy, even rhythm.[1]

These criteria for care and exhibition were understood by traditional museum practice and the professionals that assisted Connell in putting together the Cyberworlds exhibition. The object, ISAAC, could be contained, controlled and had an OH&S solution to any variance he might display (locked case: on/alive | open case: off/unresponsive). An on/off switch set to the Museum's opening/closing hours controlled when he was alive or unresponsive, making set-up and shut down of his systems for exhibition purposes easily aligned to the rest of the exhibition. ISAAC is easily categorised and referenced, and a label could easily be written up explaining his origins, what he represented and how he was meant to be interpreted and engaged with. In other words, ISAAC was easily manageable in terms of the Museum's standards of cultural significance.

Twenty-one years on, and robotic technology has progressed and developed. ISAAC was replaced in the Cyberworlds exhibition, and he is being replaced by a robot that has humanoid qualities. His name is BAXTER. He has a screen face with eyes and a mouth; he has two arms that are programmable through touch and choreography; he is not locked within a showcase, rather human approach and open engagement are encouraged. These display techniques indicate that he is safe to engage with on a regular and unrestricted basis. Human–robot interaction has become more engaged, more realistic, more touch and experience responsive. This is not for the benefit of the machine, but rather a design response to the changing needs of human beings to relate more closely to their machine companions (Fig. 16.5).

BAXTER is a prototype developed by startup group Robological,[2] made up of three engineering colleagues: Damith Herath, Christian Kroos and Zhengzhi Zheng

[1] YouTube user djobizz, 18 Feb 2009, https://www.youtube.com/watch?v=ek9xrR4FGZI, accessed 21 November 2022.

[2] https://robological.com, accessed 17 November 2022.

Fig. 16.5 Robological, BAXTER, 2022

of *The Thinking Head Project*. Unlike ISAAC, the screen face has eyes and eyebrows, but their functionality is false. These features are only there to provide familiarity and communication for the human user. The real communication devices are the 360-degree camera that is mounted on his head above the screen, the programmable arms and the recording system that remembers what users ask of the arms and then repeat the function. In this choreography, machine and robotic interactions have become more accessible to humans. BAXTER represents the ability to intercede with, interrupt or disrupt robotic function as it happens without the need for excessive coding or safety precautions on the part of the users. In industry, this represents the ability for humans and robots to work more closely together, rather than in a strictly action/response kind of way. If a human or machine worker notices an error in a product or in a packaging or production line, through the techniques that BAXTER represents, humans have access to correcting mistakes through an easier interface. No advanced engineering or computer science degrees are necessary at the engagement level.

Again, the BAXTER object may not have been so easily displayed at the time when Cyberworlds were first unveiled in the early 2000s. Since then, several things have happened culturally to allow for this research, reflection and display cycle to take place in the third space of a museum floor. For the social aspects of BAXTER's interface to become acceptable in the minds and actions of the Museum's audience, people needed to become more comfortable with machine functionality in their own lives. From robotic vacuums to smart televisions, responsive dishwashers and clothes washers, to remote air conditioning and alarm systems controlled through smart phones, technology has infiltrated our lives rapidly. In this new century, with the adage of networked systems, autonomously controlled devices have become more mainstream to the point where life without them is considered somewhat compromised, if not lacking. This born-digital shift in human–machine and human–robot interactions allows designers to begin conceiving functional and engaging systems that also allow for a more connected, even empathetic, relationship to machines.

There are two main characterisations of robot form: mechanistic/industrial and humanoid robots. ISAAC and BAXTER demonstrate a shift away from those characterisations being dichotomous, and to becoming more layered, more engaged, more empathetic. For robotics and culture to come to this point, there is a history of human technology that the artist Stelarc would class as cyborgism. Popular culture, from comic books to novels and films would have us believe that cyborgs are a sophisticated hybrid of human and machine that think, feel and live close enough to humans to pose a significant threat to our authentic experience as humans. Stelarc, on the other hand, would consider any augmentation to our human experience to be an aspect of cyborg culture. This would extend from corrective eyewear to microscopes and telescopes.

16.5 Engagement Over Aesthetics: *The Articulated Head* Over *the Thinking Head*

> There are also other things that robots do that relates specifically to their ability to enhance or pose questions about our culture and the nature of humanity…there is a creation complex that exists in us somewhere.
>
> (Matthew Connell, from Turnbull Tillman 2015)

The Powerhouse Museum, and Connell in particular, have had a long association with the performance artist Stelarc, and the research group he collaborates with through the University of Western Sydney (UWS) called the MARCS Institute for Brain, Behaviour and Development.[3] Stelarc is well known for melding technology with his body to enhance and augment the human experience in highly experimental ways, from probing and revealing his body with "medical instruments, prosthetics, robotics, virtual reality systems, the Internet and biotechnology, to explor[ing] alternate, intimate and involuntary interfaces with the body." Stelarc was a Senior Research Fellow and visiting artist at MARCS, which specialises in the psychology of brain development in its many forms, particularly artificial intelligence(Fig. 16.6).

Previously, Turnbull Tillman and Connell have written about the way that Stelarc's artwork *The Articulated Head* (2009–10) came to the Powerhouse Museum through the Engineering Excellence competition and award. This platform was previously discussed as a funding model for artists to garner institutional support in order to exhibit and evaluate their prototypes in an exhibition setting called The Museum Model (Turnbull and Connell, 2014). The predecessor of *The Articulated Head*, called *The Thinking Head*, was a chatbot designed by Stelarc that was projected onto

[3] https://www.westernsydney.edu.au/marcs, accessed 17 November 2022.

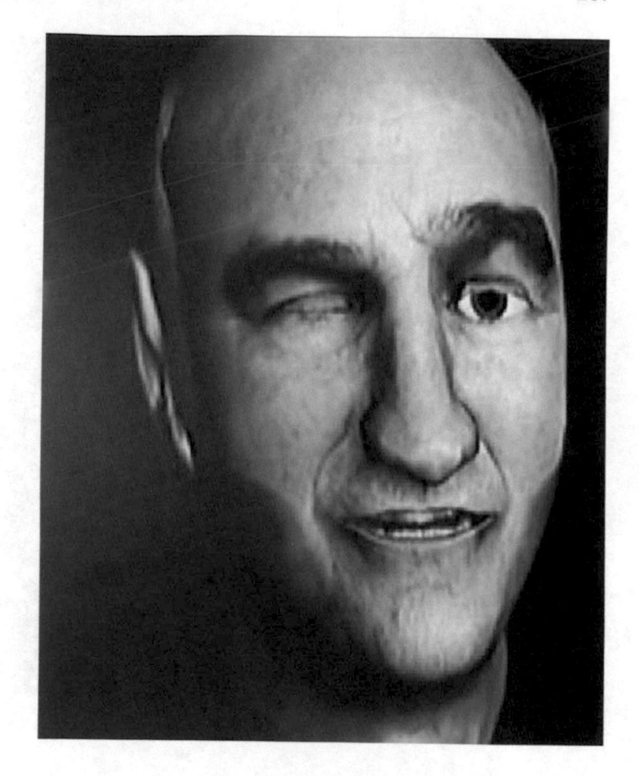

Fig. 16.6 Stelarc, *The Thinking Head*, 2003

the wall in the early stages of the Thinking Head Project (funded by the Australia Council for the Arts), which also gave rise to *The Prosthetic Head* and *The Walking Head*, which were developments that came about as a desire of the artist to provide embodiment for *The Thinking Head* (Fig. 16.7).

This work was first and foremost conceived as an artwork, though it consisted of machine parts and an artificially intelligent architecture. It utilised a computer database, a keyboard, a projector (and projection surface) and encoded software to enable engagement with humans. What made the work particularly interesting was the characterisation of Stelarc's personality within the work. The image of the onscreen face matched Stelarc's, and the conversation topics loaded into the database were topics that Stelarc was interested in and liked to think and converse about.

In 2009, *The Articulated Head* won the Research Award in the Engineering Excellence annual competition. Facilitated by Engineers Australia, the exhibition collaboration with the Powerhouse Museum always included the top two prizes in the competition, the Bradfield Award and the Research Award, and a few others that exemplified engineering in a fascinating or innovative way. The year that Stelarc's *Articulated Head* won the Research Award also saw innovations in architecture, health and safety, distance engineering techniques and renewable resources. Where the connection to its predecessor, *The Thinking Head*, made the decision to exhibit Stelarc's piece an easy one, this time the work was an example of a display object existing in a transdisciplinary environment. Here, an iterative artwork incorporated

Fig. 16.7 Stelarc, *The Articulated Head*, 2009–11

elements of engineering, computer science and audience engagement in a way that not only celebrated their interconnectedness but engaged the third space of the museum environment that incorporated the audience's response (Fig. 16.8).

Through the Beta_space platform and the ethics committee at the University of Technology Sydney's Creativity and Cognition Studios (CCS),[4] and her research initiative, New Media Curation, Turnbull Tillman collaborated with the Powerhouse to produce a performance incorporating the platforms across which Stelarc was experimenting as a conceptual artist. In an artist talk and performance on 29 May 2011, Stelarc, MARCS, the Powerhouse Museum and New Media Curation worked together to produce a performance incorporating *The Articulated Head* and Stelarc's *Second Life* avatar in collaboration with artist Daniel Mounsey, the artist who created Stelarc's *Second Life* site, CYBORGS and ZOMBIES. Mounsey collaborates online as Pyewacket Kazyanenzo. This prototype performance, titled *CLONE*, was later the first of four performances for the Ultimo Science Festival in August of 2011 and featured in two conferences at the Museum over November and December 2011.[5] The research element of the performance happened in the form of audience evaluation by survey, in which the MARCS researchers and Turnbull Tillman posited research

[4] https://www.creativityandcognition.com/, accessed 17 November 2022.

[5] https://debturnbulltillman.wixsite.com/newmediacuration/past-1/Beta_space—accessed 17 November 2022.

Fig. 16.8 *CLONE* performance at *The Articulated Head* exhibition site, Powerhouse Museum, Sydney (MAAS), 2011

queries for them to answer during the performance and turn in at its conclusion. This data was collated by MARCS and utilised towards the next iteration/performance of *The Articulated Head*.

In this display object, conceived by one artist, realised, researched and exhibited in a research capacity, existing in a cyber-space realised by another artist and arriving on the Museum floor through an engineering sponsored competition, there is no doubt that *The Articulated Head* is not any one thing, nor was its development instantaneous. It developed iteratively, over time, with the support and funding of and across various institutions, bodies and platforms, including human audience engagement with art, technology, science and robotics. The Museum was delighted to have it for these reasons, in particular to further the development of the Cyberworlds exhibition and the ideas it encapsulates. As such, its exhibition on the Powerhouse Museum floor was extended for a year so that the researchers could gather further data. This had never happened before, and certainly instilled in the Museum's operative ethos that art was a useful portal for engaging with ideas of science, design and technology.

16.6 Performing Audiences: Materiality and Interactive Art in ISEA2013 and *SHErobots*

> Sometimes as cross-disciplinary artists we collaborate with established scientific platforms to make use of them as a place for art to develop, become louder, and then eventually speak on its own.
>
> (Mari Velonaki, in Turnbull Tillman 2015)

During her two-year contract with the Curatorial Department at the Powerhouse Museum, Turnbull Tillman was responsible for annual exhibitions such as Design-TECH and the Australian Design Awards (later rebranded Good Design). One of the first exhibitions Turnbull Tillman was invited to curate and produce on her own was at the behest of the former Director, Dawn Casey, who wrote an email requesting that she "take care of the ISEA business." This business involved working with the UK-based international symposium brand ISEA (International Symposium of Electronic Art)[6] for the 2013 instalment across Sydney. It was to be managed by the Adelaide-based Australian Network for Art and Technology (ANAT)[7] and directed by Vicki Sowry. Turnbull Tillman's task was to produce a selection of works from those shortlisted by Sowry, and appropriate to the Museum's mandate of science, design and technology. In collaboration with Principal Curator Connell and Sowry, and later ISEA's 2013 Executive Creative Producer, Alessio Cavallaro, Turnbull Tillman selected and directed the installation of two floors of some of the most engaging, enticing, automated, biological, robotic and performance-based artworks to be produced in Australia over the last 20 years.

The three exhibitions selected by the Museum's curatorial team were touring exhibitions that would fit into the production requirements for a museum (rather than an art gallery). They would preferably be research-based, robust, and engaging to a range of audiences rather than a specific, singular or specialised audience. Conceived of as art-focused, these exhibitions were also considered research projects whose next iteration was commissioned by the funding body they exhibited with. These three exhibitions were a selection of ANAT's *Synapse* residency programme, Symbiotica's *Semipermeable* (+) and Experimenta's *Speak to me…* Artists such as Helen Pynor, Keith Armstrong, Oron Catts, Nigel Helyer and Wade Marynowsky featured across two temporary exhibition spaces, positing experimental ideas to do with medicine, light (and dark), molecular biology focused on the membrane, digital international relations and robotics. These ideas were realised in sculpture, film, machinery, interactive engagement, autonomously interactive machine parts controlled by computers, petri dishes, inkjet printers and performances.

When reflecting on works incorporating robotics, both contemporary and historical, Connell spoke of Wade Marynowsky's *Acconci Robot*[8] as a standout work

---

[6] https://www.isea2013.org/, accessed 17 November 2022.

[7] https://www.anat.org.au, accessed 17 November 2022.

[8] https://wademarynowsky.art/Acconci.html, accessed 17 November 2022.

Fig. 16.9 Wade Marynowsky, *Acconci Robot*, 2012

for him from ISEA2013. Based on the 1969 performance *Following Piece* by Vito Acconci, where he followed unknowing participants in the streets for as long as he could, Marynowsky designed an innocuous looking robot constructed of and resembling a wooden packing crate. It was fitted with image recognition sensors and software at eye, waist and ankle levels and a set of low, hidden wheels in order to move about when triggered. When an audience member approached or engaged with the works, the robot was unresponsive and still. As the audience member gave up and retreated, the robot would soundlessly begin to follow them (Fig. 16.9).

Housed within a low walled platform on the third floor the Powerhouse amongst the other Experimenta *Speak to me…* works, the *Acconci Robot* encouraged visitors to cross the divide between object and audience, to become more of a *thing*. Often to the delight of braver audience members willing to cross this divide physically, they were rewarded with the surprise of being followed so closely that they were frightened and immediately hopped the barrier back to the audience side while the robot tried to follow, often clumsily hitting the barrier and then turning to scan his articulated space for more unwitting participants. This was an amazing work because the joy of engaging with it was found within the experience. It was contemplative and intriguing, but again, one didn't need a degree in art history, or even an appreciation of art, to understand and enjoy the work (Fig. 16.10).

The pioneering work *Diamandini* (2011–), by author Velonaki, has roots in the ISEA universe. First exhibited in ISEA Istanbul 2011, where there were multiple themes and platforms to engage with, Velonaki and *Diamandini* exhibited in *Uncontainable: Signs of Life: Robot Incubator* (14 September to 7 October 2011 at Taksim Cumhuriyet Art Gallery/Maksem). In the words of curator Kathy Cleland:

Fig. 16.10 Mari Velonaki, *Diamandini*, 2011

> In the *Signs of Life: Robot Incubator* exhibition there are robots that look like machines but display human-like psychological behaviours; a humanoid robot that looks like a sculpture come to life, a doll-like robotic automaton performer and interactive modular robots that display hybrid machinic/biomorphic characteristics. (Cleland 2011, p. 8)

With the purpose of an arts incubator being to test new ideas in an environment supportive of the specific industry of the artwork, ISEA was the perfect setting in which to premiere *Diamandini*. In a community of artists who experiment with the hybridity of electronic art en masse, Velonaki was able to discuss the subject of robots moving through human spaces. Her kinetic agent could move through the gallery, surprising people in much the same way as Marynowsky's *Acconci Robot* did two years later, but in the guise of a drifting, elongated girl, seemingly searching for her space in society. Her movements weren't as restricted as Marynowsky's shipping container, largely because she looks and behaves somewhat human. She approaches visitors to the gallery, with her key purpose being to negotiate the space she inhabits in relation to the audience. Everything about *Diamandini* provides a strong metaphor for both the original and the new patriarchal 'other', the first being women, the second being robots.

When conceiving *SHErobots*, curators Dagmar Reinhardt, Lian Loke and Author Turnbull Tillman initially invited Velonaki to exhibit *Diamandini*. Unfortunately, the timing was off, and 'Dia' was scheduled to be on loan to the emerging National Communications Museum in Melbourne. When Velonaki instead offered *Fish-Bird*, the removal of gender or even a humanoid appearance offered an alternate intrigue to displaying a female presence. If the environment created was intended to be wholly female, having the secondary other of robotics might expand the definition of 'other.' When attempting to de-silo the roles that women play in relation to robotics research, this enquiry extended into the roles that the exhibitors were curious about. Themes of touch, intimacy, domestic labour, child and elder care, the performative presentation of self, construction of home, material and meaning making, and gender (non) expression all come to the fore. In this way, *SHErobots* pays homage to women working across all forms of robotics, conceiving and expressing ways that robots can work collaboratively with them so they can engage in society with more equity and visibility than previously. The fact that this happens on an art gallery floor as a social probe, part of a larger social experiment, shows we have a way to go.

16.7 Conclusion

This chapter sets out to discuss the display of robots as cultural objects in Museum and Gallery settings. In presenting case studies from the Powerhouse Museum and Tin Sheds Gallery, authors Turnbull Tillman and Velonaki have occupied and analysed a third space in order to conceive and exhibit transdisciplinary objects that may not fit tidily into a research stream. Where the Museum exhibitions focused more on the making and the materials of the robots, the Gallery floor tends to be a place to experiment and de-silo the taxonomies that history constructs. Conversations with fellow curators Matthew Connell, Dagmar Reinhardt and Lian Loke, and artist and Author 2 Velonaki, brought forward the different intentions and outcomes of considering robots social as cultural entities, and how audiences both respond to and dictate these tropes. In a broader social context, questions around who is designing, making, defining and displaying robotics and in what contexts (history vs care concerns) are left for the reader to consider. More personally, people may also be left considering what it means to be alive, have agency, be assigned gendered tasks and possess humanity.

Acknowledgements This chapter has been seven years in the making. It derives from two important interviews listed in the References below. The first with Director of Curatorial, Powerhouse Museum (MAAS) Matthew Connell in 2015, and the second with Director of the Creative Robotics Lab and National Facility for HRI (UNSW), Mari Velonaki in 2020. Enormous thanks to both for their time and careful consideration of the topic.

References

ANAT: Australian Network for Art and Technology: https://www.anat.org.au. Accessed 17 Nov 2022

Beta_Space: New Media Curation. https://debturnbulltillman.wixsite.com/newmediacuration/past-1/Beta_space. Accessed 17 Nov 2022

Cleland K (2011) In: Aceti L (ed) ISEA ISTANBUL 2011: Uncontainable. Conference and exhibition program 14–21 Sept 2011. https://www.isea-archives.org/docs/2011/program/ISEA2011_Istanbul_Program.pdf. Accessed 15 Nov 2011

Creativity and Cognition Studios, University of Technology, Sydney. https://www.creativityandcognition.com/. Accessed 17 Nov 2022

ISEA2013: Resistance is futile. https://www.isea2013.org/. Accessed 17 Nov 2022

Knox E (2022) Pathetic fallacy. http://www.elenaknox.com/pathetic-fallacy.html. Accessed 21 Nov 2022

Kroos C, Herath D, Stelarc (2012) Evoking agency: attention model and behaviour control in a robotic art installation. Leonardo 45(5):401–407 (2012). https://doi.org/10.1162/LEON_a_00435

Latour B (2004) Why has critique run out of steam? Crit Inq 30(2):225–248

MAAS: Powerhouse Museum (2022) Robots and machines event. https://www.maas.museum/program/robots-robots-and-machines/. Accessed 21 Nov 2022

MacDorman KF, Ishiguro H (2006) The uncanny advantage of using androids in cognitive and social science research. Interact Stud 7(3):297–337

Magnusson J (2013) Objects versus things. ResArc, Sweden: Dr Hélène Frichot, critical studies in architecture, KTH Stockholm. Published 11 Mar 2013. http://philosophiesresarc.net/2013/03/11/objects-vs-things/. Accessed 9 Oct 2015

MARCS Institute for Brain, Behaviour and Development. University of Western Sydney. https://www.westernsydney.edu.au/marcs. Accessed 17 Nov 2022

Marynowsky W (2012) The acconci robot. https://wademarynowsky.art/Acconci.html (2012). Accessed 17 Nov 2022

Muller L, Bennett J, Froggett L, Bartlett V (2015) Understanding third space: evaluating art-science collaboration. In: Proceedings of the 21st international symposium on electronic art (ISEA 2015). Vancouver, BC, Canada. http://isea2015.org/proceeding/submissions/ISEA2015_submission_332.pdf. Accessed 30 Sept 2015

Powers D et al (2008) PETA: a pedagogical embodied teaching agent. In: Proceedings of the 1st ACM international conference on pervasive technologies related to assistive environments, PETRA 2008, Athens, Greece, July 16–18, 2008. https://www.researchgate.net/publication/221410488_PETA_a_pedagogical_embodied_teaching_agent. Accessed 21 Nov 2022

Reinhardt D, Loke L, Tillman D (2022) SHErobots: tool : toy : companion. Tin Sheds Gallery, University of Sydney, 20 Oct 2022–10 Dec 2022, Australia. ISBN: 978-0-6455400-5-5

Robological (2022). https://robological.com. Accessed 17 Nov 2022

Turnbull D, Connell M (2011) Prototyping places: the museum. In: Candy L, Edmonds E (eds) Interacting: art, research and the creative practitioner. Libri Publications, Faringdon

Turnbull D, Connell M (2014) Curating digital public art. In: Candy L, Ferguson S (eds) Interactive experience in the digital age. Springer, London, pp 221–241

Turnbull Tillman D (2015) Robots and culture: an interview with Matthew Connell, 11 Sept 2015. New Media Curation. http://www.newmediacuration.com/blog/ (2015). Accessed 21 Nov 2022

Turnbull Tillman D, Velonaki M (2020) Shifting conversations in robotics with culture. New Media Curation. https://debturnbulltillman.wixsite.com/newmediacuration/post/shifting-conversations-in-robotics-with-culture (2020). Accessed 15 Nov 2022

Turnbull Tillman D, Velonaki M, Gemeinboeck P (2015) Authenticating experience: curating digital interactive art. In: Tangible, Embedd and Embodied Interaction (TEI'15), 16–19 Jan 2015, Stanford, pp 429–432

Velonaki M (2022) Fish-bird. MV Studio Website. https://mvstudio.org/work/fish-bird-cicle-b-movement-b/ (2002/3). Accessed 21 Nov 2022

Velonaki M (2009) The woman. MV Studio Website: https://mvstudio.org/work/the-woman/ (2009). Accessed 14 Nov 2022

Velonaki M, Diamandini MV (2022) Studio Website: https://mvstudio.org/work/diamandini-2011-2013/ (2009). Accessed 15 Nov 2022

Index

A
Academic mindset, 204
Action research, 207
Active learning, 194
Actuators, 46, 192–194, 196–198
Agency, 2, 5, 33, 68, 69, 71, 73, 105, 122, 123, 156, 164, 165, 172, 182, 186, 206, 219, 220, 226, 240, 247–249, 251–254, 258, 259, 273
Arduino, 6, 181, 197, 199
Art-based research, 207
Artificial Intelligence (AI), 1, 3, 6, 7, 67–69, 74, 76, 89, 90, 94–96, 101, 113, 116, 117, 124, 125, 164, 170, 183, 220, 229, 232, 239, 266
Artistic methods, 194, 206
Auditory feedback, 55
Automata, 14–17, 25, 67–71, 92, 258
Autonomous, 7, 8, 27, 46, 69, 165, 166, 171, 191, 204, 219, 226, 230, 232, 260

B
Blockchain, 7–9, 89–94, 96

C
Care industry, 201, 202
Choreography, 182, 219, 220, 222, 226, 239, 240, 242, 252, 253, 264, 265
Co-creation, 209
Co-design, 119, 140, 168, 207
Complexity, 6, 26, 27, 54, 77, 94, 101, 103, 114, 141, 154, 190, 206, 245
Complexity science, 210
Computational approaches, 194
Computational design, 85, 181, 183, 195
Constructive design, 195
Constructivist learning, 191
Cradle to cradle, 191
Cryptocurrency, 8, 89, 96
Cultural robotics, 1, 2, 5–8, 45, 90, 96, 101, 179, 180, 205, 219, 220

D
Danish Emergency Management Agency (DEMA), 164, 167, 168, 170–174
Design-based research, 189, 196, 207, 211
Designerly methods, 206
Digital fabrication, 76–78, 81, 84, 86–88, 157, 190, 203
Disciplinary perspective, 206
Discipline-specific jargon, 206
Disposable technologies, 193
Disruptive feminine, 238, 254
Drone, 4, 107, 163–174

E
Ecosystem, 3, 6, 7, 89, 90, 94–96
Embodied interaction, 102, 190, 222
Embodied knowledge, 194
Emergency, 4, 163–165, 167–170, 173, 174
Emotional sustainability, 203
Experiential learning, 191, 193, 194, 205
Expressive movement, 53, 190

F
Fibonnaci spiral, 203

© The Editor(s) (if applicable) and The Author(s), under exclusive license to Springer Nature Switzerland AG 2023
B. J. Dunstan et al. (eds.), *Cultural Robotics: Social Robots and Their Emergent Cultural Ecologies*, Springer Series on Cultural Computing,
https://doi.org/10.1007/978-3-031-28138-9

G
Green transition, 190, 208
Greenwashing, 200
Gripper, 196, 201, 226, 227

H
Health data, 75, 79, 88
Human-drone interaction, 165
Human-robot interaction, 2, 3, 27, 35–37, 45, 47–49, 53–55, 128, 132, 133, 179, 181, 192, 194, 196, 203, 205, 207, 222, 233, 239, 241, 250–252, 264, 266
Human-robot relations, 237, 240

I
Industrial robot, 183, 219, 226, 229, 237, 238, 240, 242, 250, 253
Integration, 3, 14, 33, 113, 116, 125, 165, 181, 189, 193, 205, 207, 240, 251, 263

K
Knowledge silos, 205

L
Life cycle assessment, 191

M
Materials, 3, 5, 20, 23, 27, 41, 46, 48, 50, 51, 53, 67, 69, 70, 75, 76–79, 81–88, 130, 179–185, 187, 189–194, 196, 197, 200, 201, 204, 206, 208, 209, 212, 222, 225–228, 230, 241, 242, 248–250, 257–259, 273
Microbots, 192
Microcontroller, 71, 197
Movement, 3, 16, 19–22, 35–37, 46–56, 71, 75, 79, 80, 83–86, 103, 104, 149, 166, 167, 172, 181, 192, 194, 195, 197–199, 201, 205, 206, 208, 209, 219, 223, 228–232, 240–247, 251–253, 262, 272

N
Nature-based solutions, 196
Non-fungible token, 3, 7, 9, 89–91, 93, 96

O
Optimisation, 206, 209
Outcome-based exploration, 209

P
Paradox theory, 210, 211
Participatory research, 153
Problem-based learning, 193, 196
Process-based exploration, 209, 211
Programmatic art, 7
Prototyping, 6, 25, 31, 192, 197, 199, 209
Public perception, 164, 165, 169, 173, 174

R
Robot bodies, 196, 198, 199, 246, 254
Robot fabrication, 75, 87
Robotic gripper, 192, 214
Robots and sustainability, 189

S
Search And Rescue (SAR), 4, 107, 163–167, 169–174
Sensation, 238, 239, 241–246, 251–253
Siloed thinking, 205
Situational awareness, 48, 164–166, 170, 171, 174
Smart contract, 89, 91, 93, 96
Social justice, 210
Social robotics, 1–8, 15, 26–29, 55, 103, 127, 128, 130, 132, 181–183, 185, 186, 190, 237–239, 250, 251
Societal factors, 193
Soft robotics, 4, 179, 189–198, 200, 204, 206, 208–211
Somaesthetics, 197, 203
Sonic interaction design, 3
Staying with the trouble, 206, 211
STEM, 190
Sustainability as a value, 209
Sustainability by form, 196, 200, 201
Sustainability by function, 196, 200
Sustainability-driven development, 191
Sustainable design cards, 198
Swarm, 4, 46, 69, 81, 163–174
Systems thinking, 190, 210, 211

T
Tame problem, 208
Team culture, 164, 165, 169, 172, 174
Technological narratives, 33
Technology-driven development, 191

Thought collective, 207
Thought style, 207
Transdisciplinary, 1, 4, 7, 179, 181, 189–195, 204–207, 209–211, 257, 258, 267, 273

U
Universal female, 237, 238, 253
Unmanned Aerial Vehicles (UAVs), 163, 165, 166
Unsustainability, 208
User interface, 165, 166, 168, 169, 173

V
Visualisation, 75–78, 81, 164, 168–170, 174

W
Ways of knowing, 206
Wicked problem, 200, 207, 208, 212
World Cafe, 199, 200, 203

Printed in the United States
by Baker & Taylor Publisher Services